M. Fourcroy, Thomas Elliot

Elementary Lectures on Chemistry and Natural History

M. Fourcroy, Thomas Elliot

Elementary Lectures on Chemistry and Natural History

ISBN/EAN: 9783743432925

Manufactured in Europe, USA, Canada, Australia, Japa

Cover: Foto ©berggeist007 / pixelio.de

Manufactured and distributed by brebook publishing software (www.brebook.com)

M. Fourcroy, Thomas Elliot

Elementary Lectures on Chemistry and Natural History

ELEMENTARY LECTURES

ON

CHEMISTRY

AND

NATURAL HISTORY.

CONTAINING

A METHODICAL ABRIDGEMENT

OF ALL THE

CHEMICAL KNOWLEDGE

ACQUIRED TO THE PRESENT TIME;

WITH

A COMPARATIVE VIEW OF THE DOCTRINE OF STAHL,

AND OF THAT OF SEVERAL MODERN CHEMISTS:

THE WHOLE FORMING A

COMPLETE COURSE OF THOSE TWO SCIENCES.

Tranflated from the French of

M. FOURCROY,

Doctor of the Faculty of Medicine of Paris, and of the Royal Society of Medicine.

BY THOMAS ELLIOT.

WITH MANY ADDITIONS, NOTES, AND ILLUSTRATIONS, BY THE TRANSLATOR.

IN TWO VOLUMES.

VOL. II.

EDINBURGH:

Printed for C. ELLIOT; G. ROBINSON, London;
And W. GILBERT, Dublin.

M,DCC,LXXXV.

CONTENTS

OF THE

SECOND VOLUME.

LECT. XXXVI. XXXVII. XXXVIII. *Iron* - Page 9

Essaying of iron ores - - 15

Martial vitriol - - - 23

Prussian blue - - - - 29

Martial nitre - - - - 39

Alkaline martial tincture of Stahl - - 40

Martial æthiops - - - 41

——— *marine salt* - - - - 43

Artificial pyrites - - - - 48

——— ——— *effects* - - - ibid.

Preparation of iron in Britain - - Note 54

steel - - Note 60

Qualities and theory of the different kinds of iron Note 62

cold short iron - Note 63

Solutions of the different kinds of iron in each of the acids - - - - Note 64

LECT. XXXIX. *Copper* - - 66

Essaying of copper ores - - - 69

Copper in volatile alkali - - Note 74

Vitriol of copper - - - 75

brass - - - Note 82

Specific gravity of the alloys of copper - Note 84

ECT. XL. *Silver* - - - 87

Essaying of silver ores - - 91

Cupellation of silver - - - Note 93

Lunar nitre - - - - - 96

its use in bleaching - - Note 97

Lapis infernalis, lunar caustic - - - 98

Nitrous acid obtained pure by means of a solution of silver 99

Arbor Dianæ, or arbo philosophorum - - 100

Luna cornea - - 102

CONTENTS.

Phenomena - - - - - Page 304
Different substances susceptible of the spirituous fermentation - - - - 305
Product of the spiritous fermentation - - 308

LECT. LIII. LIV. *Aquavitæ or ardent spirit* - 310
Rendered pure by distillation - - ibid
Methods of ascertaining its purity - - 311
Action of fixed alkali upon it - - 314
Acrid tincture of tartar - - - ibid.
Lily of Paracelsus - - - - 315
Ether, vitriolic - - - - ibid.
—— *nitrous* - - - 319
—— *curious phenomenon* - - Note 323
—— *Dr Black's process* - - Note 324
—— *marine* - - - - ibid.
—— —— *good* - - - Note 326

LECT. LV. *Tartar* - - - 331
Tartar sal de Seignette - - - 336
Tartarous acid - - - - 334
Stibiated or emetic tartar - - - 338
Note 340
LECT. LVI. *The acid fermentation of vinegar* - 344
Note 346
Terra foliata tartari - - 349
—— —— *mineralis, or acetous mineral salt* - 350
Spiritus Mindereri - - - ibid.
Terra foliata mercurialis - - 352
Saccharum saturni, Goulard's extract - - 352
Verdigris - - - - - 354
Radical vinegar - - - ibid.
—————— *its inflammability* - - 355
Acetous ether - - - 356
Putrid fermentation - - - 358

LECT. LVII. *Animal Kingdom* - - 362
Natural history of animals - - ibid
Class 1. *Of quadrupeds.* Zoology - - 364
Linnæus's method - - - 366
Klein's - - - 367
Brisson's - - - 369
Class 2. *Cetacea* - - - ibid.

LECT.

LECT. LVIII. Clafs 3. *Birds.* Ornithology - 370
Clafs 4. *Amphibious animals* - 371

LECT. LIX. Clafs 5. *Fifhes.* Ichthyology - 372

Clafs 6. *Infects.* Entymology - - 378

LECT. LX. Clafs 7. *Worms* - - 384
Clafs 8. *Polypi* - - - 389

LECT. LXI, LXII. *Functions of animals* - 393
Digeftion - - - 396
——— - - Note 398

LECT. LXIII. *Chemical analyfis of animal fubftances* 408
Blood - - - 409
—— *coagulation* - - Note 411

LECT. LXIV. *Milk* - - 417
Cheefe - - - 420
Butter - - - ibid

LECT. LXV. *Of the fat* - - 422
Of the bile - - - 426
Of the biliary calculi or ftones - - 430
Of the faliva and pancreatic juice - - 431
Gaftric liquor - - - Note 432
Of the feminal fluid - - - 433

LECT. LXVI. *Of urine* - - - 433
Of the phofphoric ammoniacal falt - - 439
Of the fubftance difcovered in the fufible falt by M. Prouft 441
Of the fufible falt with bafe of natrum extracted from urine - - - 443
Microcofmic falt - - Note 446

LECT. LXVII. *Kunckel's phofphorus* - - 447
Of the acid of phofphorus - - 455
Of the urinary calculi - - - 460

LECT. LXVIII. *Of the infenfible perfpiration and fweat* ibid
Of the folid excrements of animals - - 461
Of the foft and white parts of animals - 462
Of the flefh or mufcles of animals - - 464
Of the bones of animals - - - 467

LECT. LXIX. *Of the different substances used in medicine, and the oils which are extracted from different animals* 471
Castor - - - 473
Musk - - - 474
Hartshorn - - - ibid
Spermaceti - - - 475
Eggs - - - 477
Tortoise, frog, and viper - - 478
LECT. LXX. *Cantharides, ants, millepedes* - 480
Honey and wax - - 483
Resin Lac - - - 485
Kermes and cochineal - - ibid
Crab's eyes and coral - - 486
Coralline - - 487
Of the chemical analysis of animals and vegetables 488

ELEMENTARY LECTURES

ON

CHEMISTRY

AND

NATURAL HISTORY.

LECTURES XXXVI. XXXVII. and XXXVIII.

Species 11. Iron.

IRON, called *Mars* by the alchemifts, is an imperfect metal of a white colour, blueifh, and inclining to grey, difpofed in fmall facets. It is fufceptible of a very fine polifh and a great brilliancy. By its hardnefs and elafticity it is capable of deftroying the aggregation of all the other metals.

Iron has a fmell, particularly when it is heated or rubbed: it alfo has a very remarkable ftyptic tafte, which has a ftrong effect upon the animal-œconomy.

Iron is the lighteft of the metallic fubftances except tin. A cubic foot of this metal, forged, weighs 580 pounds. It lengthens by the ftroke of a hammer; but as it is very hard, and as it ftiffens greatly, laminated leaves cannot be formed with it: its ductility in wire-drawing is much more remarkable; it is drawn into very fine threads, with which harpfichords are made. This property appears to depend on its tenacity: iron is indeed the moft tenacious of all the metals except gold: a thread of iron, one-tenth of an inch in diameter, fupports, without breaking, 450 pounds weight.

Pure iron has a cryſtalline form particular to itſelf. In the furnaces where this metal has been ſlowly cooled, it has been found in quadrangular pyramids, articulated and branched, formed of octagons laid on one another. We owe this obſervation to M. Grignon, forgemaſter at Bayard in Champagne. Beſides all the properties of which iron partakes in common with the other metallic ſubſtances, it exhibits ſtill three which are peculiar to itſelf alone: one is magnetiſm, or the property of being attracted by the magnet, and its becoming itſelf a very good magnet, either when it remains a long time in an elevated poſition, or in a direction from ſouth to north; or when it has ſerved for conducting the electric fire of thunder, as ſeveral facts atteſt; or when two pieces of iron are ſtrongly rubbed againſt one another. The ſecond property is, its taking fire and fuſing ſuddenly by the ſtroke of flint; a phenomenon to which the poets unanimouſly aſcribe the diſcovery of fire by our firſt parents. The third property which diſtinguiſhes iron, is, its being the only metallic ſubſtance which is found in plants and animals, a part of whoſe humours are coloured by it. It is even probable that theſe organic ſubſtances form this metal themſelves; for the plants raiſed in pure water contain it, which may be extracted from their aſhes.

Iron is a very copious metal in nature, being found, independently of what is contained in plants and animals, in almoſt all the coloured ſtones, in the bitumens, and in the greateſt number of the metallic ores. But we ſhall talk here only of the mineral ſubſtances which contain a great quantity of this metal, which may be ſmelted for its extraction. In theſe ores, which are very numerous, the iron is either in a metallic or in a calcined ſtate, or mineralized by different ſubſtances.

1. Native iron is diſtinguiſhed by its colour and malleability. It is very rare, and found only accidentally in iron mines. M. Margraaf found ſome of it in ridges at Ebenſteek in Saxony. Dr Pallas diſcovered ſome of it

in Siberia; and M. Adanson ſays that it is common at Senegal.

2. Iron is very often, in the ſtate of ruſt, more or leſs calcined. It then forms the bog iron ores. It is diſtinguiſhed into rich and poor iron, fuſible and dry iron. The rich is iron only in a ſmall degree ruſty, and containing only a very ſmall quantity of earth. The fuſible iron fuſes eaſily, and is of a good quality: the metal is combined only with ſeveral ſtones of eaſy fuſion. The dry iron is more calcined, and intermixed with very refractory ſubſtances. All the bog-iron is diſpoſed generally in beds, in the ſame manner as ſtones are; and it has the appearance of having been depoſited by water. It is formed often into kinds of helmets or ſpherical bodies, flattened and irregular. It is not rare to find in it organic ſubſtances, ſuch as wood, leaves, barks, ſhells, changed into iron. It is neceſſary to obſerve, that organized ſubſtances are never found converted into any other metal but iron; and it appears that this converſion greatly depends upon the analogy which ſubſiſts between this metal and organic ſubſtances. In the wood of Boulogne, near Auteuil, there is a mine of bog iron, in which the vegetable ſubſtances change into iron almoſt before our eyes.

3. The eagle-ſtone, or ætites, is a variety of the bog-iron. Theſe ſtones are of different forms, commonly oval or polygonal, formed of concentric layers depoſited around a nucleus, which is frequently moveable in the centre of the ſtone. This ſtone has received its name from the belief that eagles depoſite it in their neſts, and that it has the property of facilitating their hatching: hence it has been concluded, that it acted ſtrongly upon the fœtus contained in its mother's womb; and ſome authors have even aſſerted, that it was poſſible to expedite the delivery of a woman in labour, by the application of the eagle-ſtone to her, or to retard it by applying it to her arm.

4. The hematites is a kind of bog-iron, which ſeems

to be formed in the manner of ſtalactites. Its name comes from its colour, which is generally red or of a blood colour; although, however, this colour varies. The hematires is compoſed generally of layers, which cover one another, and are themſelves formed of converging needles. On the external part of the ore are many tubercles or nipples. The hematites is diſtinguiſhed not only by the colour, but alſo by the form. Of this kind is the hematites in needles which is found in Lorrain, and the nippled hematites which is in the form of raiſin-ſtalks, or the botryoid hematites. Theſe ores are found very often along with bog-iron.

5. The magnet is only a bog-ore of iron. It is known by its property of attracting iron-filings. It is found in Auvergne, in Spain, in the Bay of Biſcay. The varieties of it are diſtinguiſhed by the colour.

6. The emerald, ſmyris, is a grey or reddiſh ore of iron, which ſeveral mineralogiſts take for a kind of hematites. It is very hard and refractory; there is great abundance of it in the iſlands of Jerſey and Guernſey. In mills it is reduced to powder, which is uſed for the poliſhing of glaſs and of metals.

7. The ſparry iron is a calx of iron, combined with the chalky acid, and carried along by water. It is generally of a white colour: it is to be found, however, of all kinds of hues; of grey, yellow, and red. It is always diſpoſed in greater or leſs laminæ, ſemitranſparent like ſpar: it is very heavy, and often regularly cryſtallized: it is found in conſiderable quarries, mixed frequently with pyrites, like that of Allevard in Dauphiné; ſometimes with the grey ore of ſilver, like the iron of Baigorry; or with manganeſe, like that of Styrie. Some mineralogiſts think that this is a ſpar, in which the metallic calx has been depoſited. The ſparry iron is entirely decompoſed, without addition, in cloſe veſſels, and gives over the chalky acid. Some iron remains in a black powder, which is very readily attracted by the magnet, and which fuſes eaſily with a ſtrong fire.

8. Na-

8. Nature alſo preſents iron in a ſaline ſtate, united with the vitriolic acid, forming martial vitriol or green copperas. This vitriol is found in the galleries of the mines of iron, particularly of thoſe which contain pyrites. It is found ſometimes in green cryſtals, or in the form of beautiful ſtalactites; at other times it is not ſo pure, and has ſuffered ſome alteration. If it has loſt only the water of its cryſtallization, it is of a whitiſh or greyiſh colour: it is called *ſori*. When it has undergone a little ſtronger calcination, it is yellow, and called *miſſy*. If the calcination has been ſo ſtrong as to carry off a conſiderable portion of the acid, the vitriol will be red, and called *colcothar* or *natural chalcite;* when mixed with ſome inflammable matter, it is called *melanteri*, on account of its black colour. All theſe different matters have received the name of *inkſtones*, on account of their property of making ink like vitriol of iron.

9. Iron is found frequently united with ſulphur: it then forms martial vitriol. This kind of ore has received the name of *pyrites*, becauſe it is ſo hard as to give many ſparks when it is ſtruck with ſteel. The martial pyrites are commonly in ſmall round, ſometimes regular, maſſes; for the moſt part, they are ſpherical, cubic, or twelve-ſided. Their form is very various; as is evident from reading the Pyritology of Kunckel. Externally ſome are brown, and of the colour of iron; others are yellow, and reſemble ores of copper, even at their ſurface: internally all of them are yellow, and ſeemingly coppery; and are formed for the moſt part of needles or pyramids of ſeveral ſides, whoſe tops converge to a common centre. Generally the pyrites are found in the neighbourhood of iron mines, and diſperſed through the clays and in coal pits: the upper layer of the latter conſiſts almoſt always of pyrites. All the pyrites are eaſily decompoſed. A very weak degree of heat ſuffices to carry off their ſulphur. Almoſt all of them undergo alteration when expoſed to the air, and

particularly in a moist place: they swell, burst, lose their colour, and are covered with a greenish white efflorescence, which is only some martial vitriol. It seems that this alteration, called *vitriolisation of the pyrites*, depends upon the united action of the air and water upon the sulphur. There is formed some vitriolic acid, which dissolves the iron, and is raised above the pyrites, like a kind of vegetation, by gradually removing the small pyramids which compose this mineral. All the pyrites do not equally effloresce. The globular pyrites, whose colour is very pale, and whose texture is not very close, are very quickly vitriolised. Those which are of a brilliant yellow, of a coppery colour, and are formed of small laminæ closely adhering to one another, effloresce but very difficultly, and ought to be carefully distinguished from the first, since they differ from them in their colour, form, texture, and properties.

10. Iron is found combined also with arsenic. The ores of iron, combined with arsenic, are called *wolfram* or *spuma lupi*: they are of a violet colour, more or less, or red or blackish, very much resembling that of the ores of tin, to which they approach also in their considerable weight. The arsenic may be separated from it by roasting. They are distinguished from the true arsenical pyrites or mispikel by the colour. This last is white, frequently crystallised in thick cubes, and contains but very little iron. The other resembles the ore of tin; and several mineralogists have thought that it contained some of it. The wolfram is found in Franche Compté, in Vosges, in Saxony, and other places.

11. Black iron is known by its colour, by the property which it possesses of being attracted by the magnet, and of being quite soluble in the acids. This iron is sometimes crystallised in form of polygons or in rounded laminæ, and presents very brilliantly the various colours of the rainbow: of this sort is that of the Elbe. This iron forms a considerable mountain, which is smelted in the open air. The ore of Sweden is also black

black iron, but is not cryſtalliſed: it is in maſſes more or leſs ſolid, mixed with ſome quartz, ſpar, aſbeſtus, &c. It is often ſo hard as to receive a poliſh, and its ſurface appears to be like a mirror. It has alſo got, as well as the preceding, the name of *ſpecular ore*: conſiderable quarries of it are to be found. This iron varies in the intenſity of its colour: there is ſome of it perfectly black, which the magnet readily attracts; ſome blueiſh, which is leſs attracted; and ſome grey, which is very little. The iron of Norway is alſo black iron; but it is generally in ſmall ſcales like mica, frequently mixed with grenate and ſchorl. Black iron aſſumes ſometimes the form of grains. It is alſo cryſtallized in cubes; which has cauſed it to be named by ſome naturaliſts, *galena of iron*, or *eiſeng lants*. When the ore of iron, in the form of mica, is of a black colour, it is called *eiſen-mann*, particularly if the ſcales are very large: when theſe ſcales are red, and when the powder which covers them is of the ſame colour, it is called *eiſen-ram*. The ore of iron in eight-ſided black cryſtals, very regular, and diſperſed in a kind of colubrine or hard ſteatite, which comes to us from Sweden, Corſe, and other places, ſeems to belong to this claſs of ores of iron. It is very brittle, and attracted by the magnet.

The ores of iron are eſſayed in the following manner: After they are reduced into powder, they are mixed with double their weight of pounded glaſs, and one part of calcined borax: the mixture is carefully rubbed; it is put into a crucible, and a little marine ſalt added; the crucible is covered, and the fire applied till it fuſes. When the whole is cooled very ſlowly, the malleable iron is generally found in a ſmall ſpherical button, frequently cryſtalliſed at its ſurface.

The treatment of the ores of iron varies according to the ſtate in which the metal is found. There are ores which have no need of any preparation before they are fuſed; others ought to be pounded and waſhed, ſometimes even roaſted, in order to be made more tender and fuſible.

The bog and ſparry iron are ſmelted in the ſame way, by fuſing them acroſs coals. The furnaces in which the iron is fuſed vary in height, which is from 12 to 15 feet. Their cavity repreſents two quadrilateral pyramids, which join by their baſe towards the middle of the furnace; this place bears the name of *etallage*. A hole is made at the bottom of the furnace, to give an exit to the fuſed metal: this hole, which is ſtopped with earth, anſwers to a triangular canal, hollowed in the ſand, and deſtined to receive the fuſed iron. Firſt ſome firebrands lighted are thrown into the bottom of the furnace, then charcoal, and after that the ore and ſome fuſing ſubſtances: for the moſt part theſe ſubſtances are calcareous ſtones, called *caſtine*, and ſome argillaceous ſtones called *arbuc*, ſometimes ſome quartz or flints: the ore, the ſtones, and charcoal, are thrown alternately into the furnace, care being taken to cover the whole with a layer of charcoal, which ought to reach the ſuperior opening of the furnace. The fuſion is effected by the aid of two bellows. The iron fuſes, being reduced by the charcoal. The ſtony matters which are added to the ore becoming fuſed and vitrefied, facilitate the fuſion of the iron, which begins at the top of the etallage of the furnace; the fuſed metal is collected at the bottom of the furnace, in the part called the *crucible*; it is made to run by the anterior opening of the furnace into the canal hollowed in the ſand; it forms what is called the *font*. After the iron, there paſſes a vitreous matter called *ſlag*: it is formed by the vetrification of the ſtones which had been added to facilitate the fuſion of the iron: it is of a green whitiſh or blue colour, which is communicated to it by a portion of a fuſed calx of iron. The font is brittle, and has not the ductility of iron. Metallurgiſts are not agreed [illegible] the cauſe of this property of the font Some [illegible]ſe that it is owing to the preſence of a portion of

ſlag;

flag; others afcribe it to this, that the iron is not fully reduced, and contains a portion of calx. Brandt fuppofed that it was owing to arfenic, and M. Sage thinks that it is zinc which renders it brittle. M. Bucquet confidered the font as an iron not properly reduced, which ftill contains a portion of metallic calx interpofed between its parts. The metallurgifts diftinguifh feveral kinds of fonts, the white, the grey, the black, &c. They give the name of fpeckled font to that which, upon a grey furface, has black ftains. The white font is the worft; it approaches to the character of the femimetals. The grey holds the middle place between the firft and the black; which is the beft, and furnifhes iron of a good quality.

The fufed iron is carried to the refining furnace. This is a forge a little hollowed, in which a mafs of font is put, which is covered with a good quantity of charcoal. The fire is blown till the iron begin to foften. When it is in this ftate, it is *baked* feveral times. This kneading makes it prefent more furface; fo that the portion of iron, in the ftate of a calx, may be reduced. The metal is feparated alfo from a portion of flags, which remain in it. It is then fubjected to the hammer, in order to be reduced into bars. The hammering, by approximating the parts of the iron, facilitates the feparation of the little flag and of the portion of calx which it may ftill contain; confequently it finifhes what the fufion had been unable to do for want of being fufficiently complete. The iron is heated and beat feveral times, till it be as perfect as we wifh to have it. Forged iron is diftinguifhed into foft iron and hard iron or fteel.

Steel is the beft and hardeft iron, of the moft fine and compact grain. The qualities of foft iron very much approach to thofe of fteel: its grain, however, is lefs compact; and when it is broken by bending, it appears to be compofed of filaments or fibres. This is what is called *nervous* or *tough iron*. But this appearance is pro-

produced only accidentally; for if the ſofteſt iron is broken clean, and with one ſtroke, it does not turn out of that tough appearance; whilſt by breaking the worſt iron with precaution, it may be made to appear ſo. It is rather more expedient to attend to the grain of the metal when we are to judge of its quality. The hard iron is more brittle; its grain is coarſe, and ſeems to be formed of ſcales: it is diſtinguiſhed into iron cracking with heat, and into iron cracking with cold. This diſtinction is founded upon experience; the cauſe of which is not well known. It is not uncommon to find, in the ſame bar, brittle iron, ſoft iron, and ſteel.

The ſteel formed by the forge is rarely perfect ſteel; beſides, it is in ſmall quantity. By art iron may be converted into ſteel. For this purpoſe ſhort iron bars are made uſe of: theſe are put into an earthen box full of a cement, compoſed generally of very combuſtible matters, as the ſoot of chimneys, or the charcoal of animal ſubſtances: to theſe are added frequently cinders, calcined bones, marine ſalt, or ſal ammoniac. The box being well filled, it is heated ten or twelve hours till the bars be very white and beginning to fuſe. In this operation the iron is purified and completely reduced by means of the combuſtible matters which ſurround it on all ſides. The portions which were not quite in a metallic ſtate are metallized. With reſpect to the ſaline and earthy matters which are added, what purpoſes they ſerve is not well underſtood. The ſteel prepared in this way is called *ſteel of cementation*. This ſeems to be the pureſt iron.

Steel may be converted into iron, if it is treated by cementation with dry matters, and particularly with calcareous matters and lime, which ſeem proper to calcine a part of it.

It is evident that all the preparations to which iron is ſubmitted, are neceſſary on account only of its being more difficultly fuſed, and of its being never perfectly purified by a ſingle fuſion.

The

The ores of iron, and particularly the black iron, like that of the Iſle of Elbe, contain ſo much of the metal, and ſo little altered, that they do not require to be fuſed. It is only ſoftened among the coals in the refining furnace, and ſubmitted to the hammer. This is called the *Catalonian method:* it can be employed only in the caſe of ores, which contain few ſtrange matters ſuſceptible of converſion into ſlag.

The chemical properties of iron are very numerous; and for rightly underſtanding them, we muſt examine them in very pure ſteel.

Steel, expoſed to a fire incapable of making it red hot, varies in its colours; it whitens, becomes yellow, orange red, violet, and at length blue: it remains a long time of this laſt colour; but if it receives a ſtronger heat, it changes into a diſagreeable water-colour. Not a conſiderable heat turns ſteel red and bright: at firſt it is of a cherry-red, and at length it is white and ſhining, and it burns with a very ſenſible flame. It does not fuſe but with an extreme heat. If ſteel-filings are thrown into the middle of a fire, or even acroſs the flame of a candle, it ſuddenly takes fire, and produces very vivid ſparks: ſuch alſo are thoſe occaſioned by the flint of a firelock. The ſteel collected upon white paper is found to be fuſed, and to reſemble ſcoriæ or droſs. Ordinary iron, expoſed to the focus of M. de Trudaine's lens, ſuddenly emits burning ſparks. M. Macquer, who fuſed ſteel and iron with this lens, obſerved, that the ſteel was moſt fuſible, owing certainly to the metal's being pure and homogeneous. Fuſed iron, which cools ſlowly, aſſumes a particular cryſtalline form, as has been already obſerved. M. Mongez calls it a pyramid of three or four ſides. Steel, though very hard and refractory, is very eaſily calcined: as ſoon as it begins to redden, it combines with the air, and burns without an apparent flame. A bar of iron, kept red for a long time, preſents ſcales at its ſurface, which may be ſtruck off with a hammer: the

metal

metal in this case is only in part calcined, being attracted by the magnet. A more perfect calx of iron may be made, by exposing some steel-filings under a muffle: it is converted into a powder of a reddish brown, not attracted by the magnet, called *crocus martis astringens*. This martial calx differs according to the state of the iron, and to the degree of calcination which it undergoes. There are some croci martis of a yellow brown; others are of a chesnut colour; and others again are of a finer red, and resemble carmine. Crocus martis astringens, exposed to a very strong heat, fuses into a blackish and porous glass. It is in part reduced by being heated slowly in close vessels: in its reduction it affords a certain quantity of chalky acid; which would seem to indicate, that the iron seizes upon this acid in its calcination, or rather that the pure air which is united to the metallic calx finds a principle with which it constitutes the chalky acid. In another place, I have offered a conjecture that this acid is a compound of pure air and inflammable gas. Were this conjecture demonstrated, it would be easy to explain what passes here. It is known that iron, heated in a pneumato-chemical apparatus, affords inflammable gas. This fact has been demonstrated by Dr Priestley. This gas, combined with the air which constitutes the calx of iron, forms aërial acid. Whatever be the theory of this singular experiment, we must agree that it is contrary to the new doctrine of calcination, but at the same time that it does not in a greater degree favour the theory of phlogiston. The part of an historian which I have taken, requires that I should represent what is contrary, as well as what is favourable, both to the theory of the gases and of Stahl.

Crocus martis astringens is very easily reduced by means of combustible matters. Upon mixing it with a little oil, and slightly heating it in a crucible, it becomes black and easily attracted by the magnet; by

this

this procefs may be made a very good fpecies of martial æthiops.

The pureft iron, expofed to moift air, very foon parts with its metallic brillancy: it becomes covered with a powdery cruft, yellowifh, and brighter than the crocus martis aftringens. This matter is called *ruft*. Ordinary iron is much more fubject to ruft than fteel is. The more it is divided, the more rapid is its alteration by the air. It is in this way that the medicine is prepared in pharmacy, which is called *crocus martis aperiens*. Steel-filings are expofed to the air, and moiftened with water · by this means it very quickly becomes rufty. It is made ftill more quickly with iron in the ftate of an æthiops treated in the fame way. In this alteration the metal agglutinates, and forms maffes, which undergo porphyrifation before it is employed in medicine. We are not yet pofitively certain of the change produced in the iron by ruft. Some of my own experiments lead me to confider the ruft, or the crocus martis aperiens, as a combination of iron and the chalky acid. I have diftilled this faffron of mars in a pneumato-chemical apparatus, and I obtained from it a great quantity of chalky acid. The iron was changed into a black powder, eafily attracted by the magnet. M. Joffe, a Paris apothecary, has communicated a fimilar procefs to the Royal Society of Medicine, for obtaining readily martial æthiops. He recommends reddening crocus martis aperiens in a retort, to which a balloon with a fmall hole in it is adapted without being luted: by this means the heat difengages the chalky acid, which M. Joffe allows to efcape by the hole in the balloon, and the iron remains pure. By this means I have feveral times cryftallized the cauftic vegetable fixed alkali, with which I had wetted the fides of the balloon adapted to the retort. By the tranfmiffion of the aërial acid from the crocus martis aperiens upon the alkali, a kind of neutral falt, which with M. Bucquet I have named *tartar with the chalky acid*. I have made many other experiments

periments upon the ruſt of iron, which I reſerve for a particular memoir: they all have convinced me, that that ſubſtance is a true neutral ſalt, formed by iron and the chalky acid. I think it ſhould be called *creta martialis*, to diſtinguiſh it from the true calx of this metal. This ſalt is abſolutely the ſame with what M. Bergman calls *ferrum aëratum*. This theory was alſo adopted by M. Macquer; it very well explains why the iron readily ruſts in a humid and impure air; why this alteration is ſo quick and ſo deep in a place whoſe air is ſpoiled by the reſpiration of animals, by combuſtion, by the vapours of animal matters, as in ſtables, ſtalls, privies. Iron, of all the metallic ſubſtances, undergoes the greateſt alteration by the contact of the air: and this alteration is not confined to the ſurface; frequently very thick bars of iron are ruſted into their centre.

Water has a great action upon iron: it divides it, and even diſſolves a part of it, according to the experiments of M. Monnet. It takes up the more of it the purer the iron is, and the greater the quantity of air it contains. When iron is agitated ſome time in water, it is by little and little very minutely divided; and by pouring off the water, a little troubled, a very black and ſubtile powder of iron ſubſides, which has been called *martial æthiops of Lemery*. It is carefully dried with a gentle heat in a cloſe veſſel, as an alembic, leſt the contact of air ſhould ruſt it. The magnet attracts this martial æthiops very well: it is no more than iron reduced to a very fine powder. As this operation is very long and delicate, ſeveral chemiſts have found means to ſimplify it. M. Roeulle employed Count Garaye's mill, and by this means obtained a very fine æthiops in much leſs time than Lemery's proceſs requires. I think that of M. Joſſe might be ſubſtituted to it with advantage, which is much more expeditious. Several other proceſſes will be found afterwards as proper for the preparation of martial æthiops.

Steel in bars, heated to a certain degree, and ſuddenly plunged into cold water, aquires a very conſiderable hard-

hardneſs, and becomes very brittle. Theſe qualities are the more perceptible the ſtronger the heat and the colder the liquor was into which it was immerſed. This operation is called *tempering*. The hardneſs of the ſteel may be varied at pleaſure: it can be alſo ſoftened eaſily, by heating it to the ſame degree at which it was before tempering, and by leaving it to cool ſlowly. This effect ſeems to depend upon this, that the ſudden cooling of the ſteel, changes the diſpoſition of its parts, and prevents its cryſtallization. All the metals are ſuſceptible of acquiring hardneſs by being tempered: but this quality increaſes with the infuſibility of the metal: on this account iron poſſeſſes it in ſo eminent a degree.

Iron, in its metallic ſtate, does not unite with the earthy matters; but the calx of iron facilitates the vitrification of all the ſtones, and gives them a green colour.

Lime, magneſia, and the cauſtic fixed alkalis, have no remarkable action upon iron: it is weakly attacked by the volatile alkali; digeſted a few days, it becomes turbid, and allows a little æthiops to precipitate. This experiment of the academicians of Dijon proves, that the volatile alkali divides iron in the ſame way as water.

Iron is ſoluble in all the acids. M. Monnet has obſerved, that only boiling oil of vitriol acts upon iron: by diſtilling this mixture to dryneſs, flowers of ſulphur are found ſublimed in the retort, and a white vitriolic maſs ſoluble in part in water, but which cannot give cryſtals, becauſe the heat has decompoſed it. This acid, diluted with two parts of water, diſſolves the filings of iron very well in the cold: the ſolution is accompanied with the diſengagement of a great quantity of inflammable gas. A loud detonation may be produced by approaching a lighted candle to the mouth of the matraſs, after having been ſhut with the hand for ſome time. This gas burns with a reddiſh flame, and frequently gives very ſmall ſparks, reſembling thoſe of the filings of iron. M. Macquer thinks, that in this combination the vitriolic

olic acid disengages a great quantity of the phlogiston of the iron, and that the inflammable gas belongs entirely to this metal. This opinion rests upon this, that this gas may be disengaged from it without any intermedium, by the sole action of the fire: and likewise upon this, that M. de Lassone has procured it from it by the action of the caustic, fixed, and volatile alkalis. The partisans of the new doctrine suppose, that inflammable gas is a modification of sulphur, and belongs to the vitriolic acid, as we have explained in treating of this acid. But the facts which I am going to mention favour more strongly the doctrine of M. Macquer than the pneumatic theory. According as the vitriolic acid acts upon the iron, a portion of this metal is divided, and forms a black powder, taken by Stahl for some sulphur, and which M. Monnet found to be some martial æthiops. It seems that this portion of iron suffered only a mechanical division, resembling that produced by the water. As soon as a part of the iron is united with a part of the acid, although this last be far from being saturated, the solution stops, and no longer acts upon the metal. M. Monnet, who made this observation, says, that upon pouring some water into the mixture, the action of the acid recommences. This phenomenon arises from the water of the spirit of vitriol being combined with the martial vitriol already formed; and the portion of the acid which is not saturated, requires to be diluted with a new quantity of water in order to dissolve the iron. The vitriolic acid dissolves more than the half of its weight of iron. This solution filtrated and evaporated by cooling, gives a transparent salt of a beautiful green colour crystallized in rhomboids. This is the martial vitriol or green copperas.

It is not worth while to make martial vitriol, since nature furnishes abundance of it, and art with no difficulty extracts it from the martial pyrites. It is quite sufficient to leave these pyrites exposed to the air for some length of time: moisture facilitates their decomposition:

position: they become covered with a white efflorescence, which requires merely solution and evaporation for the production of the martial vitriol. This decomposition of the pyrites, according to Stahl, depends on the double affinities. Sulphur is composed of phlogiston and vitriolic acid: neither water nor iron, separately, is able to decompose it; but by the united action of these two bodies, the iron seizes upon the phlogiston of the sulphur, the acid of the sulphur unites with the water, and dissolves the metal; the pyrites, which are less susceptible of efflorescing, as those which are brilliant, being roasted, in order to dissipate a portion of sulphur which they contain, and afterwards exposed to the air, effloresce readily: the vitriol is separated by washing. The solution of this salt deposites afterwards a certain quantity of iron in the state of ochre: it is only when this deposition happens that the liquor is evaporated and crystallised. The partisans of the pneumatic doctrine think, that in the vitriolisation of pyrites, the sulphur which is divided in them, as in the combinations with the alkaline substances, combines with a portion of pure air, and forms oil of vitriol; which diluted with the water of the atmosphere in the state of vapour, dissolves and unites with the iron, at the same time producing heat. The necessity of the contact of the air to favour the efflorescence of the pyrites, gives a certain degree of force to this opinion, as we have observed upon Combustion: but we ought not to forget, that the theory of M. Macquer unites these two doctrines, inflammable gas being disengaged in the vitriolisation in great quantity. We shall take more notice of this fact when we come to speak of the combination of iron with sulphur.

Martial vitriol has a green colour like that of the emerald, and a very strong astringent taste. It sometimes reddens the syrup of violets; but this effect is not constant. Its crystals, according to the researches of Kunckel and M. Monnet, contain more than the half of their weight of water. If it is briſkly heated, martial vitriol liquefies, as all the ſoluble ſalts do, which are more ſoluble in hot than in cold water: by drying, it becomes of a whitiſh grey. If it is heated with a more violent fire, it allows a portion of its acid to eſcape in the form of ſulphureous gas, and it aſſumes a red colour; in this ſtate it is called *colcothar*. Vitriol, calcined to redneſs, very ſenſibly attracts humidity from the air, owing to a portion of vitriolic acid which it contains. Martial vitriol, diſtilled in a retort in a reverberatory furnace, at firſt gives over ſome water ſlightly acid, called *ros vitrioli*. The balloon is changed, to obtain ſeparately the oil of vitriol, which, when the fire is violent, paſſes over black, and emits a ſuffocating odour of volatile ſulphureous acid. Theſe circumſtances depend either, according to Stahl, upon a portion of phlogiſton, which it ſeparates from the iron; or, according to the doctrine of the gaſes, upon this, that it is deprived of a part of its air, which becomes fixed in the iron. Towards the end of the operation, the acid which diſtills aſſumes a concrete and cryſtalline form; it is called *glacial oil of vitriol*. This experiment, deſcribed by M. Hellot, did not ſucceed with M. Baumé; but its certainty is allowed by chemiſts. By diſtilling glacial oil of vitriol in a ſmall retort, ſome ſulphureous gas comes over, and the oil paſſes white and fluid. Should it owe its concrete ſtate to the preſence of this gas? It unites with water, cauſing a great noiſe and heat, leaving the ſulphureous gas to diſſipate. The reſiduum of the diſtilled martial vitriol is red, and called *colcothar*: by waſhing it with water, a white ſalt, little known, may be ſeparated, called *ſalt of colcothar*, or *fixed ſalt of vitriol*. An inſipid red earth remains, which is a pure calx of iron, called *ſweet earth of vitriol*. Martial vitriol, expoſed to the air, grows a little yellow, and is covered with ruſt. Cold water diſſolves half its weight of this ſalt; hot water more; but when it is ſaturated it appears turbid, by means of a greater or a leſs quantity of ochre. This ochre is ſeparated by filtration; and

and the ſolution allowed to cool, produces rhomboidal cryſtals, of a pale and tranſparent green. The ſupernatant liquor ſubmitted to evaporation, affords a new ſet of cryſtals by cooling; and when cryſtalliſation is at an end, a mother-water remains of a blackiſh green or a brown yellow: this, when evaporated with a ſtrong heat, and allowed to cool, forms a ſoft unctuous maſs, which ſtrongly attracts the air's humidity. This maſs, completely dried, yields a powder of a greeniſh yellow. According to M. Monnet, the mother-water of martial vitriol contains ſome iron in the ſtate of a calx. This chemiſt convinced himſelf of this, by diſſolving immediately by means of heat a calx of iron in this acid: this ſolution is brown, and does not cryſtalliſe. The calx of iron may be ſeparated from its acid, not only by the aluminous earth, but alſo by copper and the filings of iron; which does not happen to perfect martial vitriol. A very abundant ſolution of perfect martial vitriol expoſed to the air, in time is changed into a vitriolic mother-water ſimilar to the preceding.

Martial vitriol may be decompoſed by lime and the alkalis. Lime-water poured into a ſolution of this ſalt, forms a precipitate in tufts of a deep olive green: a portion of this precipitate is rediſſolved in lime-water, and communicates to it a reddiſh colour. *Annis* 1777 and 1778, I read two memoirs to the Academy upon the martial precipitates, obtained by the mild or cauſtic alkalis; in which I accurately deſcribed the phenomena of theſe precipitations, and the ſtate of the iron in theſe different circumſtances. I ſhall give the prinpal reſults relative to vitriol. The cauſtic fixed alkali precipitates the ſolution of green vitriol in flocci of a deep green, which in part rediſſolve in the alkali, and form a kind of a martial tincture of a very beautiful red. When we add leſs of this alkali, the precipitate may be collected and obtained in a blackiſh æthiops, if we dry it rapidly and in cloſe veſſels. Without theſe two precautions, the iron very quickly rufts, being divided and moift. The vegetable alkali, faturated with fixed air, forms a precipitate of a greenifh white, which does not rediffolve in the alkali: this difference is owing to the prefence of the chalky acid, which unites with the iron as it feparates from the alkali in its union with the vitriolic acid. The pure or cauftic volatile alkali feparates from the folution of martial vitriol a green precipitate, fo deep that it appears black, which does not rediffolve in the volatile-alkali: we may by fudden exficcation, without the contact of air, obtain it black, and attractible by the magnet. The precipitate formed by the concrete volatile alkali, or by the volatile united with the chalky acid, is of a greenifh grey: it rediffolves in part in this falt, and communicates to it a red colour; which is the reverfe of what happens to the precipitations by the fixed alkali, fince this laft cauftic falt very quickly diffolves the martial precipitate; whilft the fame falt, united with the chalky acid, diffolves it but with difficulty.

The vegetable aftringent matters, as the gall-nut, fumach, the rind of pomegranates, walnut-fhells, Peruvian bark, Cyprian nuts, logwood, &c. have the property of precipitating martial vitriol black. This precipitate, which cannot but be taken for iron, is fo extremely divided, that it remains fufpended in the liquor. When fome gum arabic is added to the mixture, the fufpenfion of the precipitated iron is permanent, and there refults a black liquor, known by the name of *ink*. It is not yet fully known what paffes in this experiment. M. Macquer, M. Monnet, and the greateft part of chemifts, confider the precipitate of the ink to be united with a principle in the gall-nut, which difengages it from the acid. They feem induced to believe, that this principle is in an oily ftate. The gentlemen of the Academy of Dijon, obferving that this precipitation by the galls does not happen in an acid folution, and this precipitate difappears by the addition of an acid, think, that the

the aſtringents direct their action to the vitriolic acid and precipitate the iron pure: becauſe the aſtringent principle has more affinity with this ſalt than the metal has. M. Gioanetti, phyſician at Turin, has made ſeveral experiments upon the iron precipitated from its ſolutions by aſtringents. From a view of his reſearches, given in his analyſis of the waters of St Vincent, that this metal is not attracted by the magnet; that it acquires this property when heated in a cloſe veſſel; that it diſſolves in the acids, but without effervescence; that the ſolutions are not blackened by the gall-nut, which ſhows that the iron is united with the aſtringent principle; and that it is in the ſtate of a kind of neutral ſalt. In the third volume of the Elements of Chemiſtry of the Academy of Dijon, a ſet of experiments is found upon the aſtringent principle of vegetables, which ſeem to aſſimilate this ſubſtance to the acids. Indeed, it reddens the blue colour of vegetables; it unites with the alkalis; it decompoſes the livers of ſulphur; it diſſolves and appears to neutralize the metals; it decompoſes all metallic ſolutions with particular phenomena; it is elevated in diſtillation, without loſing its action upon the metals; and it preſents a great number of other properties, which the order that we have laid down obliges us to paſs over *.

A phenomenon ſtill more difficult to explain than the action of the gall-nut upon martial vitriol, is the decompoſition of this ſalt by an alkali calcined with ox-blood.

 By

* We cannot do better than read and conſider the excellent reſearches of the academicians of Dijon upon the principle of aſtringency. They elucidate the labours of Meſſrs Macquer, Monnet, and Gioanette, upon this important ſubject. However, the ſubject is not yet exhauſted; and it requires a more complete examination, in particular to diſcover the nature of this ſingular principle, which is found in all aſtringent vegetable ſubſtances, and which ſeems to be ſoluble in a great number of menſtrua, ſuch as water, acids, alkalis, oils, ſpirit of wine, æther, &c. See *The Elements of Theoretical and Practical Chemiſtry, &c.* calculated to ſerve for the public courſes of the academy of Dijon, vol. iii. p. 403, to 425.

By this means we get a precipitate of a fine blue colour, insoluble in the acids. This precipitate is called *Prussian* or *Berlin blue*, because it was discovered in that city. Stahl relates, that a chemist, called *Diesbach*, having borrowed of Dippel some fixed alkali to precipitate a solution of cochineal, mixed with a little alum and martial vitriol, this last gave him an alkali, upon which he had distilled his animal oil. This salt precipitated blue the solution of Diesbach. Dippel inquired into the cause of this precipitate, and by a less complicated process prepared Prussian blue, which was announced *anno* 1710 in the Berlin Transactions, but without any explanation of the operation. The chemists, thro' envy, laboured to succeed in the operation; and at last discovered it. It was not till *anno* 1724, that Mr Woodward published a process for the preparation of this colouring substance in the Philosophical Transactions. Four ounces of nitre fixed by tartar, and as much dried ox-blood, are mixed together: this mixture is calcined in a crucible till it be in the form of charcoal, and produces no more flame: it is washed with water in order to dissolve all the saline matter, and the solution is called *phlogisticated alkali:* this ley is concentrated by evaporation: then two ounces of martial vitriol and four of alum are dissolved in a pint of water: the solution of these salts is mixed with the alkaline ley; a blueish deposition takes place; the liquor is to be filtrated, and some marine acid is to be poured upon the precipitate. The precipitate then becomes of a more beautiful and of a deeper blue: it is dried with a gentle heat, or in the open air. Since Mr Woodward, many chemists have studied both the preparation and the theory of Prussian blue. With regard to its preparation, it is now known, that a great number of substances are capable of giving the alkali the property of precipitating the iron blue. Geoffroy, in the Memoirs of the Academy *anno* 1725, has said, that all the coals of animal matters possess this property. M. Baumé asserts, that the phlogisticated alkali

kali may be prepared alſo with the coals of vegetable ſubſtances by means of a more violent heat. M. Spielman has prepared it with the bitumens: Meſſrs Brandt and Monnet with ſoot. The manufacturers of Pruſſian blue are numerous; and each of them, it would appear, employ different matters for its preparation. M. Baunach informs us, that in Germany they uſe the claws of birds, and the horns and ſkin of oxen. All animal matters, however, do not appear to be proper for the phlogiſtication of the alkali. I have in vain attempted to prepare it with ox-bile, by a proceſs reſembling that which is employed with the blood. I obtained an alkali which precipitated vitriol only in a greeniſh white, and the precipitate diſſolved entirely in the marine acid.

Chemiſts have differed greatly about the theory of Pruſſian blue. Meſſ. Brown and Geoffroy conſidered it as the phlogiſticated part of the iron diſengaged by the blood-ley, and intermixed with the earth of alum. The Abbé Menon (firſt volume *des Savans Etrangers*) thought that it was the iron pure, and freed of every foreign ſubſtance by the phlogiſticated alkali. M. Macquer has refuted the opinions of theſe authors in a memoir, which has juſtly merited the name of Maſter-piece from all the chemiſts, and which is inſerted in the volume of the Academy for the year 1752. He thinks that the Pruſſian blue is iron only combined with an exceſs of phlogiſton, furniſhed by the phlogiſticated alkali which received it from the ox-blood. He obſerves, 1. That Pruſſian blue, expoſed to the fire, loſes its colour, and becomes ſimple iron. 2. That it is quite inſoluble in the acids, even the moſt powerful. 3. That the alkalis can diſſolve and become ſaturated with its colouring matter. An alkaline ley requires only to be heated upon it till the alkali acquires no more colour. This alkali, when ſaturated with the colouring matter of the Pruſſian blue, has loſt the moſt of its properties. It is no longer cauſtic; it does not efferveſce with the acids.; it does not decompoſe the ſalts with earthy baſe, but it precipitates all the metallic ſalts; and it ſeems that this decompoſition happens in conſequence of a double affinity of the acid with the alkali, and of the colouring part united with the alkali. The alkali may thus deprive the twentieth of its weight of Pruſſian blue of its colouring principle: it then arrives at its point of ſaturation. The acids diſengage a ſmall quantity of a blue feculent matter from it; and it inſtantly precipitates martial vitriol in the form of perfect Pruſſian blue.

With regard to the alkali prepared in the ordinary way, M. Macquer obſerves, that it is far from being completely ſaturated with the colouring part, and that on this account it throws down at firſt a green precipitate from martial vitriol. The fact is, the portion of alkali which is ſaturated precipitates it blue; but the portion which is not ſaturated precipitates it in the ſtate of ochre, which renders the blue precipitate green by means of the mixture of this laſt colour with the yellow. According to this ingenious theory, the acid which is poured upon the precipitate ſerves to diſſolve and to increaſe the livelineſs of colour of the portion which is not in the ſtate of Pruſſian blue. The alum which is added to the ſolution of vitriol ſaturates the alkali, which is not charged with colouring matter; and the earth of this ſalt, depoſited with the Pruſſian blue, brightens its ſhady colours. As it is neceſſary to pour the acid upon the precipitate of the martial vitriol, in order to brighten the blue, this acid may be added to the alkali previous to its being uſed for precipitation; becauſe the acid, in ſaturating the portion of pure alkali, does not unite with that portion which is united with the colouring part; and thus it may inſtantly form a beautiful blue precipitate. The phlogiſticated alkali may likewiſe be ſaturated with this colouring principle, by digeſtion upon Pruſſian blue till it ceaſe to deprive it of colour. M. Macquer had given this alkali, ſatu-

ſaturated with acid, as a good teſt of the preſence of iron in the mineral waters: But M. Baumé has obſerved, that this liquor itſelf contained a certain quantity of Pruſſian blue; which circumſtance might lead us into miſtakes. He therefore propoſes ſetting it ſome time to digeſt along with a little vinegar, with a gentle heat, in order that it may depoſite all the blue matter which it contains. Such was the excellent treatiſe of M. Macquer upon the Pruſſian blue: but this celebrated chemiſt well perceived a deficiency, particularly with regard to the nature of the colouring ſubſtance. He could no be perſuaded that this was pure phlogiſton; becauſe, under that ſuppoſition, he did not conceive how iron, overcharged with this principle, entirely loſt at the ſame time the property of being attracted by the magnet, and of being ſoluble in the acids; properties, according to Stahl, owing to the preſence of phlogiſton in the metal. M. de Morveau is the firſt who, in his excellent diſſertation upon phlogiſton, has inquired into the nature of the colouring part of Pruſſian blue. From two gros of this compound, he extracted by diſtillation 22 grains of a yellow empyreumatic liquor, which efferveſced with the mild alkalis, reddened blue paper; and of which Meſſ. Geoffroy and Macquer, although they alſo diſtilled Pruſſian plue, have made no mention.

M. Sage, anno 1772, ſent to the Electoral Academy of Mayence a memoir upon the phlogiſticated alkali, which he calls *animal ſalt*. M. Sage thinks that the fixed alkali, treated with blood, and ſaturated by digeſtion upon Pruſſian blue, in M. Macquer's way, is a neutral ſalt, formed by the animal acid and the fixed alkali. By inſenſible evaporation it gives cubic eightſided cryſtals, or priſms with four ſides, terminated by pyramids alſo with four ſides. This ſalt decrepitates upon coals: it fuſes with a violent heat into a ſemitranſparent maſs, ſoluble in water, and proper to make Pruſſian blue. M. Sage thinks that the acid which

neu-

neutralises the alkali in this neutral salt, is the phosphoric acid, because the mixture of alkali and ox-blood, when urged with a strong fire, fuses, emits an acrid vapour, accompanied with white and brilliant sparks; which, according to him, are only some burning phosphorus. This opinion upon the acid of the Prussian alkali would have been corroborated, if, on the one side, by its distillation with charcoal, phosphorus were obtained, which would also take place by making use of Prussian blue; and if, on the other part, Prussian blue were formed by the combination of the fusible or phosphoric salt with base of vegetable fixed alkali with a solution of iron. As M. Sage in his memoir has made no mention of such experiments, his theory cannot be admitted.

Messieurs the chemists of the Academy of Dijon have adopted a part of this doctrine in their elements. They consider the phlogisticated ley as a solution of a neutral salt: they direct the crystallisation of it by evaporation, instead of purifying it by means of vinegar in M. Baumé's way. This salt, say they, is very pure. When it is thrown upon fused nitre, a detonation takes place. They have said nothing upon its decompositions and the nature of its principles; they called it *crystallised Prussian alkali.*

M. Bucquet, upon precipitating by means of the marine acid, and filtrating the ley prepared for the Prussian blue, observed that this alkali, though very clear, and apparently deprived of all the Prussian blue which it contained, allowed, however, a blue powder to precipitate. After having filtrated it more than twenty times in the space of two years, in order to free it of the portion of Prussian blue, which precipitated after every filtration, he at last found that no more was produced by the solution of martial vitriol. I have had in my possession, more than these five years, a small portion of this ley: it has deposited nothing these two years but a slight blueish covering upon the sides

of

of the veſſel which contained it, and it has preſerved a ſimilar colour. I have had occaſion to obſerve this phenomenon twice; ſince which I have underſtood that it was taken notice of by M. Bucquet in his courſes, and I believe that it is conſtant. M. le Duc de Chaulnes ſhowed to M. Macquer a phlogiſticated ley, which gave no Pruſſian blue when it was previouſly mixed with an acid. This chemiſt thinks that the cauſe of this is, that this ley was prepared in metal veſſels. M. Bucquet imagined, from the obſervation related above, 1. That the Pruſſian blue is all contained in the alkali employed to prepare it: 2. That the acids alone are able to precipitate it: 3. That when this alkali, in a longer or ſhorter time, has depoſited all the colouring part which it contains, it at length ceaſes to produce Pruſſian blue.

The Phyſical Journal, April 1778, contains obſervations upon Pruſſian blue, by M. Baunach apothecary at Metz, which greatly favour M. Bucquet's opinion. After having deſcribed the proceſs which is employed in the manufactures of Germany for its preparation, M. Baunach aſſures us, that the ley made in this manufacture by the fuſion of the alkali and of the charcoal of birds claws, and of oxens horns and ſkin, precipitates all the metals, and even the calcareous earth, blue. This alkali diſſolves the metals, after having precipitated them; and they may be ſeparated from it of a very beautiful blue colour by the marine acid. The ſingular facts announced in this Memoir; ſuch as the diſtillation of the Pruſſian blue, produced by this ley, which gives no oil nor volatile alkali; the ſolubility of the blue precipitate, formed by the marine acid when poured upon this ley, in the nitrous acid; the calcareous earth found diſſolved in this laſt-mentioned acid, which deprived the blue of colour; a particular and phlogiſticated earth which it could not diſſolve; do not theſe ſeem to announce, that this blue is not of the ſame nature with that which is precipitated from the ordinary

ordinary phlogiſticated ley, in which M. Macquer found ſome iron, which could come from nothing but from the blood.

From all theſe facts I am of opinion we may conclude, that a great many experiments ſtill remain to be made upon Pruſſian blue, and that our knowledge of it is far from being complete. To what we have ſaid upon this ſubject I ſhall only add a few obſervations, with which an acquaintance is very requiſite.

1. Pruſſian blue, diſtilled with a naked fire, gives a very great quantity of inflammable gas; and at the ſame time ſome oil, concrete volatile alkali, and a little acid phlegm. This gas burns blue like that of marſhes: it has an empyreumatic odour: lime-water gives it the property of burning red and detonating with the air, becauſe it abſorbs the aërial acid which is united with it: it is found in the phlogiſticated alkali. M. de Laſſone looks upon the gas of Pruſſian blue as a particular inflammable gas.

2. Pruſſian blue, after this analyſis, puts on the form of a blackiſh powder, which the magnet attracts. Before it aſſumes this colour, it has an orange one, as M. de Morveau obſerved. This chemiſt was alſo of the opinion, that the Pruſſian blue, which the heat converted into an orange, might be of uſe in painting.

3. The volatile alkali, heated above Pruſſian blue, decompoſes it, ſeizing upon the colouring ſubſtance, and leaving the iron in the ſtate of ochre. M. Macquer, *anno* 1752, announced this fact. Meyer, who followed him, gave the name of *colouring liquor* to the volatile alkali ſaturated with the colouring part of the blue, and adviſed its uſe in analyſing mineral waters. I have obſerved, that when the volatile cauſtic alkali is diſtilled upon Pruſſian blue, the liquor which paſſes over has not the property of giving a blue colour to martial ſolutions; and conſequently the colouring principle is not ſo volatile as the alkali. When only a part of this ſalt has come over by diſtillation, the reſiduum is of an olive-

olive-green: by dilution with diſtilled water and filtration, this liquor is charged with the colouring part, and gives a very lively Pruſſian with martial vitriol.

4. I diſcovered that lime-water, digeſted with Pruſſian blue, diſſolved the colouring matter by means of a little heat. The combination is very rapid. The lime-water becomes coloured, and the blue aſſumes the colour of ruſt. The lime-water, filtrated, is of a clear yellow colour: it does not turn the ſyrup of violets green: it has no longer an alkaline taſte, and the aërial acid does not occaſion any precipitation: it does not unite with the other acids: in a word, it is neutraliſed by the colouring matter of the blue; and by being poured into a ſolution of martial vitriol, it gives an elegant blue, requiring no acid in order to brighten it.

5. The cauſtic fixed alkali, in the cold, quickly deprives the blue of its colour. I have obſerved, that a very briſk heat takes place in theſe experiments; that the alkalis, in their ſtate of purity, attract the colouring matter from a much greater quantity of Pruſſian blue than when they are ſaturated with the chalky acid; and with the martial ſolutions afford a much greater quantity of blue than when in the ſtate of ſaturation with that acid.

6. I have found that magneſia likewiſe has the property of taking the colouring matter from the blue; but its power is much weaker than that of lime-water *.

7. Pruſſian blue, pounded, and projected upon nitre in fuſion, produces ſome ſparks, which denote that it contains ſomething combuſtible.

8. Pruſſian blue, prepared without alum, acquires the property of being attracted by the magnet by means of

* I give here the general reſults only of ſeveral important facts which the decompoſition of the Pruſſian blue, by means of the alkaline ſubſtances, preſented to me: I reſerve a more complete deſcription of them for particular memoirs.

of a ſlight calcination; but that in commerce never acquires this property by the action of fire.

Martial vitriol very eaſily decompoſes nitre. This decompoſition is owing in part to the vitriolic acid, which, uniting with the alkali of the nitre, drives off the nitrous acid; but it is in a great meaſure occaſioned by the reaction of the iron upon the nitrous acid. If for this purpoſe we uſe martial vitriol moderately dried, a very great quantity of very red fuming nitrous acid is obtained; the reſiduum, waſhed, furniſhes vitriolated tartar and ſome fixed alkali; and a ſweet earth of vitriol remains upon the filtre: but if a vitriol ſtrongly calcined has been employed, we get a very ſmall product. This product ſeems to be formed of two liquors; one of which, being of a dull and almoſt black colour, ſwims above another one, which is red and weighty, as an oil would do above water: on this account M. Baumé conſidered this liquor as a kind of oil. A white ſaline maſs paſſes into the neck of the retort, which attracts humidity from the air, diſſolves in water with rapidity and heat, emitting a ſtrong odour of ſpirit of nitre and very thick vapours. This ſolution, ſaturated with fixed alkali, gives vitriolated tartar; hence the white maſs is only oil of vitriol, rendered concrete by a portion of nitrous gas. The weighty liquor of the balloon does not appear to differ in any reſpect from Glauber's ſpirit of nitre; but the light liquor which ſwims above it, being mixed with oil of vitriol, produces a briſk effervеſcence, and even a dangerous exploſion: almoſt all the nitrous acid is diſſipated, and the oil of vitriol remains, but in a concrete and cryſtalliſed form. M. Bucquet, who communicated this diſcovery to the academy, had at firſt obſerved that the concrete oil of vitriol obtained in this diſtillation emits red nitrous vapours when it diſſolves in water. He thought that this acid owed its ſolidity to the preſence of the nitrous gas; and in order to be convinced of this, he tried to mix the blackiſh brown nitrous acid, which ſwam above the red, with very concentrated oil

 of

of vitriol: but in the very moment of the mixture of these two substances, so rapid a commotion took place in the mixture, that the nitrous acid, which was poured upon the vitriolic, was driven to a great distance with considerable noise: the person who made the mixture was covered with it; a great number of red and inflamed pimples immediately rose upon his face, which suppurated like the small-pox. The oil of vitriol became concrete, and entirely similar to that obtained in the distillation just now described. It appears that this acid may owe its concrete state to gases of a different nature; and perhaps that which M. Hellot obtained from martial vitriol, distilled without any intermedium, is it rendered glacial by the presence of sulphureous gas. What remains after the distillation of nitre with martial vitriol calcined to redness, is only a kind of scoriæ of iron, from which we can obtain but very little vitriolated tartar by washing.

We do not know if the solution of martial vitriol might be precipitated by inflammable gas: it seems that in this experiment the iron assumed that which it lost in its solution by the vitriolic acid; since, from M. Monnet's observation, the gas of liver of sulphur, precipitated by an acid, gave the property of furnishing crystals to a vitriolic mother-water.

Liver of sulphur precipitates martial vitriol of a blackish colour. This precipitate is a kind of combination of iron with sulphur.

The nitrous acid is rapidly decomposed by iron, which disengages from it a great quantity of nitrous gas, particularly if the acid made use of is concentrated, and if the iron is divided. In this case, the metal is almost entirely calcined by the air, which it takes from the nitrous acid. The solution is of a brown red; in a certain time it allows the martial ochre to be deposited, because it contains a vercy calcined iron. If, in this state, more iron is put into it, the acid dissolves it, as Stahl has shown, and the calx of iron which it held

held diſſolved is inſtantly precipitated. However, by employing a weak nitrous acid and iron in ſmall bits, a more permanent ſolution may be obtained, in which the metal is more adhering to the acid. This laſt combination is greeniſh, and ſometimes of a clear yellow. Both theſe ſolutions, when evaporated, grow turbid, and depoſite the martial ochre of a red brown. If it is ſtrongly evaporated, inſtead of affording cryſtals, it turns into a reddiſh yellow, which is only in part ſoluble in water, and of which the greateſt portion precipitates. By continuing to heat the martial nitre, many red vapours are diſengaged, the maſs dries, and gives a calx of a red like that of bricks. This maſs diſtilled in a retort, furniſhes a little fuming nitrous acid, much nitrous gas, and ſome aërial acid: no pure nor dephlogiſticated air can be obtained; which is undoubtedly owing to the calx of iron, which is very well calcined in this ſolution, not being reducible. The laſt gas which I obtained, goes to ſupport the conjecture which I have given about the preſence of the chalky acid in the acid of nitre. The calx which remains after the diſtillation of martial nitre, is of a lively red, and might furniſh a fine colour for painting. A ſolution of martial nitre, however copious, did not appear to me to be precipitated by the affuſion of water. The alkalis decompoſe it with phenomena, differing according to their nature. The cauſtic fixed alkali precipitates it in a clear brown; the mixture very ſoon changes to a blackiſh brown, and much deeper than the colour of the firſt ſolution. This phenomenon ariſes from this, that a portion of the precipitate is diſſolved in the alkali, though in very ſmall quantity. The mild fixed alkali throws down a yellowiſh calx, which very quickly becomes of a fine orange red. If the mixture is agitated whilſt the effervefcence goes on, much more of the precipitate diſſolves than of that by the cauſtic alkali. M. Monnet has taken notice of this phenomenon, and attributes it to the gas which is produced. This ſolution

tion of iron by the fixed alkali is called *alkaline martial tincture of Stahl*. It is of a very fine red. For its preparation, M. Baumé recommends taking a nitrous solution of iron, containing but a moderate portion of iron. On the contrary, Stahl advised a very saturated solution. M. Monnet observed, that a yellow solution gave a great deal of precipitate, almost none of which redissolves in the alkali, and which does not colour it as the martial tincture ought to be; whilst a very red solution instantly makes one with the same alkali.

The alkaline martial tincture of Stahl loses its colour in a certain time, and allows the calx of iron which it contains to be deposited. It may be decomposed by means of an acid. The acid of nitre throws down a calx of a red colour like that of bricks, which is soluble in the acids, and called *Stahl's crocus martis aperiens*. The pure or caustic volatile alkali precipitates the nitrous solution of iron of a deep green and almost blackish. Aërated sal ammoniac redissolves the iron which it has separated from the acid, and produces a still more lively red than the tincture of Stahl. This solution of iron might be of great advantage in cases where a powerful tonic and deobstruent are both indicated.

The copious and red martial solution gave me always but very little true Prussian blue by means of the alkali saturated with the colouring matter of this compound. I had only a blackish precipitate, which is redissolved by the marine acid; the liquor had then a green colour.

M. Maret, secretary to the Academy of Dijon, has communicated to the Royal Society of Medicine a very speedy process for making martial æthiops: it consists in precipitating the nitrous solution of iron by the caustic volatile alkali washing and drying the precipitate quickly. M. d'Arcet, appointed by this company to examine M. Maret's process, did not obtain constantly the same result with this physician. In my Memoirs upon the Martial Precipitates, presented to the Academy anno 1776-77, I have determined the cafes in which M. Maret's experiment fucceeded, and in which it did not. For this purpofe, 1. The folution of iron muft be frefh made, and it muft have been made in the cold very flowly with a weak acid, and the iron muft be very minutely divided: 2. That the volatile alkali be recently prepared, very cauftic, and, in particular, by repofe deprived of the fmall portion of calcareous earth and blackifh combuftible matters which ufually rife from the fal ammoniac and lime: 3. That the precipitate be inftantly feparated from the liquor, and rapidly dried in clofe veffels. In fpite of all thefe precautions, this precipitate is fometimes not very black; it gets a light brown colour: its furface may be taken off in fcales, the inferior furface of which is blackifh; which proves that the contact of the air flightly rufts the upper furface. I have obtained an æthiops more beautiful and more certain by precipitating the marine and acetous folution of iron by means of the cauftic fixed and volatile alkalis, and by rapidly drying thefe precipitates after they were well wafhed in clofe veffels. But notwithftanding all this, I think that thefe kinds of æthiops, however pure they are fuppofed, always retain a fmall portion of their precipitants and firft folvents, as M. Bayen has obferved with regard to the precipitates of mercury; and that they ought not to be employed in medicine with as much confidence as thofe which I fpoke of before. M. d'Arcet, in his report to the Royal Society of Medicine concerning M. Maret's procefs, has communicated another of M. Croharé's for making martial æthiops. This apothecary, celebrated for feveral well-executed chemical works, prepares this medicine by boiling water, acidulated with a little nitrous acid, upon iron-filings. The metal is inftantly very much divided, and affords a great quantity of æthiops; but as it might be fufpected to retain a fmall quantity of acid, would it not be furer to prefer

M.

M. Joffe's procefs, which is of very eafy execution, and whofe ufe can infpire no fuch apprehenfion?

The marine acid diffolves iron with rapidity; the folution fends off a great quantity of an inflammable gas, which feems to be produced in greater quantity than from the folution of this metal in vitriolic acid. It likewife has a fmell fomewhat different from that of the vitriolic inflammable gas. Does not this phenomenon feem to prove that one of the principles of the marine acid is a combuftible body? M. Lavoifier inclines to this laft opinion. He thinks that the inflammable gas differs according to the acid employed; and that that which is difengaged from the folution of iron in the marine acid is one of its principles. The marine folution of iron produces a great deal of heat: the heat continues till the acid be faturated: a portion of iron is precipitated in form of a true æthiops, as in all the other folutions.. After it is filtrated, it is of a green colour, inclining to yellow: it is much more ftable than the two preceding. If it is put into a well-corked bottle, which is filled with it, only a very fmall quantity precipitates. I have kept fome of it thefe four years in this ftate, which has fuffered but a fmall depofition of a pale yellow: if, on the contrary, it is expofed to the air, in fome months almoft all the iron which it contains is precipitated of a deeper colour than that of the preceding; and by keeping it in a tight bottle, but which is half empty, it depofites more than the firft, and lefs than the fecond. This experiment, which I made with great accuracy, proves that the air has great fhare in the decompofition of all the metallic folutions. This phenomenon happens in all the folutions of iron, and it is a general obfervation with refpect to thofe of almoft all the metals. Stahl announced, that the marine acid, in its combination with iron, affumed the characters of fpirit of nitre: but this fact has not been obferved by any chemift: it feems that Stahl referred only to the yellow colour of this folution, and to the ſmell which it emitted; a ſmell indeed a little different from that of the pure marine acid. The marine martial ſolution evaporated does not cryſtalliſe regularly. M. Monnet obſerved, that if it is left to cool when it is in a ſyrupy conſiſtence, it forms a kind of maſs, in which cryſtals are ſeen flattened and in form of needles, which powerfully attract humidity. This maſs fuſes with a very gentle heat, and ſeems to merit the name of butter of iron: by a ſtronger heat it is decompoſed, but not ſo eaſily as martial nitre, and it aſſumes a ruſty colour when it is dry. The marine acid ſeparates from it, and may be obtained by diſtillation; which, according to M. Brandt's remark, carries along with it a little calx of iron. M. le Duc d'Ayen, in one of the four excellent memoirs which he has given to the Academy concerning the combinations of the acids with the metals, has at great length examined what paſſes in this decompoſition of the martial marine ſalt in a retort. This operation afforded him very ſingular products: at firſt an acidulated phlegm came over with a gentle heat; the marine acid is then concentrated, and its gas, much more volatile than the water, was in part fixed by the iron. With a ſtronger fire, a little of the acid was raiſed along with ſome iron; and ſome cryſtals, not deliqueſcent, are formed in the balloon. At the ſame time cryſtals are ſublimed along the ſides of the retort, which are tranſparent, very light, and in form of razor-blades, which decompoſed the light like the beſt priſms, and exhibited very beautiful ſhades of red, yellow, green, and blue. At the bottom of the retort a ſtyptic deliqueſcent ſalt remained of a brilliant colour and leafy form, which exactly reſembled the ſpecies of mica in large laminæ, which are improperly called *talc* or *Muſcovy glaſs*. This laſt ſalt, when expoſed to a violent heat in a retort of free-ſtone, is decompoſed, and furniſhes a ſublimation ſtill more aſtoniſhing than the former products. It was an opaque matter truly metallic, which, when examined by the microſcope, preſented

presented regular crystals or edges of hexagonal prisms, which M. le Duc d'Ayen compares to the ornaments in the cielings of rooms. These crystals were as brilliant as the most polished steel, and the magnet attracted them very strongly: this, we see, was iron reduced and volatilised *.

From these accounts, it is evident how fully stored chemistry is with singular phenomena, and how great discoveries this beautiful science promises to those who would incline to make as accurate and complete experiments as M. le Duc d'Ayen has described in his researches.

Let us not forget to observe, that this reduction of the iron favours the pneumatic doctrine; and perhaps similar products might be obtained from many other metallic solutions when treated in the same way.

The solution of martial marine salt is decomposed by lime-water and the alkalis, like all the martial solutions: but these precipitates are less altered, and may be very easily reduced, particularly those which are reduced by the caustic alkalis. I have already observed that this combination gave the purest æthiops by precipitation. Liver of sulphur, hepatic gas, and inflammable gas, decompose it as they do the other two.

Water, charged with aërial acid, easily dissolves iron. For this purpose, it is sufficient to put some filings into this acid spirit, and leave the mixture in digestion for some hours. This liquor, when filtrated, has a sharp and a little styptic taste. Mess. Lane and Rouelle acknowledged this property in the chalky acid. M. Bergman, who calls this combination *ferrum aëratum*, says, that when exposed to the air, it becomes covered with

* I have in my cabinet a black ore of iron, which presents very small brilliant laminæ half a line in breadth, whose form greatly approaches to the crystals obtained by M. le Duc d'Ayen. They are small very thin scales of a very shining iron gray, laid horizontally, which cross one another in all directions, and which are scattered on a reddish opaque quartz, or a kind of coarse jasper. This pretty specimen came from Lorraine.

a pellicle of the colours of the rainbow, which the pure alkalis decompofe, but when combined with fixed air have no fuch effect. This folution turns the fyrup of violets green, and gives Pruffian blue with the phlogifticated alkali; fome martial ochre precipitates, when it is left expofed to the air, or when it is evaporated. I think it fhould be called *creta martialis*. Iron has a a great tendency to unite with the chalky acid. Nature prefents it to us very frequently in this ftate: bog iron ores, fparry iron, appear to be entirely formed of this combination. The martial mineral waters contain the iron frequently in the ftate of creta martialis. This falt, feparated from water, and dry, has but little tendency to diffolve in this fluid: but it diffolves in great quantity in the fpirit of acid of chalk; from which it precipitates in proportion to the volatilifation of the acid.

Iron makes nitre detonate: by projecting into a very red crucible a mixture of equal parts of filings of fteel and very dry nitre, in a little time a very rapid motion commences; many very brilliant fparks are thrown out of the crucible. When the detonation is finifhed, the crucible contains a reddifh calx of iron; a fmall portion of which is combined with the alkali: by wafhing, the water diffolves the alkali, and the martial calx remains upon the filter. It is called *crocus martis of Zwelfer*. It is of a reddifh yellow, not very foluble in the acids. The alkali which is feparated by the wafhing is cauftic, according to the moft of the chemifts, who think that moft of the metallic calces act like pure lime upon this falt charged with aërial acid *.

Iron eafily decompofes fal ammoniac. Two gros of filings of fteel, triturated with a gros of fal ammoniac, expel

* It muft be obferved, that fince the theory of Black upon caufticity, the neceffary experiments have not been made to determine this parity of action between lime and the metallic calces. Nothing complete can be faid upon this point, till experience evince the truth of the fact.

expel all the alkaline gas. M. Bucquet, who diſtilled this mixture in a pneumato-chemical apparatus with mercury, obtained fifty-four cubic inches of an aëriform fluid, the one half of which was alkaline gas, and the other inflammable gas. Four ounces of the ſame filings, and two ounces of ſal ammoniac, diſtilled in a retort with an ordinary recipient, gave about two gros of alkaline ſpirit charged with a little iron, which it very ſoon allows to depoſite in the ſtate of ochre. The reſiduum of theſe operations is a martial marine ſalt. The decompoſition of ſal ammoniac by iron is founded upon this, that this metal unites very well with the marine acid; which is proved by the diſengagement of the inflammable gas which is obſerved in this experiment. In pharmacy, a medicine is prepared with ſal ammoniac and iron, called *flores martiales ſalis ammoniaci*, or *ens martis*. A pound of ſal ammoniac in powder, and one ounce of iron-filings, are mixed together: this mixture is expoſed in an earthen veſſel covered with another of ſimilar materials to a fire capable of reddening the inferior part of this apparatus. In five or ſix hours a yellow matter is ſublimed, which is to be kept in a vial. Theſe are the flores martiales. This ſubſtance is formed in a great meaſure of ſal ammoniac ſublimed with a little calx of iron. As this metal very eaſily decompoſes ſal ammoniac, we muſt uſe but a ſmall quantity, in order that the greateſt part of the ſal ammoniac may be ſublimed without alteration. The portion of martial calx, which is volatiliſed by means of the marine acid and the volatile alkali, gives a colour to the ſal ammoniac.

The calx of iron decompoſes this ſalt better than the metal itſelf does, diſengaging the volatile alkali even without heat. What is obtained by diſtillation is very fluid and ſufficiently cauſtic. I have got volatile alkali which efferveſced with acids, by diſtilling ſal ammoniac with half its weight of crocus martis aperiens. In this experiment the aërial acid diſengaged from the iron unites with the volatile alkali, which is thus rendered effervescent.

Iron is acted upon by inflammable gas: but this alteration, which is very sensible from the colour, has not been examined anent the other properties of this metal.

Sulphur rapidly combines with iron. A mixture of filings of iron and of sulphur in powder, moistened with a small quantity of water, grows hot in a few hours: then it swells, agglutinates, absorbs water, splits with a noise or evident crackling, and emits many aqueous vapours, accompanied with a very evident hepatic odour, somewhat similar to that of inflammable gas. If the mixture is made of a great size, it inflames in twenty-four or thirty hours, and as soon as the watery vapours have ceased. Towards the mutual action of these substances, the heat augments with great rapidity, and the inflammation very soon takes place. The smell is then much more exalted, and seems to be owing to the inflammable gas produced by the reaction of the sulphur and iron. This smell is mixed with that of liver of sulphur and that of pure inflammable gas: and it is undoubtedly this inflammable gas which is disengaged in great quantity, that causes the inflammation observed in this experiment, since the flame is much brighter than that of sulphur. It is driven a foot upwards, according to M. Baumé, who observed this phenomenon from a hundred pounds of filings and as much sulphur; it lasted only two or three minutes. The mixture remained inflamed or red for forty hours. M. Baumé explains this phenomenon by the disengagement of the phlogiston into free fire. Lemery the father called this an artificial volcano; and he imagined, that the fires which are kindled in the interior parts of our globe, and which, raising the surface of it, produce earthquakes and volcanos, were owing to a combination similar to that of the pyrites heaped together and moistened. These terrible effects, as the ſame chemiſt ſays, may be imitated, by burying in the earth a mixture of ſulphur in powder and iron filings reduced into a paſte with water, and covering it up with earth ſtrongly preſſed together. This experiment did not ſucceed with M. Bucquet, who repeated it with great exactneſs: but from Dr Prieſtley's experiments we may conceive the reaſon. This philoſopher has obſerved, that the mixture of iron and ſulphur moiſtened, abſorbed a certain quaintity of air, which is undoubtedly neceſſary to its inflammation. This laſt fact agrees very well with M. Lavoiſier's theory. Accordingly it ſeems, that the ſulphur very much divided by the reaction of the moiſtened iron, attracts a portion of the pure air of the atmoſphere, that it may change into vitriolic acid; then the iron is diſſolved, and produces inflammable gas, which kindles by the heat occaſioned in the mixture. On this account a great quantity of martial vitriol can be extracted from the mixture. There is a great analogy between this combination of the iron with the ſulphur by the humid way and the efloreſcence of the pyrites.

By fuſion ſulphur very eaſily combines with iron, and compoſes a pyritous matter diſpoſed in needles. As the ſulphur in this caſe greatly increaſes the fuſibility of the iron, this metal may be quickly fuſed by means of this combuſtible body. For this purpoſe we muſt paſs a ſmall bar of this metal, heated to whiteneſs, into a roll of ſulphur, and receive the fuſed matter in water. In the water we find black brittle globules ſimilar to the pyrites, and formed like thoſe of ſmall, very elongated, and concentric pyramids.

Arſenic combined with iron gives a coarſe, brittle, and very little known alloy: with cobalt it conſtitutes a mixed metal in ſmall compact grains, hard, and very difficult to break. It is preſumed not to unite with biſmuth.

Combined with antimony, it forms a hard alloy in ſmall facets, which the hammer but ſlightly flattens.

Iron

Iron has more affinity with the ſulphur than with the regulus; conſequently it has the power of decompoſing the antimony. Five ounces of Marechal nails are made red-hot in a crucible, and a pound of powdered antimony thrown in: a ſtrong quick heat is applied in order to fuſe the mixture: when it is well fuſed, an ounce of powdered nitre is projected, in order to facilitate by a proper fuſion the ſeparation of the ſcoriæ from the regulus: the materials are left to cool, and a regulus of antimony is found which does not contain any iron. If one part of iron to two of antimony be uſed, there will be a martial regulus. The ſcoriæ which are found above this laſt regulus, prepared with nitre and tartar, have a yellowiſh colour ſimilar to that of amber: for that reaſon Stahl has called them *ſcoriæ ſuccinatæ*. He preſcribes reducing them into powder, and boiling them in water. This fluid carries along with it the moſt divided part; it is then decanted, filtrated, and the powder remaining upon the filtre detonated thrice with nitre. It is waſhed and dried: and this is Stahl's crocus martis aperiens antimoniatus.

It is ſtill uncertain if zinc can unite with iron. M. Malouin, in his memoir upon Zinc, (Academy 1742), has ſhown that this ſemimetal may be applied, like tin, to the ſurface of iron, and defend it from the contact of the air; which is an indication that theſe two metals are ſuſceptible of combination.

Nickel appears to be capable of a very intimate combination with iron; theſe two metallic ſubſtances, according to M. Arvidſon, never admitting of an entire ſeparation.

Mercury in its metallic ſtate forms no union with iron. Their immediate combination has been in vain attempted; but they unite in the form of a calx. M. Navier has obſerved, that a ſnowy white precipitate was obtained from the admixture of a ſolution of iron and of mercury by the vitriolic acid; the mixture evaporated forms flat very light cryſtals reſembling ſedative ſalt, falt. M. Navier has afcertained that thefe cryftals are combinations of iron and mercury.

Lead contracts no union with iron.

It is not known if iron and tin can unite by fufion. The covering the furface of iron with a coat of tin, or the preparation of white iron, feems to fhow that this combination takes place, as we are going to fee. To tin the iron, the furface requires to be very convenient and brilliant; for which purpofe it is fubjected to the action of an acid or of a file, or of fal ammoniac; then it is plunged vertically into a veffel full of fufed tin: it is turned up to increafe the thicknefs of the coat; and when it is fufficiently tinned, it is taken out and rubbed with faw-duft or bran, to carry off the greafe or pitch with which the fufed tin was covered, and which had adhered to the furface of the tin-iron. If the iron to be tinned is reduced into fmall plates, the tin will not only adhere to the furface, but will penetrate it, and combine with all its parts; and upon cutting it, the fame white colour will be obferved in its middle as at its furface; which fhows that white iron, well prepared, is a true chemical combination: befides, it is more malleable than iron; and of it veffels are made of a form which this metal in its pure ftate could not be made to affume.

The ufe of iron is fo extenfive and fo well known, that it were ufelefs to infift upon it. It is fufficient only to know, that no art can difpenfe with it; and, as M. Macquer expreffes it, it is the foul of all the arts. The different modifications it can affume make it a fubftance very adequate for the multiplicity of ufes to which it is deftined. The font ferves to fufe utenfils which are more or lefs folid, or have more or lefs refiftance, as neceffity requires. The hardnefs and tenacity of forged iron very properly fuit the various purpofes to which it is applied. The fame may be faid of the different forts of fteel: the finenefs of the grain and the temper conftitute many kinds of it; all of which are made ufe of in the different arts to which they are adapted. The calces of iron ferve to give a red or brown colour to the porcelains, the earthen wares, the enamels, and other fubftances. They are employed alfo in the preparation of the artificial precious ftones, and combined with oil in painting.

In medicine iron is an important remedy, productive often of the moft falutary effects. It is the only metal which contains nothing hurtful, and whofe effects need not be dreaded. It has even fuch an analogy, as we have already feen, with the organic matters, that it feems to make a part of them, and frequently to owe its production to the animal œconomy or to vegetation. The effects of iron upon the animal œconomy are fufficiently numerous. It ftimulates the fibres of the membranous vifcera, and feems to act in a fpecial manner upon thofe of the mufcles, whofe tone it augments. It ftrengthens the nerves, and gives the weakened machine a remarkable energy and vigour. It excites feveral fecretions, particularly thofe of the urine and fanguine evacuations. It provokes natural hemorrhagies, as the menftrual and hemorrhoidal fluxes. It increafes and multiplies the contractions of the heart, and confequently the force and quicknefs of the pulfe. It acts upon the fluids in no lefs degree. It eafily paffes into the circulation, and combines with the blood, to which it gives denfity, confiftence, colour, and a greater concrefcibility; at the fame time it communicates to it fuch an activity, as eafily to pafs into the moft minute veffels, ftimulating the fides of the canals which contain it, fupporting the energy and life. The beautiful experiments of M. Menghini, publifhed in the Memoirs of the Inftitutes of Boulogne, have proved, that the blood of perfons who have made ufe of iron is more coloured, and contains a greater quantity of iron, than it contains naturally. M. Lorry, who in the practice of medicine has cultivated this nicety of obfervation, and the pretty remarks which characterife the knowing phyfician and philo-

philosopher, obferved, that the urine of a patient to whom he adminiftered iron in a very divided ftate, was manifeftly coloured with the gall-nut. This metal, then, is a tonic, corroborant, ftomachic, diuretic, alterant, and an incifive remedy, whofe action comprehends the properties of a great number of medicines. Like aftringents, it braces the fibres, and increafes their ofcillation; and is preferable to other remedies which enjoy the fame virtue, on account of the greater certainty and durability in its effects; becaufe it combines with the organs themfelves by the intermedium of the fluids, which ferve for their nutrition. On thefe accounts, it is proper in all the cafes where the fibres of the vifcera, of the mufcles, and even of the nerves, have their action weakened; in the weaknefs of the ftomach and inactivity of the inteftines, and in the difeafes arifing from thefe caufes; and, laftly, in all cafes where the fluids have not the proper confiftence and concrefcibility, but are too watery, as in pale habits, and in thofe difpofed to dropfy, &c. It is employed under many different forms; as the porphyrifed limatura martis, martial æthiops, crocus martis aperiens and aftringens, tinctura martialis alkalina Stahlii, flores martiales. To thefe medicines might be added, perhaps, the iron precipitated from the acids and rediffolved by the volatile alkali, fal martialis fedativus, and the Pruffian blue recommended by the academy of Dijon. Externally, martial vitriol is ufed as a ftyptic in hemorrhagies and the like.

Iron, poffeffing the power of magnetifm, or the artificial magnet, produces alfo fingular effects upon the animal œconomy. Applied to the fkin, it relieves pains, allays convulfions, excites rednefs, fweat, and frequently alfo an eruption of fmall pimples: it feems alfo to impede the acceffion of epileptic fits. Left in water for a dozen of hours, it is faid to communicate to it a purgative property. Although all thefe facts require to be confirmed by many experiments, it cannot be doubted that

that the magnet has a very ſenſible virtue. M. Thouret phyſician of the Faculty of Paris, and of the Royal Society of Medicine, has communicated, in the firſt volume of this laſt company's hiſtory, a beautiful obſervation relative to this ſubject. A patient at Rouen complained of a fixed pain in the different branches of the ſeventh pair of nerves diſperſed upon the cheek, by applying a magnet to the different parts of this region, the ſkin ſeemed to make reſiſtance to the magnet. There is no queſtion that new obſervations will confirm theſe diſcoveries, and elucidate the practice of phyſic; which ſome perſons have been inclined to render more intricate by making it appear myſterious *.

* The iron of Great Britain is made from three different kinds of minerals. From iron-ore, commonly called *Lancaſhire ore*, from the name of the county where it is found in greateſt abundance. This ore is very heavy, of a fibrous or lamellated texture: the colour is dark purple, approaching to a ſhinning black, and when reduced to powder it becomes of a deep red: it lies in veins like the ores of moſt other metals, from which it is dug in a ſimilar manner. The bog-ore reſembles a deep yellow ochry clay, and ſeems to be the depoſition of ſome ferruginous rivulet, whoſe courſe had formerly been over the ſurface of a flat marſhy plain. It lies in beds of an irregular thickneſs, commonly from twelve to twenty inches, and is very various in its breadth from ſide to ſide; ſo that it never forms a uniform extended bed of any great dimenſion. Miners, who are accuſtomed to ſearch for bog-ore, know where to dig, by examining the nature of the plants which grow upon the ſurface, as they always indicate the quality of the ſoil which is beneath them. The iron-ſtone has no regular appearance, and does not in the leaſt reſemble a metal externally. It lies in beds of great extent like other ſtony matters, and is commonly ſtratified with ſeams of pit-coal, forming alternate layers; though theſe are not always rich enough to repay the expence of working. Such are the chief differences of the three principal minerals from which iron is obtained; although there is an infinite variety in the quality and appearance of the iron-ore and iron-ſtone.

The ores of iron are in general roaſted before they are put into the furnace for fuſion. This operation is not always performed in order to free the minerals from ſulphureous and other volatile matters, as many of the ores contain no ſuch matters. But then the proceſs of roaſting calcines the adhering matrix, makes the ſiliceous earth crack, and renders the whole maſs ſo friable, that it may be

eaſily

eaſily broke into fragments of a convenient ſize for melting. After the mineral is duly prepared, it is next ſubjected to the moſt intenſe heat which the force of fuel can raiſe in a furnace; and if the quality of the mineral requires an addition, a certain portion of limeſtone is added to promote and facilitate the fuſion of the metal. The furnaces which are erected for melting iron, conſiſt of a large capacious cavity, from ſixteen to twenty-five feet high, and from ten to fourteen wide: the moſt approved ſhape nearly reſembles that of a hen's egg, with the largeſt end undermoſt; below which there is a ſquare cavity to contain the melted metal, and at top a very ſhort vent about twenty inches in diameter. The inner wall is built of fire-ſtone, which indures a ſtrong heat with little riſk of melting, and has all the joints cemented with mortar compoſed of ſand and clay. This is ſurrounded with more building, which deviates more and more from a circular form, and the whole is incompaſſed with a ſquare building, which is about twenty feet at the baſe, and gradually converges to the top. This outer wall conſiſts of very large ſtones, and is ſometimes bound round with large beams of wood to give additional ſtrength, and to fortify the whole building. The foundation of ſo large a maſs muſt be quite ſolid. But under the hearth a void ſpace is left, by means of an arch, to prevent any watery moiſture from cooling the bottom of the melted metal. The hearth is compoſed of a large flat ſtone; and upon every ſide of it a ſtone is ſet on edge about eighteen inches deep, which forms the ſquare cavity made at the bottom of the furnace to collect the metal. One of the ſtones is ſomewhat ſmaller than the reſt, which leaves an opening above it in order to give vent to the ſcoriæ after the metal has riſen to a certain height. A little above the edge of this cavity, ſoon after the body of the furnace begins to ſwell, a hole is made in the wall to permit the noſel of the bellows to enter, and a proper vacuity is left in the ſame ſide of the building to lodge the body of the bellows. It is a matter of ſome nicety to proportion all the parts to each other ſo that the furnace may work well: and the principle upon which the perfection of a furnace depends is ſtill ſo little underſtood, that the conſtruction of iron furnaces is varied in different countries, more from caprice than from judgment and experience.

After the whole ſtructure is completed, the great cavity is filled with charcoal, which is kindled at the top, to heat the body of the furnace gradually. When the charcoal has burnt down a certain length, the founder begins to charge the furnace with proper materials for working. This charge conſiſts of charcoal, limeſtone, and the irony mineral mixed together in a juſt quantity. The founder muſt be extremely attentive that none of the materials exceed the proportion which it ſhould bear to the reſt. If the quantity of charcoal be too large, the exceſs burns in pure loſs, the iron is too much ſcorified, and does not poſſeſs the ſame excellent quality it would have done, had the proportions been more juſt. If the

char-

charcoal be added in too ſmall quantity, then the iron is not ſuſſiciently purged from the heterogeneous matters which vitiate the quality, the matrix is ſtill left crude, full of large brilliant laminæ and ſulphureous particles. Beſides, when the furnace has been once overcharged with the mineral, it never can afterwards be brought to conſume ſo large a quantity as it would otherwiſe have done. For this reaſon the founder muſt attend the furnace with aſſiduity and care during the firſt days of working, in order to learn the trim; and he ſhould all the while be attentive to augment the proportion of the mineral gradually, ſo that the furnace may never be overcharged by the injudicious addition of too large a quantity at once.

The moſt intenſe heat is excited oppoſite to the blaſt of the bellows; and as the mineral deſcends gradually from the top of the furnace, it is prepared for fuſion before it reaches the blaſt. The chief part of the operation, the converſion of the ore into metal, goes on at this point: and as ſome eſtimate of the perfection of the proceſs may be formed from inſpection, the founder often looks through the hole where the bellows enters, to obſerve what paſſes within. He there perceives the metal falling down in drops like rain, which collect in the baſon at the bottom of the furnace. The pure metal ſinks below from its ſuperior gravity, while the ſlag or ſcoriæ, which conſiſt of the earthy matters, converted almoſt into a vitreous maſs by the force of the heat, float atop of it. At certain intervals, the metal collected in the baſon will become agitated with great inteſtine commotions, boil and bubble up, and emit vivid ſparks, which ſeem to deflagrate in the air. Thoſe fits of ebullition will continue for half an hour at a time, and return at ſtated periods with wonderful regularity. They are ſuppoſed to purify the metal, though the cauſe of the phenomenon is not well underſtood. Sometimes the furnace will take a ſcouring, and melt away, till a large excavation is formed on the ſide of it, without any perſon being able to aſſign a reaſon why it ſhould begin, or why it ſhould ceaſe. The excavation moſt commonly occurs in the loweſt part of the wall oppoſite to the blaſt, or on the higher part of the wall on the ſame ſide with the bellows.

An intelligent workman can judge, with great correctneſs, of the perfection of his *font* (the term uſed for the quantity of metal run from the furnace at once), from obſerving the appearance of the flame which iſſues from the top of the furnace, and from inſpecting the colour and conſiſtence of the ſcoriæ which flow from the ſurface of the metal. When the flame riſes to a great height above the furnace with clouds of thick ſmoke, attended with a number of ſparks ſhooting to a diſtance in the air, we may expect an ebullition on the hearth. When the anterior wall is covered with a brown and ſmoaky colour, we may be aſſured that the minerals are not completely fuſed, and that the pure, heavy, metallic parts are not tho-

thoroughly separated from the light impure scoriæ: but, on the contrary, a green tint, verging to a white, denotes an excellent state of the fusion. With regard to the appearance of the scoriæ, if they seem full of brilliant particles or scales, if they run white from the hearth, especially towards their extremities, or if the scoriæ be very light and limpid as water, with a small degree of tenacity, and if they harden immediately upon exposure to air; all those circumstances mark a heat too much concentrated, which must be tempered by the addition of more mineral. If the scoriæ are opaque and black, there has been a deficiency of charcoal; but if they be black, resembling the colour of iron, and at the same time light, we may conclude that a just proportion of charcoal and mineral has been employed. Yet, after all, the most expert workmen agree, that the nature and quality of the metal is affected by the occurrence of a thousand unknown circumstances, which we do not thoroughly understand, and which it is by no means easy to investigate.

A good-going furnace will often yield from three to four tons of iron in the course of twenty-four hours. The metal is let out commonly at three intervals, by driving a hole with a punch contiguous to the small stone of the bason; and when the whole has run out, the hole is well shut with clay until there is occasion to open it again. This operation is called *tapping the furnace*. The metal is run into moulds of whatever shape and size the workman wishes. The bellows is stopped during the flow of the metal, and opened immediately after, as the furnace is continually supplied with a fresh charge when any part of the former is exhausted. In this manner it is kept working perpetually, until part of it fall into decay by use.

Iron that flows from the furnace is called *cast iron*, *pig iron*, or *yetlin*. It is very hard and brittle, does not admit of extension under a hammer, and commonly receives but little impression from the teeth of a file. Those properties are directly contrary to the properties which malleable iron possesses: for malleable iron is neither brittle nor hard, admits of great extension under a hammer, and may be shaped into any form by means of a file; but is altogether infusible in the greatest heat which it is in the power of fuel to excite, however long it be exposed to its action.

The difficulty of making good cast iron is so very great, that the most sagacious workman cannot always promise to run every font of the same quality: there is even a difference in the quality of the different parts of one font; and this is sometimes so remarkable, that one half of a kettle proves hard, while the other half proves soft, although the whole metal be run out at once from the furnace. Variations, so great as this, are attended with great practical inconveniences; so that it becomes an object of consequence to correct the fault completely.

In order to ascertain how long iron retained its softness, M. Reaumur kept ſome ſoft iron in fuſion, taking out a ſmall portion at different intervals to try the quality. What was taken out firſt, proved ſoft when cold; after a quarter of an hour it ſtill yielded to the file, but with more difficulty; after remaining fuſed another quarter of an hour, it proved hard, and did not yield to the file at all. M. Reaumur found in the trials which he had made for this purpoſe, that caſt iron, remelted in a crucible with animal bones and a ſmall quantity of charcoal-duſt, was rendered ſoft and uniform in its quality. He found, at the ſame time, when kept in fuſion without addition, the contrary effect took place, and that ſoft iron became even brittle by this proceſs. In the courſe of examining the beſt method of ſoftening caſt iron, M. Reaumur at laſt diſcovered, that the mere application of heat was alone ſufficient, provided the requiſite degree was employed. This was ſomewhat nice to regulate; for if the heat did not exceed a cherry red, no length of expoſure was of the ſmalleſt ſervice: The operation began as the heat increaſed, and ſucceeded beſt when it approached to the fuſing point; for thoſe plates were moſt completely ſoftened whoſe edges had begun to melt. In theſe experiments there was an opportunity to obſerve the progreſſion of the change, which began at the ſurface, and proceeded gradually towards the centre. The outſide was ſometimes fairly converted into ſoft malleable iron, while the internal parts retained their original texture. The diſtinction is ſufficiently marked. Malleable iron is of a lamellated, caſt iron of a granulated, texture. The good or bad quality of the caſt iron is determined from the colour, ſize, and diſpoſition of the grains. What is in general moſt eſteemed has the grains fine, diſtinct, apart from one another, and approaches nearly to the appearance of ſteel tempered with a gentle heat: and with regard to colour, the browneſt are the moſt eaſily kept ſoft. That which has a black colour with a fine diſtinct grain is the moſt malleable of all. Of two different fonts, one ſhall be of a lighter grey, and better grained; the other ſhall be of a deeper black, with larger and leſs diſtinct grains. In this caſe we prefer that which has the moſt perfect granulation: for in common we lay much more ſtreſs upon the texture of the grain than upon the tint of colour. After all, we muſt confeſs, that thoſe marks often prove fallacious, this metal is ſo various in its nature.

Caſt iron is extremely valuable from the extent of its uſe to the purpoſes of life. By the improvement of the manufacture, it is formed into a great variety of utenſils. It is employed for ſmall or large work, from the leaſt-ſized tacket up to the largeſt piece of ordinance weighing many tons.

Caſt iron, as was before obſerved, is not malleable. In order to make it ſo, the Swedes, who are reputed to make the beſt bar-iron, purify the font by repeated fuſions. The firſt part of the proceſs is to place the iron upon a mixture of charcoal-duſt, and the ſcoriæ of former meltings, before the blaſt of a bellows driven by water. The

The blaſt plays into a cavity about eighteen inches deep, and is directed ſo as to ſtrike the angle where the bottom and oppoſite ſide join. The cavity is filled till within one third of the top with the mixture of charcoal-duſt and ſcoriæ, upon which the maſs of iron is placed obliquely, in order that it may be parallel to the direction of the blaſt, which plays chiefly upon the under ſide; and the metal ſtands about ſix inches diſtant from the nozel of the bellows. The whole is then covered with charcoal, which is kindled below, and the bellows is ſet a-working. The ſcoriæ promote the fuſion of the iron, which melts at the under ſide; and as it runs down the workmen puſh it nearer the bellows, that the melting ſurface may be all the while retained at the ſame diſtance from the blaſt. When all the metal is melted, it begins to ſwell and bubble at the ſurface, and will ſometimes riſe above the edge of the cavity. The tenuity of the fuſion muſt be ſuch as will permit the foreign matters to ſeparate from the metal. This kind of boiling is continued about half an hour, prolonged at the diſcretion of the workman, if he judge longer time requiſite from the quality of the metal. Once the metal is ſo cool as to fix, it is ſubjected to the ſame operation, repeated almoſt in a ſimilar manner. In this ſecond operation the workman is attentive to expoſe the upper ſurface of the metal firſt to the influence of the fire, and takes care to turn the iron, ſo that the whole maſs may melt uniformly. After the maſs melts, it throws off ignited charcoal-duſt and ſcoriæ, forming whirls of fire all over the forge. This is ſuppoſed to be a true purification of the iron, which laſts about ſeven minutes, and the whole proceſs conſumes about two hours. The next ſtep is to beat the iron with a forge hammer, when a great deal of ſcoriæ and other impurities are expelled from the metal. About 350 lbs are commonly prepared at a time, and about 24 tons of charcoal are allowed to the workmen for every 400 lbs of malleable iron. By this proceſs caſt iron is rendered fit for the uſe of the blackſmith; and if the ſucceſs has been complete, the iron is alike ductile when it is hot and when it is cold. But there are two imperfections which frequently degrade the quality of the metal. One is a great degree of fragility when hot, though it is ſufficiently ductile and tenacious when cold. The other is the converſe of this, ductile and tenacious when hot, fragile when cold. Iron with thoſe faults is called by the workmen *red ſhort* and *cold ſhort iron*; and that which is cold ſhort is accounted the worſt and moſt troubleſome fault. They ſeem ſo far to depend upon the nature of the ore, that a founder cannot always correct them by the moſt dextrous and judicious management. Faultleſs iron has been long ſought after; and of late a method has been diſcovered of converting yetlin into malleable iron with great eaſe and certainty. The improvement conſiſts in running caſt-iron into an air-furnace, where it is kept about the conſiſtence of wax by a proper regulation of heat. The furnace is open at both ends, and the flame of the fire blows over the surface of the metal. The side-walls are perforated with round holes, through which the workmen keep pockering the metal with iron rods. During this operation the iron is continually emitting a blue vapour; the nature of which has not yet been examined with care. There is no violent ebullition, nor any other remarkable appearance takes place, while the conversion is going on, which is thoroughly completed in the space of two hours with a mass of iron weighing 600 lb. The malleable iron obtained at the end of the process has all the qualities of good metal, and can be made from pigs of all kinds; which is not the case in the other ways of working. Upon what principles the change happens, is not thoroughly understood; though it seems sufficiently evident that the effect of heat is to expel some volatile matter from the cast-iron, which flies off under the form of blue vapours.

As the erection and support of a blast to fuse iron is attended with considerable expence, some improvers have tried to melt down the iron stone by the force of such a current of air as could be raised by the construction of the furnace. An attempt was lately made near Edinburgh, with great hopes of success; but, after the fire had lasted for a week, no fusible iron was procured. There was a small quantity of malleable iron at the bottom, and a large quantity of metallic matter remained in the slag. This experiment shows, that it is more easy to make malleable iron than cast iron. It is probable, that, some-time back, no other kind of iron could be made; and this conjecture is confirmed by the richness of the slags which are found near old iron-works: these are now wrought over again in many places with sufficient profit. At the furnaces in Navarre, the iron is not brought into fusion, but reduced from the ore in a kind of furnace, open at top, and carried from that directly to the forge-hammer, without the intervention of any intermediate process.

Two separate pieces of malleable iron unite perfectly into one, when made red-hot in the fire and hammered on the anvil. This operation is called *welding*, and the heat requisite is called a *welding heat*. It is customary to throw some sand into the fire along with the iron, which vitrefies the scoriæ, and keeps them so fluid, that they are squeezed out with the blows of the hammer; and thus the clean surfaces of the heated iron are permitted to approach each other. In this way anchors for large vessels are made, up to the weight of two ton. Soft iron is also used for the fabrication of the smallest instruments.

Although cast and forged iron be both of them extremely valuable, yet we find steel possess properties far superior to either. The excellence of steel consists chiefly in the great firmness of cohesion, and in the facility with which it may be tempered to any degree of hardness or of softness, without having its strength impaired. Steel of the best quality is prepared from the finest forged iron, by cementation

mentation with charcoal-duſt in a ſtrong heat. The bars of iron are imbedded in the charcoal-duſt, which lies in a ſtone-box raiſed from the floor on treſſels. Theſe boxes are placed in a kind of vaulted furnace, where a fire is kindled on the floor, which plays round them, and keeps them hot for ſome days. One end of the box has a hole in it, at which a bar can be occaſionally taken out, to obſerve what progreſs has been made in the converſion into ſteel. When that is complete, the fire is allowed to decay, ſo that the boxes cool gradually; which occupies a longer time than the cementation did. In this way forged iron is converted into ſteel. This mode of working is the moſt generally practiſed, and the moſt approved; though ſteel may be alſo made directly from crude iron by fuſing it with certain mixtures. Steel is accordingly made after this way in many parts in Germany. But it is obſerved, that all kinds of ores do not afford ſteel of equally good quality; ſo that the workmen call thoſe ores which yield the beſt ſteel, *ſteel ores*. In general, ſteel obtained by this proceſs is not ſo good as what is made in the ordinary way; ſo that we muſt be careful not to confound it with that ſort which is called *melted ſteel*, as this is prepared by fuſing ſteel made from bar iron; and the fuſion improves the quality, by favouring the eſcape of any heterogeneous matter, and by rendering every part of the grain of an equally fine texture.

After the ſteel is formed, it is made to receive a temper, by firſt rendering it perfectly hard, and then reducing it to the exact degree of ſoftneſs which the workman wiſhes it to retain. When ſteel is made red-hot in the fire, and then plunged ſuddenly into cold water, it becomes ſo exceſſively hard, that no tool is able to cut it: but if, inſtead of being inſtantly cooled, the ſteel is allowed to loſe its heat gradually, it then is found in the oppoſite extreme, and is perfectly ſoft: therefore, by combining theſe effects in different degrees, the ſteel may be brought to any temper the workman pleaſes. The firſt ſtep of the proceſs is to make the ſteel perfectly hard, and afterwards to ſoften it down to any temper we chooſe. The workman then gives the ſteel the hardeſt temper; next he returns it back to the fire, and gives it a gentle heat, if he means the hardneſs to be but little diminiſhed; if he intends to make it ſoft, he allows it to remain a longer time; and various tints of colour which heat makes ſteel to aſſume, enable the workman to judge of the temper with perfect exactneſs. In a gentle heat, ſteel acquires a yellow hue; next it becomes purple; then blue; and laſt of all, whitiſh grey, before all ſhades of colour vaniſh in a red heat. When the ſteel is removed from the fire, in the firſt ſtage it is extremely hard; in the ſecond, it is fit to be formed into tools for working upon metals; in the third, when blue, it becomes very pliant and elaſtic, and is of a proper temper to make ſprings, and to fabricate inſtruments for cutting all ſorts of ſoft materials. A bar of ſteel may be made to aſſume all theſe colours at once, by placing one

 end

end in the fire and keeping the other at a distance. As steel is the only substance from which tools can be made, it is not only highly useful, but it is most indispensably requisite, in the practice of every art; and it well deserves the name of *mother of the arts*, which it has so often obtained.

Having thus endeavoured to exhibit a general view of the manufactures of iron through all the steps of the process, it will in the next place be proper to attempt an investigation of the causes which produce the difference between cast iron, forged iron, and steel. In doing this, we shall follow chiefly the doctrine which M. Bergman delivers, in his Treatise on the Analysis of Iron.

The first observation to be made is, that if there be a piece of each of the three preparations of iron, dissolved in the vitriolic acid, the proportion of residuum left undissolved varies exceedingly in each of them; cast iron containing the largest share of insoluble matter, steel the next, and forged iron least of all; and that, of whatever foreign matters the residuum was composed, it always contained a portion of plumbago, upon whose properties the conversion of cast iron into steel and malleable iron is to be explained. Plumbago is, from the latest examination, supposed analogous to sulphur in its composition, or to be in reality nothing but a species of sulphur composed of aërial acid and phlogiston; the aërial acid combining into a solid concrete, to form plumbago, in the same way as vitriolic acid unites with phlogiston in the formation of common sulphur. The proof of this composition is derived from the analysis of plumbago by Mr Scheele, who, in all his experiments, had reason to infer, that plumbago was a compound solely of aërial acid and phlogiston, and that it was liable to a similar resolution and recomposition with vitriolic sulphur. It is therefore probable, that plumbago will be decomposed, in all cases where a third substance is presented to its action, which attracts the phlogiston more strongly than the aërial acid does; then the plumbago will be decomposed, the aërial acid will escape, and the phlogiston will attach itself to the iron. and saturate every part of it which was not thoroughly metallised before. Now, in a vast variety of trials, M. Bergman constantly found, that cast iron contained a large quantity of plumbago; steel, much less; and forged iron, scarcely any. He also found, that forged iron contained the largest share of phlogiston, cast iron the least, and steel an intermediate quantity. Here, then, we perceive, that the species of iron which contains the largest share of plumbago, contains the smallest share of phlogiston; and that the species which is fully enriched with phlogiston is quite purged of plumbago. From this circumstance M. Bergman infers, that the difference between crude and forged iron consists in the proportion of phlogiston, and in the presence or absence of plumbago: crude iron is less rich in phlogiston, and contains plumbago: forged iron is saturated with phlogiston, and free from the adulteration of plumbago.

The

The internal change which takes place in the converſion of crude iron into forged iron, is ſuppoſed to ariſe from the decompoſition of the plumbago. The heat, diminiſhing the coheſion of the conſtituent parts of the plumbago, enables the phlogiſton to unite with the metallic earth, and expels the whole aërial acid. Thus the plumbago is decompoſed, and the crude iron ſupplied with a competent ſhare of phlogiſton. Before this decompoſition, the crude iron had been deficient in its quantity of phlogiſton; in conſequence of it, it is provided with abundance, and brought to the ſame point of ſaturation with malleable iron: ſo that, according to M. Bergman, the decompoſition of plumbago, the expulſion of the aërial acid, and the adheſion of the phlogiſton to the impoveriſhed metal, changes crude iron into malleable iron.

In the formation of ſteel, this reſolution is not altogether ſo complete, though the metal has loſt part of the plumbago, and acquired a larger ſhare of phlogiſton; for we muſt remember, that ſteel is in an intermediate ſtate between caſt and forged iron. Experiments, by ſolution in acids, ſhow, that ſteel contains more phlogiſton than caſt iron, but leſs plumbago; and, on the contrary, that it contains leſs phlogiſton than forged iron does, but more plumbago. If, therefore, the proceſs of converting caſt iron into ſteel be ſtopped in the middle, we ſhall have a ſpecies of ſteel. Accordingly, caſt iron may be converted into ſteel by the mere application of heat. But as the fineſt ſteel is made from forged iron, it may be aſked, how the plumbago is to be found in the iron which contains none? To ſay that it attracted plumbago ready formed, where none exiſted before, would moſt certainly be abſurd; but to ſuppoſe that plumbago is regenerated by the union of its integrant parts, involves no contradiction. This is the conjecture employed to ſolve the difficulty. The aërial acid of the charcoal being expelled by the force of heat, combines with the redundant phlogiſton of the forged iron; and this compoſes plumbago: and the iron being thus in part robbed of its phlogiſton, and at the ſame time furniſhed with plumbago, is converted into ſteel. All the experiments upon iron correſpond with this doctrine. M. Reaumur found, that, during the proceſs of ſoftening crude iron, the outſide, which was moſt expoſed to the heat, was completely changed into the nature of forged iron; the intermediate ſtratum was converted into ſteel, while the more internal part was not affected in the leaſt.

Such is the outline of M. Bergman's doctrine, which is by far the moſt elegant and ingenious which has ever yet been offered upon the ſubject of the converſion of iron and ſteel; and as ſuch deſerves much recommendation.

But perhaps M. Bergman's inveſtigation of the cauſe of the cold ſhort quality of iron is ſtill more ſucceſsful. It had long been queſtioned, whether this imperfection aroſe from ſome inherent defect in the nature of the metal itſelf, or from the admixture of ſome heterogeneous

terogeneous

terogeneous ſubſtance which debaſed the quality. M. Bergman obſerved, that a ſolution of cold ſhort iron in vitriolic acid, largely diluted with water, began in the courſe of a few hours to depoſite a white-coloured ſediment when the ſolution was expoſed to the acceſs of air. If this ſediment was collected before any ochry precipitate followed, it was diſcovered to poſſeſs properties peculiar to itſelf. By proper treatment with reducing fluxes, it was convertible into a new metal, totally diſtinct from any metal hitherto known. To this metal M. Bergman gave the name of *ſiderum*, from its intimate connection with iron. When any portion of this ſiderum was added to pure malleable iron, it became cold ſhort iron; and when the white calx was ſeparated from cold ſhort iron, it then became perfectly ductile. He alſo tried the mixture of other metals with iron, but found that none of them produced the ſame effect. So that it clearly appears, that the cold ſhort quality of iron is entirely owing to the alloy of ſiderum. The oppoſite quality of being red ſhort is not ſo eaſy to explain: and all we ſhall obſerve here is, that every kind of iron is ſubject to a great number of varieties, which it would not be proper to enter upon, as perhaps it would not be eaſy to ſet the whole in a clear point of view.

Yetlin diſſolved in vitriolic acid is of a greeniſh colour ſomewhat reſembling green glaſs, but a good deal lighter. This acid attacks yetlin even in the cold and without dilution; but the ſolution is promoted by adding two or three times its weight of water, and by the application of a boiling heat: a good deal of plumbago remains undiſſolved at the bottom. When precipitated by the cauſtic volatile alkali, it gives a calx of a very deep dark green colour, interſperſed with ſome greyiſh matter. The cauſtic fixed alkali throws down a light green calx.

Yetlin diſſolved in the nitrous acid is of a golden yellow colour. During their action a very copious ochry-like matter precipitates, which is of a brown red colour. This ſolution has a very ſtyptic taſte. Precipitated by the fixed alkali, it yields a reddiſh orange-like precipitate. The volatile alkali throws down a browniſh matter, ſomewhat of a yellow hue.

The marine acid attacks yetlin pretty violently: the ſolution is of a very light green. During their action the metal is ſplit often into thin ſlices, which remain undiſſolved. The acid takes up a good deal of the metal. Precipitated by the fixed alkali, a greeniſh matter falls, mixed with a little greyiſh matter. The volatile alkali throws down a deep green calx. Almoſt all the fixed air of the mild fixed alkali was abſorbed, ſcarcely any efferveſcence being raiſed.

The action of the vitriolic acid upon ſteel is not ſo violent as upon yetlin; but is increaſed by dilution and a boiling heat. The ſolution here is more complete, but a ſmall quantity ſtill remains undiſſolved. It is of a clearer colour than that of yetlin. Precipitated by the cauſtic fixed alkali, a green calx is thrown down; and by the

vo-

volatile, an extremely deep green one, when viewed in the light; but when removed from the light, it appears tainted of a dark blue.

Steel is more ſoluble in the nitrous acid than yetlin is. The ſolution is of a yellow citron colour, ſomewhat inclining to red. A ſmall quantity of plumbago remains, of a blackiſh colour, as it is in the other two acids. Precipitated by the cauſtic fixed alkali, a yellowiſh citron-coloured matter is ſeparated. The volatile alkali precipitates it of a browniſh red.

Steel is not ſo eaſily acted upon by marine acid as yetlin is. The ſolution is of a lighter yellow than that of yetlin in the nitrous acid. The ſteel is likewiſe ſometimes ſplit into pieces, and there remains a good deal of a blackiſh brown matter undiſſolved. The cauſtic fixed alkali precipitates this ſolution of a moſt beautiful orange, and at the ſame time heightens the colour of the liquor. The volatile alkali cauſes a very dark brown precipitate, and the ſupernatant liquor is colourleſs.

Forged iron is the moſt ſoluble of the three kinds of iron in vitriolic acid; leaves ſcarcely any thing undiſſolved, and its ſolution is of the colour of that of ordinary green vitriol. It is not readily acted upon by the acid till it be properly diluted. When precipitated by the cauſtic fixed alkali, it gives an elegant green precipitate. The volatile alkali gives a copious calx, ſo green that it is almoſt black: ſometimes a whitiſh grey matter falls, which mixes with the green, and renders the whole of a faint blue tint.

Forged iron in the nitrous acid is of a deeper and more ochry yellow colour than ſteel in the ſame acid. It differs from that of yetlin, in its ſediment being of a browniſh ochry colour, without any appearance of red. Precipitated by the cauſtic fixed alkali, it gives a matter of a pretty orange yellow, which is at firſt preſently taken up, but by farther addition is again thrown down. The volatile cauſtic alkali throws down a ſeemingly more copious calx, which upon repoſe turned out of four different colours: the undermoſt part was of a brown colour, ſomewhat like the ochry ſediment; the next above it, black; the next, like the depoſition in the yetlin ſolution; and the uppermoſt, a dark browniſh red.

Forged iron diſſolves very well in the muriatic acid. The ſolution is of a very light green colour, inclining a little to yellow: a browniſh matter in ſmall quantity remains undiſſolved. The ſolution recipitated by volatile cauſtic alkali gives a calx almoſt wholly reſembling that ſeparated from the vitriolic ſolution, but is rather bluer. The cauſtic fixed alkali gives upon repoſe a deep green which inclines to black: a greeniſh yellow matter floats ſometimes a-top.

All theſe appearances are found to vary from the quality of the metal, and from the nature and purity of the acids and alkalis employed. All the precipitates of the different kinds of iron yield upon reduction malleable iron, and not the ſame kind of metal which was employed in ſolution.

LECTURE XXXIX.

Species 12. COPPER.

COPPER is an imperfect metal, of a brilliant enough red colour, called by the alchemists *Venus*. It has a disagreeable smell, perceptible upon friction or being heated: its taste is styptic and nauseous, less sensible, however, than that of iron. It is hard, very elastic, and sonorous. It possesses a very great degree of ductility. It is made into very thin plates and wire of great length. It loses about one-eighth or one-ninth of its weight in the hydrostatic balance. It has so vast a tenacity, that a thread of it a tenth of an inch in diameter, is capable of supporting two hundred and ninety-nine and a half pounds weight before it break. Its fracture seems to be composed of small grains. It is susceptible of a regular form. M. l'Abbé Mongez defines its crystals to be quadrangular pyramids, sometimes solid, and sometimes composed of other small ones adhering laterally.

In the earth, copper is found in different states. Its ores are very numerous, and may be all referred to the following.

1. Native copper, of a red colour, malleable, and possessing all the properties of this metal. It is distinguished into two kinds; copper of the first formation, and copper of the second formation or of cementation. Copper of the first formation is dispersed in laminæ or filaments in a matrix almost always quartzy. There are specimens of it whose crystals imitate a kind of vegetation: others are in a mass and in grains. The copper of cementation is generally in grains or laminæ, superficially lying upon stones or iron. This last seems to have been deposited in water charged with vitriol of copper and precipitated by iron. Native copper is found in several parts of Europe; at Saint-Bel in Lyonnois, at Norberg in Sweden, at Neusol in Hungary, and in several countries in America.

2. Red

2. Red copper: It is diſtinguiſhable by its red dull colour, reſembling that of the ſcales which are detached from copper reddened by the fire when it is ſtruck with a hammer. M. Monnet conſidered this ore as a natural calx of copper. It is generally mixed with native copper and mountain green. It is very rare, ſometimes cryſtalliſed in octagons or ſilky fibres, called *flowers of copper*.

3. Earthy copper, mountain green, or green chryſocolla. This ore is a true ochre of copper, of a more or leſs dull green, very light, unequally diſtributed in the matrix. It is ſometimes very pure. Three varieties of it are reckoned.

1. Simple mountain green, earthy or impure.
2. Mountain green cryſtalliſed, or ſilky copper of China. This kind, which is very common in Voſges and at Hartz, is alſo found in China: it is very pure, and cryſtalliſed in long ſilky very ſolid bundles.
3. Mountain green, in ſtalactites or malachite. This ſubſtance, which is very frequently found in Siberia, is compoſed of layers, which repreſent greater or ſmaller nipples: ſome ſpecimens are formed of needles, converging to a common centre. The different layers have not the ſame ſhades of green. Malachite is ſufficiently hard to receive a very fine poliſh. It is therefore uſed for making different toys.

4. Mountain blue, or blue chryſocolla. This is a calx of copper of a deep blue colour: it is ſometimes of a regular form, and in priſmatic rhomboidal chryſtals of a beautiful blue. In this ſtate it is called *azure copper*: at other times it preſents ſmall grains, depoſited in the cavities of different matrixes, and particularly in quartz. For the moſt part it is in layers, lying ſuperficially in the cavities of grey and yellow ores of copper. It appears that all theſe ores of copper have been precipitated from vitriolic ſolutions of copper by means

means of calcareous earths, acroſs which theſe waters run. M. Sage conſiders theſe blue ores of copper as combinations of copper with the volatile alkali: and he ſays that they differ only in inſolubility. He believes that malachite is only an alteration of this blue, which he calls *azure tranſparent ore of copper.*

5. Copper united with ſulphur: yellow ore of copper, or coppery pyrites; for the naturaliſts uſe this name indiſcriminately: however, the miners uſe to call ores of copper, all the ſpecimens of copper united with ſulphur which are rich enough in metal to be ſmelted with advantage; and they reſerve the name of *pyrites* for the ſpecimens containing a great quantity of ſulphur and little metal. The rich ores of copper are generally of a brilliant yellow, inclining more or leſs to red or green. In the earth they form ſometimes conſiderable ſtrata. Oftentimes this copper is maſſy and dull; frequently ſcaly and micaceous. This is the form of the ores of copper from Denmark, Norway, Sweden, and Sainte-Marie-aux-Mines. At other times this ore is ſcattered up and down in its matrix, as the copper of Alſace: it is then called *tyger ore of copper.* This variety is mixed with a little azure. Frequently copper ores preſent at their ſurface very brilliant blue or violet colours, which are owing to the decompoſition of their principles. If theſe colours are only at their ſurface, they get the name of *copper cats-eye, azure*, or *peacock-coloured ores*; but if the colour penetrates deeper into the heart, they are called *white* or *violet vitreous copper ores.* They contain generally a large quantity of ſulphur, a little iron, and are not very rich in copper. When theſe kinds of ores are only ſuperficially ſcattered upon their matrix, they are called *pyrites:* theſe are the Derby ores in England, ſome of thoſe of Saint-Bel in Lyonnois, and ſeveral ores of Alſace, as thoſe of Caulenbach and Feldens; beſides, they are found adhering to all ſorts of matrices, to rock cryſtal, quartz, ſpar, ſchiſtus, mica, &c.

6. Grey copper ore, or Fahlertz of the Germans, is a combination of ſulphur and arſenic with copper, iron, and ſilver. It greatly reſembles the grey ore of ſilver: it is only a little leſs brilliant, and really differs from it only in containing leſs ſilver than it does. M. Romé de Liſle alſo diſtinguiſhes a white ore of ſilver, which he ſays contains a ſmall portion of ſilver more than the grey; but this is a true ore of ſilver.

7. The hepatic or brown ore of copper. This is a combination of iron, ſulphur, and copper. M. Romé de Liſle conſiders it as a changed grey copper ore, which has parted with the arſenic and a part of its ſulphur. He obſerves, that we muſt diſtinguiſh it well from the falſe hepatic copper ore; which he ſays is nothing elſe than a brown iron ore, and contains almoſt no copper.

8. Ore of copper, black or of a dirty colour. M. Gellert calls it *ore of copper in ſcoriæ*. It is a reſiduum of the decompoſition of the yellow and grey ores of copper, which contain neither ſulphur nor arſenic, and approach to the ſtate of malachite.

In eſſaying a copper ore, after its being pounded and waſhed, it muſt be long and ſtrongly roaſted, and then fuſed with four times its weight of black flux and ſome marine ſalt. A button of metal is found, which is frequently ſtill blackened by a little adhering ſulphur: it is fuſed with four parts of lead, and ſubjected to cupellation, to ſeparate the ſilver and gold, as there is but little copper which does not contain a certain quantity of theſe precious metals.

In the great copper-works the ores are pounded and waſhed: after that, they are roaſted at firſt in the open air, and almoſt without fuel; for as ſoon as the ſulphur which they contain begins to kindle, it continues to burn of itſelf. When it is extinguiſhed, the ore is roaſted afreſh once and again along with wood: after which it is fuſed acroſs coals, to get what is called the *impure copper*.

This

This ore as yet has loſt only a portion of its ſulphur. The fuſion it is ſubjected to, is intended to give the metal new ſurfaces, in order to facilitate its roaſting. It is ſubjected to ſix or ſeven roaſtings, according to the quantity of ſulphur contained, and is then fuſed to obtain the black copper. This copper is malleable: it is however ſtill combined with a little ſulphur, which is ſeparated in obtaining the perfect metals which it contains. The black copper is fuſed with three times as much black flux, which is called the reduction of the copper; and this mixture is moulded into the form of loaves, called *loaves of liquation*.

Theſe are placed upon two plates of iron, placed in ſuch a way that they leave a furrow between them. Theſe plates terminate the top of the furnace of liquation, whoſe bottom is inclined towards the fore-part. The fire, placed below the plates, heats the loaves: the lead fuſes, and falls through the coals, carrying along the ſilver and gold, which have more affinity with it than they have with the copper. After this operation, which is called *liquation*, the loaves are found conſiderably diminiſhed, and all deformed. They are expoſed to a ſtronger fire, ſo violent as to commence the fuſion of the copper, in order to ſeparate all the lead exactly. This third operation is called *reſſuage*. The lead, charged with the perfect metals, then undergoes cupellation. With reſpect to the copper, it is refined by fuſion in a crucible; in which it is left till all the heterogeneous matter it contains is ſeparated from it in the form of ſcum. It is tried by immerſing iron roads into it, which are thus covered with a little copper; and its purity is judged of according to its more or leſs brilliant red colour. The refined copper is run into plates, or ſeparated into *en roſettes*. For this laſt purpoſe, the ſcoriæ which cover the copper in fuſion are carefully removed: then the ſurface of the metal is allowed to fix. When it ceaſes to be fluid, a moiſt beſom is applied above: the impreſſion of the cold makes

makes it contract; the portion which congeals detaches itſelf not only from the ſides of the crucible, but from the reſt of the fuſed metal, and is lifted off with pincers. Thus the greateſt part of the copper contained in the crucible is converted into *en roſettes*. The portion which remains at the bottom is called the *king*.

The pyrites of copper which contain little metal, are ſmelted only in order to extract the ſulphur and vitriol. They are roaſted and diſtilled for the ſeparation of the ſulphur. During the roaſting, a portion of vitriolic acid acts upon the metal, diſſolves it, and begins to form vitriol. The roaſted pyrites are then expoſed to the air; and when the vitrioliſation is finiſhed, the efloreſcent pyrites are waſhed, the ley filtrated; and by evaporation and cryſtalliſation, a blue rhomboidal ſalt is obtained, called *vitriol of copper*, *blue vitriol*, *blue copperas*, or *Cyprian vitriol*. It ſhall be taken notice of when we are examining the combinations of this metal.

Copper expoſed to the fire aſſumes colour in the ſame manner as ſteel does: it becomes blue, yellow, and at laſt violet. It does not fuſe till it is very red. When it is in proper fuſion, it ſeems to be covered with a green flame: it boils, and can be volatiliſed, as is obſerved in founders chimneys. Flowers of copper are alſo found in the crucibles in which it is fuſed. If this metal, in fine powder, is thrown acroſs flames, it gives them a blue and green colour. On account of this property, it is uſed in fire-works. If this metal, when fuſed, is allowed to cool ſlowly, and if, when the ſurface has become fixed, the ſtill fluid portion is decanted off, that part which ſtill adheres to the ſides of the crucible, or of the veſſel employed in this experiment, is found to be cryſtalliſed in form of pyramids; which are ſo much the larger and more regular, according to the more complete fuſion of the metal, and the management of its cooling.

The ſurface of copper which is heated with the concourſe

courſe of the air, burns and is changed into a calx of blackiſh red.

This calx is eaſily obtained, by reddening a copper plate, and then ſtriking it with a hammer: it falls off in the form of ſcales. The ſame thing happens if, after it has been made red, it is immerſed in cold water: the ſudden contraction of the parts of the metal facilitates the ſeparation of the calx covering the ſurface. This calx falls to the bottom of the water, and is called *ſcales of copper*. As this calx is not perfectly burnt, it may be calcined afreſh under a muffle: it then aſſumes a very deep brown red colour: urged with a violent fire, it fuſes into a blackiſh or brown cheſnut glaſs. The calx of copper may be decompoſed and deprived of the air, which changed its metallic properties by means of the vegetable or animal combuſtible bodies. The ſcales are in part reducible *per ſe*, ſince the founders, who purchaſe them from the braziers, think it ſufficient to throw them into great crucibles upon melted copper, with which they enter into fuſion and incorporate. They do the ſame with the filings.

The air acts upon copper the more eaſily when it is moiſt and altered: it is converted into a ruſt or green calx, which ſeems to have ſome ſaline qualities; for it has taſte, and is diſſolved by water: on this account the ancient chemiſts admitted a ſalt in copper. In this ruſt it is remarkable, that no part but the ſurface is attacked; and it ſeems even to ſerve for the preſervation of the interior parts of this metal, as we may learn from the ancient medals and ſtatues, which are very well preſerved under the coat of ruſt which covers them. The antiquaries call this cruſt *patine;* and value it greatly, becauſe it atteſts the antiquity of the pieces which it covers.

Water does not appear to attack copper unleſs it be converted into vapour: for this reaſon it is more dangerous to allow liquors to cool in copper veſſels, than to boil them in them; becauſe, as long as the liquor is

I boiling

boiling and the veſſel hot, the aqueous vapour does not attack its ſurface: it is quite the reverſe when the veſſel is cold.

Copper does not unite with the earthy matters: its calx facilitates their fuſion, and forms with them greeniſh glaſſes.

Lime and magneſia have no remarkable action upon copper; at leaſt the alterations occaſioned by them have not been much attended to.

The cauſtic fixed alkalis, digeſted in the cold with copper-filings, in a little time aſſume a very ſlight blue colour: the copper is coloured with a powder of the ſame colour. Theſe ſolutions are affected more eaſily in the cold than with heat, according to M. Monnet. It is eſſential, however, to obſerve, that this chemiſt tried theſe combinations with mild alkali, and not pure fixed alkali: this ſurely would have had a much greater action upon the copper.

The volatile alkali diſſolves this metal more rapidly. This ſalt, digeſted upon filings of copper, in a few hours gets a deep blue colour; it diſſolves, however, but very little of it. I obſerved the phenomena of this ſolution during a whole year. Into a ſmall veſſel I put ſome volatile cauſtic alkali upon copper filings: in a few months, the ſurface of the metal was covered with a blue calx; the ſides of the veſſel were covered with a pale blue calx; and the lower part of the veſſel, which contained the copper, preſented at the ſurface of the glaſs a brown calx, the top of which was yellowiſh. This liquor entirely loſes its colour when it is ſhut up: take out the cork, and it appears: it exhibits this phenomenon in a very remarkable degree only at the beginning, and when it is decanted off from above the copper. If the ſolution is old, and ſtill contains copper, its colour is a fine blue, although in cloſe veſſels; however, upon expoſure to air, it is much deeper. When this ſolution is ſlowly evaporated by means of fire, the greateſt part of

 the

the volatile alkali is diſſipated; a portion remains fixed with the calx of the metal, and is depoſited in ſoft cryſtals, as M. Monnet has obſerved. M. Sage ſays, that very fine cryſtals may be obtained from it by a ſlow evaporation; he compares them to the natural azure copper. However, this laſt does not give volatile alkali. When it is heated, it is not ſoluble in water; is not efflorefcent with the air, as that is which is prepared by art. M. Baumé ſays, that this compound forms very brilliant and beautiful blue cryſtals. This ſolution, expoſed to the air, dries very quickly, and leaves a matter of a graſs-green hue, which is juſt a calx of copper. M. Sage is of opinion that it is the origin of malachite. If an acid is poured into a ſolution of copper by the volatile alkali, no precipitate forms; but the blue colour totally diſappears, and is changed into a very ſlight pale green. This phenomenon, which has been taken notice of by Meſſrs Pott and Monnet, ſhows that there is only very little calx of copper in the volatile alkali; and that it is rediſſolved by the acid, or by the ammoniacal ſalt formed by the addition of the acid. The blue colour, however, may be reproduced by the addition of ſome volatile alkali to the mixture. The calx of copper made by means of fire, and all the other calces of this metal, are preſently diſſolved in the pure volatile alkali, which by this proceſs may be impregnated with a good quantity of metal. It quickly aſſumes the moſt beautiful blue colour; and on that account is recommended as a teſt to diſcover the ſmalleſt portion of copper in all the matters in which it is ſuſpected *.

The

* The changes of colour which the ſolution of copper exhibits from the admiſſion or excluſion of air, has attracted much notice from chemiſts on account of the beautiful appearance of the phenomenon. M. Bergman ſuppoſes, that the colour of the ſolution is intimately connected with the quantity of phlogiſton which the copper contains; and that the only effect of admitting atmoſpheric air is to carry off a portion of phlogiſton from the metal: For it is to be obſerved, that if ſome filings of copper be put into a phial full of

volatile

The vitriolic acid does not act on copper, unless concentrated and boiling. In the time of the solution, a great quantity of sulphureous gas flies off. When it is over, a brown thickish matter is found, which contains the calx of copper and a portion of this calx combined with vitriolic acid. By ablution and filtration, a blue solution is obtained; if it is evaporated to a certain degree, and left to cool, it furnishes long rhomboidal crystals of a beautiful blue colour, called *vitriol of copper*. If, instead of evaporating this solution, it is exposed a long time to the air, it affords crystals; but a green calx is precipitated, a colour which all the calces of copper assume when formed or dried by the air *. The vitriol

 of

volatile alkali, and so closely corked as to exclude the access of air, no solution will take place; because volatile alkali cannot dissolve copper perfectly saturated with phlogiston; and the phlogiston in this case has no possible means of escape, for want of the vehicle which air furnishes. But if a small quantity of air be included in the vessel along with the volatile alkali and copper, a solution is made, which proves entirely colourless; but if the air gains free admission, the solution becomes of a deep rich blue colour. We have it in our power to destroy this colour, by adding more copper and again excluding the air: and we must here remark, that the exclusion of air alone does not seem sufficient to deprive the solution of all colour; at least M. Wassenberg kept some of it, hermetically sealed in a glass tube, for nine months together, without being able to perceive any alteration. The explanation which M. Bergman offers of this change is, that the air, depriving the metal of phlogiston, reduces it more completely to the state of a calx, which exhibits a blue colour; that when only a small quantity of air is admitted, the metal, being but partially robbed of phlogiston, gives a colourless solution; and that when a deep-coloured solution is closely shut up with some fresh filings of copper, the solution becomes colourless, because the volatile alkali is capable of dissolving a larger proportion of copper when the metal contains more phlogiston; therefore the volatile alkali attacks the metallic copper, which during solution parts with a share of redundant phlogiston to what was more calcined by the action of air; and thus reduces the whole to that degree of calcination which gives an uncoloured solution.

* Blue vitriol is seldom formed by dissolving copper directly in the vitriolic acid. The blue vitriol of the shops is mostly obtained from copper pyrites. It may also be made by stratifying plates of copper

of copper has a very ſtrong ſtyptic taſte, approaching to cauſticity. When it is expoſed to the fire, it fuſes very quickly; it loſes its water of cryſtallization, and becomes of a blueiſh white. A very ſtrong heat is required for the ſeparation of its vitriolic acid, which adheres much more ſtrongly to copper than to iron. The vitriol of copper is decompoſed by magneſia and lime: the precipitate formed by theſe two ſubſtances is of a blueiſh white: if it is dried in the air, it becomes green; on this account ſome chemiſts ſay, that the precipitates of vitriol of copper are green. Abſolutely the ſame thing takes place with regard to thoſe obtained by the fixed alkalis in different ſtates: they are at firſt blueiſh, and aſſume a green colour upon drying: it is thus perhaps that mountain-green is formed. It is eſſential to obſerve, that when the vitriol of copper is precipitated by the ſolution of mild alkali, no efferveſcence is excited; which ſhows that the aërial acid combines very well with the calx of copper; a phenomenon which all metallic ſolutions do not exhibit. The volatile alkali likewiſe precipitates the ſolution of vitriol of copper of a blueiſh white colour; but the mixture very ſoon puts on a very deep blue, becauſe the volatile alkali diſſolves a portion of the copper precipitated: it alſo requires but very little of this ſalt to diſſolve all the calx of copper ſeparated from the vitriolic acid.

The nitrous acid diſſolves copper in the cold with rapidity. A great quantity of ſmoking nitrous gas flies off. It is the way which Dr Prieſtley uſed to get this gas in a very ſtrong ſtate. A portion of this metal, reduced to the ſtate of a calx, is precipitated in form of a brown powder: it is ſeparated by filtration. The ſolution, filtrated, is of a much deeper blue than that by the vitriolic acid; which ſhows that the copper is more calcined.

copper with ſulphur, and cementing them together for ſome time; when the vitriolic acid of the ſulphur, being diſengaged, attacks and corrodes the copper, forming a metallic ſalt, which, by the affuſion of water, yields perfect cryſtals of blue vitriol.

calcined. If it is evaporated with precaution, it cryſtallizes by cooling. M. Macquer is one of the firſt chemiſts who have taken notice of this property, in his memoir upon the ſolubility of ſalts in ſpirit of wine. If its cryſtals are very ſlowly formed, they preſent elongated parallelograms: if they are depoſited more quickly, they are hexaëdral priſms, whoſe point is obtuſe and irregular; and they imitate bundles of diverging needles. And, laſtly, if the ſolution is very ſtrongly evaporated, it forms a maſs deſtitute of any regular form: this has undoubtedly been the reaſon why ſome chemiſts have ſaid, that this ſolution was incapable of cryſtallization. Nitrous copper is of a very bright blue colour: it has ſo cauſtic a taſte, that it might be uſed to corrode excreſcences which grow upon the ſkin. M. Sage ſays, that it is fuſible with a temperature of twenty degrees of Reaumur's thermometer. It detonates upon burning coals; but as it contains a great deal of water, this phenomenon is not very ſenſible. When it is fuſed in a crucible, it emits a great quantity of nitrous vapours, which might be collected by uſing a diſtilling apparatus. When it is dried, its colour is green: by a greater heat it becomes brown; and in this ſtate it is only a calx of copper. I have diſtilled it with a pneumatico-chemical apparatus; and got a great deal of nitrous gas, a little chalky acid, not an atom of pure air. It was reduced in this operation to the ſtate of a brown calx. Nitrous copper attracts humidity from the air; it may, however, be kept a long time in cloſe veſſels. In warm and dry air, it becomes covered with a green effloreſcence. It is very ſoluble in water; a little more ſo in hot water than in cold. The ſolution, expoſed to the air in flat veſſels, or quickly evaporated in a dry and hot ſeaſon, leaves a green calx, as the cryſtals of this ſalt do in the ſame circumſtances.

It is precipitated by lime of a pale blue; by the fixed alkalis, of a blueiſh white; by the volatile alkali, in flocci of the ſame colour, which are very quickly diſ-

 ſolved,

ſolved, and give a very brilliant deep blue colour; by the liver of ſulphur, of a reddiſh brown without hepatic odour; by the tincture of the gall-nut, of an olive-green. Vitriolic acid alſo decompoſes it; and we get cryſtals of blue vitriol, if this acid, in its very concentrated ſtate, has been employed. Stahl announced this decompoſition: M. Monnet has ſince confirmed it; and I have had occaſion to obſerve it ſeveral times. Iron has more affinity with moſt part of the acids than copper has. When a lamina of iron is immerſed into a ſolution of nitrous copper, the copper precipitates in its metallic form, and gilds the ſurface of the iron, either by reſuming the phlogiſton of this laſt metal, according to the Stahlian doctrine; or by yielding to it the air, which converted it into a calx, according to the pneumatic theory. Vitriol of copper preſents the ſame fact; and it is a proceſs made uſe of by ſeveral perſons, to induce the belief of its converſion into copper *.

The marine acid diſſolves copper only when it is concentrated and boiling. During this ſolution, only very little gas is produced, whoſe nature is unknown: it ſeems, however, to be inflammable gas. This acid aſſumes a very deep green and almoſt brown colour. This combination forms a maſs very ſoluble in water: if it is waſhed, the water is of a fine green colour, which diſtinguiſhes this ſolution from the two preceding. By ſlow evaporation and cooling it depoſites very regular priſmatic cryſtals if care has been taken: they form on the contrary only ſharp, very ſmall, and acute needles, when the evaporation has been too rapid and the cooling too ſudden. This marine ſalt of copper is of a very agreeable

* The metallic ſalt formed by nitrous acid and copper exhibits a very beautiful appearance, when applied to the ſurface of tin-foil. If ſome of this ſalt be laid upon the tin-foil, gently wetted with water, and the tin-foil inſtantly rolled up, we ſoon perceive the action begin, firſt by the emiſſion of nitrous vapours, and then by the deflagration of the tin, which crackles and flames, and is left reduced into a grey calx. It requires celerity in rolling up the tin-foil to make the experiment ſucceed; and the cryſtals muſt at the ſame time be complete.

greeable graſs green: it has a cauſtic and very aſtringent taſte; it fuſes with a very gentle heat, and congeals into a maſs when left to cool. M. Monnet ſays, that there is a great degree of adheſion between its principles, and that the marine acid cannot be raiſed from it except by a very conſiderable heat: it ſtrongly attracts humidity from the air; it is decompoſable by the ſame intermedia with the preceding ſalts of copper. I have obſerved that the volatile alkali did not diſſolve ſo very well the calx of copper ſeparated from the marine acid, as it did that from the vitriolic and nitrous copper. The blue which it forms, is on that account not ſo lively; and there remains a portion of this calx which the volatile alkali does not diſſolve entirely. The vitriolic and nitrous acids do not decompoſe the marine ſalt of copper. The nitrous ſolution of mercury and ſilver decompoſe it, and are themſelves decompoſed in the inſtant of the mixture: a white precipitate forms by the junction of the marine acid with the mercury or ſilver, and the calx of copper unites with the nitrous acid. I have, however, obſerved, that the liquor does not aſſume the blue colour which the ſolution of copper by the nitrous acid ought to have; and that in general the calx of copper, formed by the marine acid, is very difficult of aſſuming this colour, as we have already ſeen with regard to the volatile alkali. It appeared to me, that in general the calces of copper paſs very difficultly from blue to green, and very difficultly too from green to blue. The marine acid diſſolves the calx of copper with a great deal more facility than it does the metal itſelf. This fact was accurately obſerved by M. Brandt. The ſolution is of a fine green, and cryſtalliſes as eaſily as the former; which proves, that in the metallic ſaline combinations, the metals are always in the ſtate of a calx, as we have already obſerved.

Nitre detonates with difficulty by means of copper. This ſalt muſt be fuſed, and the copper very hot, in order that the deflagration may take place; ſtill it is but very weak. This operation is peformed by throwing fine filings of copper upon nitre in fusion in a large crucible, in order that the contact may be more multiplied. When the metal is very much heated, a slight motion, accompanied with somewhat rapid flashes, is perceived. The residuum is a calx of a grey colour, a little brown, mixed with the fixed alkali: it is washed; the water takes up the alkali, which retains a little copper, and the calx of this metal remains pure. It completely fuses alone into a glass of a deep and opaque green: it is employed to colour enamels: the alkali is supposed to be rendered caustic; but there are no accurate experiments yet upon the subject.

Copper very easily decomposes sal ammoniac. M. Bucquet, who examined this decomposition with great care, obtained, by making the experiment in a pneumato-chemical apparatus above mercury, from two gros of copper-filings and one gros of sal ammoniac, fifty-eight inches of an elastic fluid; of which twenty-six inches were very good alkaline gas, twenty-six were detonating inflammable gas, and six a mephitic gas, which extinguished candles, without being absorbed by water, or precipitating lime-water. There was disengaged a little volatile alkaline spirit of a fine blue colour, which swam above the mercury. This experiment shows us that the marine acid produces inflammable gas by dissolving the copper. The residuum was a mass of a blackish green; the half of which was dissolved by water, and communicated to it a fine green colour; a distinct character of the marine salt of copper: the other half was like a kind of calx of copper formed by the marine acid. Upon repeating this decomposition in the dose of four ounces of copper with two of sal ammoniac with the ordinary apparatus of the balloon, M. Bucquet obtained two gros eighteen grains of a blue volatile alkaline spirit, which effervesced with acids, and contained about an inch of chalky acid per gros. This chemist did not know how to account for this last gas: but I am of opinion that it might come from some impuri-

ties

ties of the ſal ammoniac; for having repeated this experiment with ſal ammoniac prepared by ſublimation, I got a very cauſtic volatile alkali, not occaſioning the ſlighteſt efferveſcence with acids. The calx of copper alſo decompoſes ſal ammoniac, and gives to the volatile alkaline ſpirit which it drives over a portion of aërial acid, which renders it efferveſcent. This alkali is always blue, becauſe it carries over with it a ſmall portion of calx of copper, to which it owes this colour: however, the acids do not precipitate an atom of this metal. In pharmacy, two medicines are prepared with ſal ammoniac and copper; the firſt of which has received the name of *ammoniacal flowers of copper*, or *ens veneris*. This is nothing but ſome ſal ammoniac coloured with ſome calx of copper. A mixture of eight ounces of this ſalt with a gros of calx of copper, are made to ſublime in two earthen veſſels placed one above another. All the ſal ammoniac is volatiliſed without being decompoſed; and it attracts a little, which gives it a blueiſh colour. The ſecond preparation, called *aqua cæleſtis*, is made by digeſting ten or twelve hours a pound of lime-water with one ounce of ſal ammoniac in a copper baſon. The lime diſengages the volatile alkali, which diſſolves a little of the copper of the baſon, and which is coloured blue. Aqua celeſtis may be made in a glaſs or earthen veſſel, by adding a little filings or calx of copper to the lime-water and ſal ammoniac.

It ſeems that copper decompoſes alum: for if a ſolution of this ſalt is boiled in a copper-veſſel, a little clay is depoſited; and when this alum is precipitated by the volatile alkali, its earth aſſumes a ſmall blue colour, which detects the preſence of the copper.

The action of inflammable gas upon copper is not known. This metal unites with ſulphur very well. This combination may be made by the humid way, that is, by mixing flowers of ſulphur and filings of copper, and moiſtening the mixture with water; but it ſucceeds much better by the dry way. A mixture of equal

equal parts of ſulphur and copper-ſilings is expoſed to the fire in a crucible, which is gradually heated till it be reddened; from this combination a maſs of a blackiſh grey reſults, a kind of coarſe copper, which is brittle and more fuſible than the copper. This compoſition is prepared for dyeing and painting Indian ſtuffs, by ſtratifying in a crucible laminæ of copper and ſulphur in powder, and heating the crucible as ſaid above: the reſult is pulveriſed, and called *æs veneris*. The liver of ſulphur and hepatic gas have a remarkable action upon copper: the former diſſolves it by the dry and humid way; the ſecond ſtrongly colours its ſurface; but their effects have not as yet been examined.

Copper alloys with ſeveral metals: With arſenic it becomes white and brittle, and forms the white tombac. It unites with biſmuth, and, according to M. Gellert, forms an alloy of a reddiſh white in cubical facets. It alloys very well with regulus of antimony, and gives a coppery regulus, which is diſtinguiſhable by a pretty violet colour. It decompoſes antimony, and unites with the ſulphur which it carries off from the regulus.

It combines very eaſily with zinc. This combination may be made in two ways. 1. By fuſion, a metal is obtained of the colour of gold, which is much leſs ſuſceptible of ruſt than the pure copper, but has leſs ductility. The more its colour reſembles gold, the more brittle is the metal: beſides, it varies according to the proportion of the mixture and the precautions obſerved in fuſing it. Its varieties are, ſimilor, pinchbec, Prince Rupert's metal, and gold of Manheim. 2. If laminæ of copper are cemented with lapis calaminaris, reduced into powder and mixed with charcoal, and if the crucible is reddened with the fire, the copper unites with the zinc, and forms braſs. This is leſs apt to ruſt than copper, it is as malleable and more fuſible than it; but when ſtrongly heated, it loſes the zinc, and becomes red copper *.

Copper

* Although copper be extremely valuable on account of its ductility,

Copper is difficultly alloyed with mercury. A kind of amalgam, however, may be formed, by rubbing ſome very thin plates of copper with mercury. A plate of copper plunged into a ſolution of mercury by an acid,

tility, lightneſs, and ſtrength, it is rendered leſs uſeful on many occaſions from the difficulty of forming large maſſes of work. We cannot join heated bars of copper, by the operation of welding, as we can do thoſe of iron. Neither is it an eaſy matter to caſt copper ſolid, ſo that it ſhall retain all its properties entire: for if the heat be not ſufficiently great, the copper proves deficient in toughneſs when cold; and if the heat be raiſed too high, or continued for too great a length of time, the copper bliſters on the ſurface when caſt in a mould: ſo that the limits of fuſion at which copper will caſt well are exceedingly contracted; and from the intenſe heat neceſſary to make it flow with ſufficient tenuity, it becomes impracticable to regulate the degree of heat with exactneſs. From theſe circumſtances, pure copper is rendered leſs applicable to many different purpoſes. We find, however, that the addition of a certain proportion of zinc removes almoſt all the inconveniences, and furniſhes a mixed metal more fuſible than pure copper, extremely ductile and tenacious when cold, which does not ſo readily ſcorify in a moderate heat, and which is leſs apt to ruſt from the action of air and of moiſture. A mixture of this kind forms common braſs, which varies in its properties according to the proportion of the ingredients and the intimacy of their union. Braſs is frequently made by cementing plates of copper with calamine, when the copper imbibes one-fourth or one-fifth its weight of the zinc which riſes from the calamine. The proceſs conſiſts in mixing three parts of calamine and two of copper with charcoal-duſt in a crucible, which is expoſed to a red heat for ſome hours, and then brought into fuſion. The vapours of the calamine penetrate the heated plates of copper, and thereby add to its fuſibility. It is of great conſequence, for the ſucceſs of this proceſs, to have the copper cut into ſmall pieces, and intimately blended with the calamine. In moſt foreign founderies the copper is broken ſmall by mechanica. means with a great deal of labour; but at Briſtol, the workmen employ a very ingenious method to reduce the copper into grains with more eaſe and celerity. A pit is dug in the ground about four feet deep, the ſides of which are lined with wood, and the bottom is made of copper or braſs, and moveable by means of a chain. The top is made of braſs with a ſpace near the centre, perforated with ſmall holes, through which the melted copper runs in a number of ſtreams into the water which the pit contains. The holes are luted with clay, and the water is perpetually renewed by a freſh ſtream, to prevent it from turning warm by the influx of melted copper.

acid, is covered with a fine argentine coat, owing to the femimetal being precipitated upon the copper.

Copper and lead unite very well by fusion, as is proved by the loaves of liquation *.

It

per. As the copper falls down through the cold water, it forms itself into grains, which collect at the bottom. In this form it can be most completely mixed with the powdered calamine. The difficulty of execution arises here from the violent explosions which melted copper and water produce, unless every precaution be practised to prevent them. And even where no precaution is omitted, the works occasionally explode with a degree of violence which endangers the lives of the workmen. Brass is sometimes made in another way, by mixing the two metals directly: but in managing this process, a good deal of address is requisite, since the heat necessary to melt the copper, makes the zinc burn and flame, by which a considerable part is dissipated; so that the remaining copper is defrauded of the due proportion of zinc. If the copper be first melted by itself, and the zinc heated and plunged into it, a strong commotion ensues, though the dissipation is much less considerable than in the other case, as the zinc is quickly imbibed by the copper, and in some measure protected and retained by it. If the copper and zinc be brought separately into fusion, and one poured upon the other, an explosion happens, and a great part of the mixture is thrown about in drops, and lost. The union appears to succeed best, and with the least loss of zinc, when fluxes containing inflammable matter are added to the mixture. The flux is first brought into fusion in a crucible, the copper and zinc poured into it; and so soon as they appear thoroughly melted, they are to be well stirred together, and expeditiously poured out. The reason for pouring them out instantly, is to prevent the inflammation of the zinc, which would speedily consume, and leave nothing but red copper in the crucible. It is remarkable that either a large proportion of zinc or of copper give a metal resembling the colour of gold, while in the intermediate proportions the imitation is not so good. Five parts of copper to one of zinc prove the best with an excess of copper; though, with an excess of zinc, the resemblance to gold is still stronger. But in this case the metal is extremely brittle; and indeed no brass is made which bears extension when hot.

* When Archimedes endeavoured to determine the quantity of alloy in the crown of gold which Hiero king of Syracuse gave him to examine, his calculations all proceeded on the supposition that the specific gravity of the compound came out to be the exact mean between the gravities of the two metals of which it was composed. This was indeed the conjecture natural to form, without any previous know-

It is combined with tin in two ways; either by the application of fused tin to the copper, or by fusing these two metals together. The first operation is used in the tinning of copper; the second forms bronze. In tinning copper-vessels, they are scoured well, in order to render their surface clean and brilliant; then they are rubbed with some sal ammoniac to clean it entirely. They are heated, and some powdered resin is thrown upon them. This substance, by covering the surface of the copper, hinders it from being calcined. Then the fused tin is laid on and spread with tow.

Several persons complain, that the tinning of copper vessels is not sufficient to defend them from the action of the air, from humidity, and from salts; because these vessels are seen often covered with verdigris. It would be possible to remedy this inconvenience, by laying on a thicker coat of tin, if it were not apprehended that the superior degree of heat to that of boiling water, to which these vessels are frequently exposed, might fuse the tin, and render the surface of the copper exposed. We are justly astonished from the small quantity made use of in tinning, since M. Bayen and M. Charlard have computed, that a copper pan, nine inches in diameter and three inches three lines deep, had acquired by tinning only twenty

knowledge of the subject to direct his judgment, though it is by no means consonant to the truth of the fact, when ascertained by experiment. So far from being true, we find no mixed metal correspond accurately to the specific gravity given by computation. Some mixtures exceed the mean, others fall below it. But the most singular of all discoveries is, that the specific gravity of tin, mixed with copper, not only exceeds the mean gravity of the two metals, but even surpasses that of the heaviest of them. A mixture, consisting of ten parts of copper and one part of tin, proves heavier than pure copper. The augmentation is also sensible when the lighter metal bears a larger proportion to the whole mass. Even one part of tin to two parts of copper affords a mixture, possessing greater specific gravity than the copper itself. This circumstance has been long known, and is thoroughly established: and the whole series of facts upon the subject shows the fallacy of trusting to the specific gravity of the mixed metal, as a just test of the proportion of ingredients which enter into the mixture.

twenty-one grains. However, this ſmall quantity is ſufficient to prevent the danger ariſing from copper, when attention is paid not to allow ſubſtances capable of diſſolving the tin, to remain long in the tinned veſſels; and chiefly when the tinning is frequently renewed, which the friction of the meat, the heat, and the actions of the ſpoons with which the ſubſtances cooked in them are agitated, very readily deſtroy. There is, however, a dread which we cannot ſhake off; I mean, the tin which the braziers uſe to tin the copper pans. It is frequently alloyed with a fourth of its weight of lead; and from this circumſtance we have reaſon to fear bad effects, as we know it is very ſoluble in acids and in fats. This ſhould render it neceſſary for government to form meaſures that the braziers be not deceived in the purchaſe of their tin; and that they employ only that of Malacca or of Banca in the ſtate it comes to us from the Indies, without its being alloyed and fuſed again by the pewterers.

M. de la Folie, a citizen of Rouen, celebrated for his chemical works relative to the arts, and for the uſeful diſcoveries with which he has enriched the arts of dyeing, and of making earthen-ware, and a great number of the manufactures of Rouen, has, in order to avoid the dangers and inconveniences from tinned copper, propoſed the uſe of veſſels of beat iron covered with zinc, which has nothing dangerous in it, as we have ſeen already. Several perſons have made an advantageous trade of it: it is to be wiſhed the number of ſuch veſſels were increaſed.

When tin is fuſed with copper, a more weighty metal reſults than the ſum of the two metals employed. This alloy is the more white, brittle, and ſonorous, according to the quantity of tin. When it is very white, it is called *bell-metal.* When it contains more copper, it is yellow, and is called *airain* or *bronze:* it is uſed for the caſting of ſtatues, and to make pieces of artillery, which ſhould be very ſolid, not to ſplit with the ſlighteſt impulſe;

impulſe; not very ductile, however, that they might not be deformed by the ſtroke of the bullets.

Copper and iron are ſuſceptible of combining by fuſion and ſoddering: however, this operation does not ſucceed eaſily. When a mixture of theſe two metals is fuſed in a crucible, the iron is found frequently ſown in the copper, without having contracted a perfect union. Copper decompoſes the mother-water of martial vitriol, though iron has a ſtronger affinity with the acids than copper has.

The uſes of copper are very multiplied and very well known. A multitude of various utenſils are made with it. It is particularly the yellow copper, or its alloy with zinc, which is moſt employed on account of its ductility and beauty. As copper is a very violent poiſon, it ought never to be adminiſtered in medicine. The moſt proper remedies, in caſes of poiſon by copper reduced to a calx or verdigris, are emetics, abundance of water, livers of ſulphur, alkalis, &c.

LECTURE XI.

Species 13. SILVER.

SILVER, called *Luna* or *Diana* by the alchemiſts, is a perfect metal, of a white colour, and of the moſt vivid brilliancy. It has neither taſte nor ſmell: its ſpecific gravity is ſo great, that it loſes about an eleventh of its weight in the hydroſtatic balance. A cubic foot of it weighs 720 pounds. It poſſeſſes ſo great a degree of ductility, that ſheets of it are made as thin as paper, and it can be drawn into finer wire than the hairs of our head. A ſingle grain can be made to contain an ounce of water. By its tenacity, a ſilver wire the tenth of an inch in diameter, can ſupport 270 pounds weight without being broken. Its hardneſs and elaſticity are inferior to thoſe of copper: It is the moſt ſonorous of the metals except copper: it ſtiffens under the ham-

hammer; of which property it may be deprived by nealing. Meſſ. Tillet and Mongez have cryſtallized it. They obtained quadrilateral pyramids, ſometimes iſolated like thoſe which are found on the ſides of the crucible in which it has been fuſed, or in groups adhering laterally to each other.

Silver is found in ſeveral ſtates in nature. The principal ores of this metal may be reduced to the following.

1. Native or virgin ore of ſilver. It is diſtinguiſhable by its brilliancy and ductility: its form is very various; it is frequently in irregular maſſes of different ſizes. Sometimes it is in the form of contorted capillary filaments: and in this ſtate ſeems to borrow its formation from a decompoſed red ore of ſilver, according to the obſervation of Henckel and M. Romé de Liſle. It is alſo found in laminæ like net-work, which the Spaniards call for that reaſon *arané;* like vegetation, or in branches, formed of octagons ſticking to each other. Some of theſe ſpecimens are like fern leaves; others again preſent iſolated cubes and octagons, whoſe angles are truncated: theſe laſt are the moſt rare. Native ſilver is almoſt always diſperſed in a quartzy matrix. It is found in Peru, in Mexico; at Konſberg in Norway, at Johan-Georgenſtadt, and at Ekenfriederſdorf in Saxony; at Sainte-Maria, and other places. This metal never has been found in the ſtate of a calx.

2. The vitreous ore of ſilver, according to the majority of mineralogiſts, is formed of ſilver and ſulphur. It is of a foul grey reſembling lead. The knife cuts it as it does lead. It is often without form, ſometimes cubical, having truncated angles. M. Monnet diſtinguiſhes one variety of it, which is converted into powder inſtead of being cut. This ore gives from 72 to 84 pounds in the hundred weight.

3. The red ore of ſilver is often deep in colour, ſometimes tranſparent, cryſtalliſed in cubes, whoſe ed-

ges are truncated, or in hexaëdral pyramids, terminated by three-ſided pyramids; it is called *roſſi-clero* at Potoſi. The ſilver is united with ſulphur and arſenic: it contains perhaps ſome aërial acid, ſince, by diſtillation, and the reception of the produce in a receiver lined with fixed alkali, M. Sage got cryſtals. When it is broken, its colour is clearer within, and it ſeems formed of ſmall needles or converging priſms like ſtalactites. If it is expoſed to a well-managed fire, capable of reddening it, the ſilver is reduced, and forms capillary vegetations reſembling native ſilver. It gives from 58 to 62 pounds in the hundred weight.

4. The grey and white ore of ſilver, which differs from the copper ore called *Tablertz* only in its containing more precious metal. It is in maſſes, or rather triangular cryſtals, the ſides of which are cut *en biſeau*. The thickeſt of theſe cryſtals are not of a very ſhining colour: the ſmalleſt ſcattered upon a flat matrix have a very agreeable appearance, on account of their very vivid brilliancy. The grey ſilver gives from two to five marks of ſilver in the quintal. Sometimes the grey ſilver is introduced into the organic ſubſtances, whoſe form it imitates exactly. It is then called *figured ore of ſilver;* ſuch is that which reſembles ears of corn, and which M. Romé de Liſle took for cones and ſcales of the pine tree: of this ſort alſo wood is found mineraliſed.

5. Black ore of ſilver, called *nigrillo* by the Spaniards, according to Meſſ. Lehman and Romé de Liſle, is only a decompoſition of the red or grey ore of ſilver, and a ſort of middle ſtate between that of theſe ores and native ſilver: this laſt is frequently found in it. The laſt mineralogiſt obſerves, that what is ſolid, ſpongy, or rotten, ſprings from the red and vitreous ores, and is much richer than that which is friable and of a pitchy colour, whoſe origin is owing to the alteration of the white or grey ores of ſilver. For this reaſon it is ſubject to vary in quantity of produce. It gives in ge-

neral

neral from ſix or ſeven to near ſixty pounds in the hundred weight.

6. The horn ore of ſilver is of a dirty yellowiſh grey; ſometimes it inclines to a flax grey. It has, though rarely, a ſemitranſparency: if ſoft, may be bruiſed and eaſily cut: it fuſes with the flame of a candle. It is found cryſtalliſed in cubes, and moſt frequently it is without ſhape. It frequently contains portions of native ſilver. Mineralogiſts are not agreed about the nature of this ore. Some ſuppoſe it to contain ſulphur and arſenic. Meſſ. Lehman, Cronſtedt, and Sage, think that it is mineraliſed by the marine acid. It appears that this laſt ſentiment has been moſt generally adopted. M. Woulfe has confirmed it; and he aſſerts that ſilver and mercury are the only metals that are mineraliſed by the marine acid. M. Prouſt alſo owns this ſubſtance as a mineraliſer of theſe two metals.

7. Soft ore of ſilver of Wallerius is only native or mineraliſed ſilver, diſperſed in greater or leſs quantity in coloured earths. The gooſe-dung ore of ſilver, thus named on account of its yellow green mixed colour, is only a variety of this ſpecies.

8. Silver is found frequently combined with other metallic matters in the ores we have already deſcribed. Of this ſort is miſpickel, the grey ore of cobalt, kupfernickel, or ore of nickel; antimony, which preſents frequently the variety called *plumed ore of ſilver*; blende-galena, the martial pyrites, and white ores of copper: theſe laſt are alſo only grey ores of ſilver. All theſe ſubſtances frequently contain enough of ſilver to be ſmelted with profit; but it is eaſily conceived that they ſhould not be deſcribed as particular ores of ſilver, it being ſufficient to take notice that they are in part compoſed of this metal *.

The

* M. Daubenton has very properly diſtinguiſhed the metallic ores by the number of metals which they contain, and has given marks for aſcertaining their preſence. This new method is much more accurate, ſure, and ſatisfactory, than that of former mineralogiſts.

The eſſaying of ſilver ores ſhould vary according to their nature. Thoſe which contain native ſilver require only to be pounded and waſhed. To ſeparate exactly this metal from the ſtrange ſubſtances which alter it, it may be triturated with crude mercury. This diſſolves the ſilver, and may be afterwards volatiliſed by means of fire, to get the perfect metal. The ſulphureous ores of ſilver require roaſting, and are then fuſed with a greater or leſs quantity of flux. In this fuſion the ſilver is obtained, generally alloyed with lead, copper, iron, &c. In order to ſeparate and know exactly the quantity of precious metal which this alloy contains, an entirely chemical proceſs is made uſe of, founded upon the properties of the imperfect metals. Lead, being capable of vitrification, and in its vitrification carrying along with it the imperfect metals, ſuch as iron, copper, without touching the ſilver; from this property a proceſs is employed to ſeparate the perfect metal from thoſe which alter it. The ſilver is fuſed with lead; which is in greater or leſs quantity according as it contains foreign ſubſtances: then this alloy is put into flat and porous veſſels, made with calcined bones and water. Theſe kinds of veſſels, called *cupels*, are proper to abſorb the glaſs of lead which is found in the cupellation. The ſilver remains pure after this operation. In order to know how much imperfect metal it contained, or of what title it was, any maſs of ſilver is ſuppoſed to be compoſed of twelve parts, called *marks*, and each of theſe weigh 24 grains. If the maſs of ſilver examined has loſt a twelfth of its weight, it is ſilver of eleven marks; if it has loſt only a twenty-fourth, the ſilver is eleven marks twelve grains fine; and ſo on. The cupel, after this operation, has acquired a great deal of weight: it is impregnated with the glaſs of lead, and with the imperfect metals which were alloyed with the ſilver, and which the lead has ſeparated from it. As the lead contains almoſt always a little ſilver, it is neceſſary to cupel it previouſly *per ſe*, in order to determine the quantity which it contains: we muſt then deduct from the button, which is obtained in cupelling its ſilver, the ſmall portion which is known to be contained in the lead employed, which is called the *teſt*.

Cupellation preſents a phenomenon, from which the artiſt learns the progreſs of his operation. According as the ſilver becomes pure by the vitrification and ſeparation of the lead, it appears much more brilliant than the portion which is not ſo pure. The brilliant part increaſes by degrees; and when all the ſurface of the metal becomes pure and brilliant, the moment it paſſes to this ſtate, it exhibits a flaſh or a fulguration, which annouces the completion of the operation. The cupelled ſilver is very pure relatively to the imperfect metals which it contained before: but it may contain ſome gold; and as it always contains a certain quantity of it, another proceſs is requiſite for the ſeparation of theſe two perfect metals. As gold is much more unchangeable than ſilver is by moſt of the menſtrua, the ſilver is diſſolved by the nitrous or marine acids and ſulphur; and the gold, upon which theſe ſolvents act but weakly, remains pure. This manner of ſeparating ſilver from gold is called *parting*. We ſhall ſpeak of the different ſorts of parting after we have explained the action of each of the ſolvents of ſilver employed, and when we come to treat of its alloy with gold. The great works for the extraction of ſilver from its ores, and for obtaining it pure, are nearly ſimilar to thoſe already deſcribed for the eſſay of the ores of this metal. In great, there are in general three ways of treating ſilver. The firſt conſiſts in triturating virgin ſilver with mercury. This amalgam is waſhed to ſeparate all the earth; it is preſſed through ſhamoy leather, and diſtilled in iron retorts. The ſilver is then fuſed and melted into ingots. This proceſs cannot be followed in the caſe of the ores which contain ſulphur; for that reaſon they are roaſted and mixed with lead, in order to refine the precious metal by cupellation. Such is the proceſs uſed for the rich ores of ſil-

ſilver; as for the poor ones, a different method from the two firſt is followed. They are fuſed without a previous roaſting along with a certain quantity of pyrites. This fuſion, called *crude fuſion*, gives a maſs of copper containing ſilver, which is treated by liquation with lead: this laſt, which has carried along the ſilver during fuſion, is afterwards ſcorified by cupellation, and the perfect metal remains pure. Cupellation in great differs from that practiſed in ſmall, in this, that in the firſt the lead ſcorified is driven up from the cupel by the action of the bellows; whereas, in eſſaying, the glaſs of lead is abſorbed by the cupel *.

 Sil-

* All the ſubſtances with which ſilver is minerallifed in its ore, may be ſeparated by fuſion with lead, which promotes the ſcorification of earths and imperfect metals, while it leaves the ſilver unaltered. The firſt part of the proceſs is to put one part of the ſilver-ore, well pulveriſed, into a veſſel along with eight parts of granulated lead. This veſſel is expoſed to a fire under a muffle, and the heat gradually increaſed. In a ſhort time the lead melts, and the pounded ore riſes up and floats a-top of it. A little after, the ore grows clammy, melts, and diſperſes to the ſides of the veſſel. At this period the fire is to be kept moderate, and then to be raiſed by degrees until the whole maſs has been brought into thin fuſion, and the ſurface of the lead be covered completely over with ſcoriæ. The operation is now finiſhed. The ſilver, during the fuſion, forſakes the earthy matters to unite with the lead, which collects with the whole of the ſilver into a maſs at the bottom, while the ſcoriæ ſwim a top without a particle of ſilver in them. It is ſometimes neceſſary to ſtir the matter with an iron rod; and if the ſcorification has been ſucceſsfully conducted, we find the rod, when cold, covered with a thin, ſmooth, ſhining cruſt, equally coloured on every ſide: but if, on the contrary, the ſcoriæ ſhow a conſiderable degree of clammineſs, adhere to the rod in quantity though it be quite red hot; if they be unequally tinged, ſeem duſty, and have rough grains interſperſed; you are certain that the proceſs is imperfect: for this is a ſign that the whole of the ore has not been converted into ſcoriæ; and therefore the ſilver cannot all precipitate with the lead to the bottom of the veſſel. This precipitation of the ſilver with the lead depends upon the attraction which ſubſiſts between the two metals, which diſpoſes them to join whenever they are brought into contact while fuſed, and upon the effect which the lead has to attenuate and promote the fuſion of the ſcoriæ; ſo that every particle of the ſilver may be ſet looſe from the matrix to precipitate with the lead into a metallic maſs. If the pro-

Silver, obtained by the processes just now mentioned, is in general much less alterable than all the metals hitherto described. Light, however long applied, makes no sensible change in its properties. Heat fuses it, makes

it

process has succeeded according to our wishes, we have a distinct vitrefied mass a-top, and a regulus at bottom, consisting of lead and silver. The process is called *the precipitation of silver from its ore*, because it is hereby detached from all earthy matters, and only left united with another metal, from which it is to be fused by a subsequent operation.

Once the silver is freed from all other impurities, and only dispersed over a quantity of lead, we get clear of the lead by cupellation. The cupel must be previously heated red hot, to expel the moisture completely: then the lead containing the silver is put into it; and the air having access to the melted metal, which is placed under a muffle, the lead quickly calcines, and is converted into litharge. But as the conversion of lead into litharge is owing to the admission of air, there must be a provision for the removal of what is already formed; otherwise it would collect on the surface, and protect the rest of the lead from undergoing the same change. Now this inconvenience is provided for by making the cupels of such spongy materials as will completely imbibe all the liquid litharge as it forms: so that in the process of cupellation the litharge is perpetually entering the body of the cupel, until the whole lead be absorbed, and the button of silver left purified on the top. There is here some nicety in regulating the heat: for, if it be too low, the litharge does not sufficiently liquefy and penetrate the cupel; and if too great, the litharge penetrates the cupel so rapidly as not to have time to scorify any heterogeneous matter which may still adhere to the silver. In a successful essay, the surface of the metal should appear convex and more red than the cupel; the smoke should rise almost to the top of the muffle; the middle of the essay should be smooth, and surrounded by a small circle of litharge, which is continually imbibed by the cupel. As the cupellation advances, the fire requires to be increased, because silver is more difficult to fuse, and helps to protect the lead. When the whole lead is absorbed, we perceive a sudden corruscation of the silver, which arises from the congelation of the surface; and the workmen say that it *lightens*. At the same time, we often see efflorescences shoot from the silver in different directions, termed the *vegetation of the button*. This appearance is produced by the contraction of the solid shell of silver, which squeezes out the melted matter in the centre through little cracks and crevices. In order to prevent this accident, which sometimes occasions a loss of metal, it is necessary to increase the heat, to keep the silver still in fusion, and to allow it to cool and to consolidate gradually. The

conit boil, and volatilses it, but without alteration. A white heat is requisite for its fusion; it is more fusible than copper. When it is kept in fusion for some time, it bubbles up and emits vapours, which are nothing but volatilised silver. This fact is proved by its being found in the vents of chimneys in which great quantities are continually fused. It is confirmed by the experiment of Messrs the Paris Academicians: By exposing very pure silver to the focus of M. Trudaine's lens, these gentlemen saw this metal, when fused, emit a thick smoke, and whiten a gold plate which was exposed to it.

Silver, by cooling slowly, is capable of a very regular form, or of crystallizing into quadrangular pyramids. M. Baumé had already observed, that it assumed a symmetrical form, which was evident at its surface from filaments resembling the down-hair of feathers. I have observed, that the pure button obtained by cupellation, exhibited frequently at the surface small spaces with five or six sides put between them like windows: but the crystallization in pyramids has been observed only by Messrs Tillet and l'Abbé Mongez.

It has been for a long time believed, and some chemists still think, that silver is indestructible by the combined action of heat and air. It is so far certain, that kept in fusion with the contact of air, it does not appear to be sensibly altered. However, Juncker has advanced, that by long reverberation, according to Isaac de Hollandors's way, the silver was changed into a vitrescent calx. This experiment was confirmed by M. Macquer. This learned chemist exposed some silver, twenty-four times successively, in a porcelain crucible, to a fire which

 bakes

continuance of the heat also permits any litharge which may lie under the button to enter the cupel; so the whole is completely absorbed, and nothing but silver left in a metallic form. There are some small variations requisite when the ore contains iron, tin, or copper. The mode of working in large is conducted upon the same principles, and is almost the same in every particular.

bakes that of sèves; and from the twentieth fusion he obtained a vitriform matter of a green olive, which seemed to be a true glass of silver. This metal, when heated with the focus of a burning lens, always presented a white powdery matter at its surface, and a greenish vitreous covering on the support upon which it was placed. These two facts cannot leave a doubt of the changeability of silver: though it be much more difficult of calcination than the other metallic substances, it is nevertheless susceptible of conversion in time into a white calx, which, urged with a violent fire, gives an olive-coloured glass. Perhaps it might be possible to obtain a calx by heating this metal when reduced to very fine laminæ in a matrass, as it is done in the case of mercury.

Silver suffers no change from the atmosphere: its surface is but very little tarnished even though it be long exposed; nor has water any action upon it. The earthy substances do not combine with it: it is probable that its calx would give an olive colour to glass with which it was made to enter into fusion. The saline earthy matters and the alkalis have no sensible action upon silver; nor has great attention been paid to this subject.

The vitriolic acid dissolves silver when it is very concentrated and boiling, and when the metal is in a state of great division. A great quantity of sulphureous gas is emitted from the solution: the silver is reduced to a white substance, upon which we must pour spirit of vitriol to obtain a solution. By evaporation very small needles of vitriolum lunæ crystallise. This salt fuses with heat: it is very fixed: it is decomposable by the alkalis, iron, copper, zinc, mercury, &c. All the precipitates obtained may without addition be reduced into fine silver in close vessels.

The nitrous acid dissolves silver with rapidity, and even without the assistance of heat. This solution goes on sometimes so rapidly, that to prevent its inconveniences, the silver is obliged to be employed in a mass.

A great deal of nitrous gas is disengaged, and a more or less copious white matter precipitates, if the spirit of nitre contain any portion of vitriolic or marine acid. The nitrous acid is generally coloured blue or green: it loses this colour, and becomes transparent when the solution is finished, and pure silver has been employed: but when the silver contains copper, there is a greenish shade more or less. Frequently the purest silver that can be employed contains some gold: then as the nitrous acid has but a very little action upon gold, as it acts upon the silver, the gold is separated from it in small blackish flocci, which collect at the bottom of the vessel, and are pure only gold unaltered.

From this different action of the nitrous acid upon these two metals, it is successfully employed for their separation in the operation of parting by aqua fortis. The nitrous acid can take up more than the half of its weight of silver. This solution is very caustic: it stains the epidermis black, and entirely corrodes it *. When it is very copious, it deposites small brilliant crystals like sedative salt: by evaporation to half its quantity, it gives, by cooling, flat crystals, which are either hexagons, triangles, or squares, and which seem formed of a great number of small needles adhering sidewise. These laminæ are placed obliquely upon one another. They are transparent and very caustic; called *nitrum argenti*, *nitrum lunare*, *crystalli lunæ*. This salt is readily altered by

* Besides staining the skin, the solution of silver in nitrous acid stains v getable matters, and many kinds of stone. It is now used to mark linen for the bleachfield, where the process of bleaching, instead of effacing the colour, increases the deepness of the shade. This additional strength of colour proceeds from the influence of the sun's light upon the metal in the salt, to which it furnishes phlogiston; and if any part be shaded from the sun, it will remain colourless. When agates and other stones are stained, they must also be exposed to the light before the solution makes any change. It is clear that the silver acquires phlogiston, as in some cases metallic particles are revived merely by exposure to the rays of the sun. From the corrosion of the skin by lunar caustic, the detached phlogiston adheres to the silver and gives the deep colour.

by the contact of the light, and blackened by combustible vapours. If it is put upon a burning coal, it detonates, and leaves a white powder, which is pure ſilver. It is very fuſible. If it is expoſed to the fire in a crucible, at firſt it bubbles up upon loſing the water of its cryſtals; then it remains in tranquil fuſion. If in this ſtate it is left to cool, it forms into a grey maſs, having ſlightly the appearance of needles, and forms a preparation known in pharmacy and ſurgery by the name of *lapis infernalis*. It is not neceſſary to uſe cryſtalli lunæ, which are very tedious to make and very expenſive; it is ſufficient to evaporate to dryneſs a ſolution by the nitrous acid, to put the reſiduum into a crucible or ſilver veſſel, as M. Baumé directed, and to heat it ſlowly till it be in perfect fuſion; then it is run into ingot molds, to give it the form of ſmall cylinders. If the ſticks of lapis infernalis are broken, it is obſerved that they are formed of needles, which divide in rays from the centre, ſeeming to terminate at the circumference. The nitrum lunare, in the preparation of the lapis infernalis, muſt not be long heated: without this precaution, a part of it is decompoſed, and a button of ſilver is found at the bottom of the crucible.

In order to obſerve what paſſes in the operation, I diſtilled cryſtalli lunæ in a pneumato-chemical apparatus. They gave me ſome nitrous gas, and a great quantity of dephlogiſticated air, the pureſt that I know. In my matraſs I found the ſilver entirely reduced; the glaſs had put on the opacity of enamel, and was of a fine marrone brown colour. This brown colour is ſurely owing to mangeneſe or ſome other ſubſtance contained in the glaſs; for the colour of the glaſs, formed by the calx of ſilver, inclined to an olive green, as has been already obſerved.

Nitrum lunare expoſed to the air does not attract humidity; it very eaſily diſſolves in water, and may be cryſtalliſed by a ſlow evaporation.

The nitrous ſolution of ſilver is decompoſed by the

ſa-

ſaline earthy and alkaline ſubſtances; but with very different phenomena, according to their ſtate. Lime-water forms a very copious olive precipitate. The aërated fixed alkalis precipitate it white; the cauſtic volatile alkali, of a grey inclining to an olive-green. Although the nitrous acid has manifeſtly the ſtrongeſt action upon ſilver, it has not the greateſt adheſion and affinity with it: the vitriolic and marine acids are capable of carrying off the ſilver with which it is combined. For that reaſon, by pouring a few drops of theſe acids into a nitrous ſolution of ſilver, a precipitate in a white powder is formed when the vitriolic acid is uſed; and in thick flocci, like a coagulum, by the uſe of the marine. In the firſt caſe, vitriolum argenti is formed; in the ſecond, a marine ſalt of ſilver. Theſe two ſalts, not being very ſoluble, are precipitated. It is not neceſſary to uſe the vitriolic and marine acids, uncombined, to effect theſe decompoſitions; for this purpoſe, the neutral ſalts, reſulting from their union with the alkalis and earths, may be employed: hence a double decompoſition and combination takes place; the nitrous acid ſeparating from the ſilver, unites with the baſe of the vitriolic and marine ſalts.

Upon this difference of relation between the acids and ſilver, a proceſs is formed, which is uſed to obtain a very pure nitrous acid, exempt from the mixture of other acids, and ſuch as is required in ſeveral metallurgical operations, and in moſt part of chemical inveſtigations. As, in diſtilling nitrous acid, it rarely happens that a certain part of the vitriolic or marine acid does not come over along with it, chemiſts have found methods of ſeparating theſe foreign fluids; and for this purpoſe make uſe of the nitrous ſolution of ſilver with great advantage. A nitrous ſolution of ſilver is poured into the impure nitrous acid, till no farther precipitation is occaſioned: the precipitated vitriol of ſilver, or horn-ſilver, is allowed to ſubſide; the acid is poured off, and diſſolved with a gentle heat, to ſeparate the

ſmall

ſmall portion of lunar ſalts which it may contain. What is obtained is very pure nitrous acid; it is called *aqua-fortis præcipitata*.

The greateſt part of the metals are capable of decompoſing the nitrous ſolution of ſilver, in conſequence of their ſuperior affinity with the nitrous acid. The arſenical neutral ſalt, diſſolved in water, produces, in a nitrous ſolution, a reddiſh precipitate, formed by the union of the ſilver with the arſenic. This precipitate imitates the red ore of ſilver. Silver is obtainable in its metallic ſtate by precipitation with moſt of the metals and ſemimetals; but as the ſeparation only of this perfect metal by means of mercury and copper is of any importance, it is to be conſidered here on account of the phenomena of the firſt and the utility of the ſecond.

Silver, ſeparated from the nitrous acid by means of mercury, is in its metallic ſtate; and the ſlowneſs of its precipitation gives riſe to a particular ſymmetrical arrangement, called *arbor Dianæ*, or *arbor philoſophorum*. There are ſeveral proceſſes for obtaining this cryſtallization. Lemery directed an ounce of fine ſilver to be diſſolved in moderately ſtrong nitrous acid; which ſolution is to be diluted with about twenty ounces of diſtilled water; then two ounces of mercury are to be added. In forty days, a very beautiful vegetation forms. Homberg has given a much ſhorter proceſs. An amalgam, made in the cold with four gros of ſilver in thin plates, and two gros of mercury, is diſſolved in a ſufficient quantity of nitrous acid; to this ſolution a pound and a half of diſtilled water is added. Into an ounce of this liquor a ſmall ball of a ſoft amalgam of ſilver is immerſed, and the precipitation of the ſilver almoſt inſtantly takes place. The precipitated ſilver, united with a portion of mercury, falls in filaments, as if priſmatic, upon the ſurface of the amalgam; other filaments lay themſelves upon the firſt, in ſuch a manner as to imitate a vegetation in form of trees. Laſtly, M. Baumé has deſcribed a method of obtaining Diana's tree; in which there is a little difference from Hom-

berg's

berg's proceſs, and which ſucceeds more certainly. He directs mixing ſix gros of a ſolution of ſilver, and four gros of a ſolution of mercury, by the nitrous acid, both ſaturated, and to add to them five ounces of diſtilled water, and to pour them into a glaſs veſſel upon ſix gros of an amalgam, made with ſeven parts of mercury and one of ſilver. Theſe two methods ſucceed with much more readineſs than thoſe of Lemery, by the reciprocal action and relation between the metallic ſubſtances. The mercury contained in the ſolution attracts that of the amalgam; the ſilver contained in this laſt alſo acts upon that contained in the ſolution; and from theſe attractions a more ready precipitation of the ſilver takes place. The mercury which makes part of the amalgam being more abundant than would be neceſſary to precipitate the ſilver from the ſolution, produces a third effect of very important conſideration: it is that which attracts the ſilver, by the affinity and tendency which it has to combine with this metal: it effectually combines with it, Diana's tree being only a brittle cryſtalliſed amalgam. This cryſtallization ſucceeds much better in conical veſſels, as of glaſs, than in round or wide veſſels, ſuch as the cucurbit recommended by M. Baumé. It is found alſo neceſſary to keep the veſſel in which the experiment is made, from agitation, which might oppoſe the ſymmetrical arrangement of the amalgam.

Copper, put into a ſolution of ſilver, likewiſe precipitates this metal in its brilliant and metallic form. This proceſs is generally employed to ſeparate the ſilver from its ſolvent, after the parting is finiſhed. Copper plates are ſteeped in the ſolution, or rather the ſolution is put into copper veſſels: the ſilver inſtantly ſeparates in tufts of a whitiſh grey colour. The liquor is decanted when it is blue, and no more ſilver precipitates. The ſilver is waſhed with ſeveral waters: it is fuſed in crucibles, and then ſubmitted to cupellation with lead, to ſeparate a portion of copper to which it is united in the precipitation.

tion. The ſilver thus procured is the pureſt of all: it is twelve marks fine. From theſe two precipitations by mercury and copper, it is clear, that the metals, ſeparated from their menſtrua by metallic ſubſtances, are precipitated with all their properties *.

The marine acid, in its ſtate of fluidity, does not immediately diſſolve ſilver; but it appears very ſuſceptible of diſſolving it when in the ſtate of a gas; as the operation of the concentrated parting ſhows, which conſiſts in expoſing to the fire golden laminæ alloyed with ſilver, cemented with a mixture of martial vitriol and common ſalt: the vitriolic acid diſengages the ſpirit of ſalt, which directs its action to the ſilver, and diſſolves it. The following is a much more ready and more eaſy proceſs for combining the marine acid with ſilver. Some ſpirit of ſalt is poured into a nitrous ſolution of ſilver: the very copious precipitate which is inſtantly formed, is the combination of the marine acid and ſilver, which has a greater affinity with it than with that of nitre, and which conſequently quits this laſt to unite with the marine. The ſame combination is obtainable by pouring ſome marine acid into a ſolution of vitriol of ſilver; becauſe this acid has a greater affinity with the metal than the vitriolic has. The marine acid may alſo be combined with it, by heating it upon a calx of this metal precipitated from the nitrous acid by fixed alkali.

The marine ſalt of ſilver has ſeveral properties which it is of importance to know. It is remarkably fuſible. When it is expoſed in a phial to a gentle fire, as upon

* When the ſilver is precipitated from acid of nitre by means of copper, we find that one part of copper is ſufficient to precipitate three parts of ſilver, and to ſaturate all the acid which held the ſilver diſſolved. In this caſe, moſt authors obſerve, that the ſilver is not quite pure, but contains a ſmall quantity of copper; as indeed, in almoſt all precipitations, ſome of the precipitant adheres to the precipitate. The ſilver is found, in a metallic ſtate, divided into extremely minute particles; and upon this account employed in cold ſilvering, where the ſilver is applied to the ſurface of other metals by means of detergent ſalts, which clear them of all impurities.

it fuses into a grey semitransparent sub-
milar to horn; and for that reason it is
nea. If it is run upon a porphyry, it fixes
ubstance, as if crystallised into argentine
is long heated with the contact of air, it
: it easily passes through the crucibles;
lised, and another is reduced into metals,
ules of silver, set in the portion of luna
ompofed. Luna cornea, exposed to the
hite colour, and very soon turns brown.
water but in very small quantity; since
ling distilled water takes up three or four
ng to experiments of Monnet. The al-
le of decomposing luna cornea dissolved
reated with heat, along with these salts:
employed for obtaining the purest and
ich we know. A mixture of four parts of
d one of luna cornea, is put into a cru-
l: when it is in proper fusion, it is taken
left to cool, and broken: the silver is se-
is found below the febrifugal salt formation, and the superabundant alkali em-
aumé, the author of this process, says,
y of alkali which he prescribes prevents
a from passing through the crucible, by
l its parts, which it decomposes at once.
as observed, that the concrete volatile
ixed with luna cornea, does not decom-
at it unites without precipitating the sil-
bination seems to resemble the sal alem-
g to M. Macquer. The Berlin chemist
oned has given another process for the
na cornea, and for obtaining the silver
per-

o call to mind here, that the most part of the me-
s with marine acid exhibit this property. Many
fusible enough to merit the name of *butters*, which
n by chemists; such are the butters of arsenic,
r, zinc, and tin, &c.

perfectly pure. Five gros sixteen grains are triturated in a mortar, with an ounce and a half of concrete volatile alkali, making an addition of as much water as will form a paste: the mixture is stirred till the effervescence cease; then three ounces of very purified mercury are added; and the whole triturated till we perceive a fine amalgam of silver: it is washed with water, then triturated, and the washing renewed, till the water come off clear, and the amalgam be very brilliant; then this last substance is dried, and distilled in a retort, till the vessel be of a white red: the mercury passes over into the receiver, and the silver is found pure at the bottom of the retort. In this way silver is obtained in perfect purity, and without any sensible loss. It is this silver which should be used for delicate chemical experiments. The water made use of for washing the mixture has taken along with it two substances; a certain quantity of sal ammoniac, which it holds in solution, and a white powder, which it cannot dissolve. When this powder is sublimed, a small quantity of silver is found at the bottom of the subliming vessel. This experiment proves, that luna cornea is not completely decomposed except by means of a double affinity; and in fact, in Margraaf's process, the volatile alkali unites with the marine acid, only because the silver combines upon its part with the mercury, which attracts and solicits it to quit the acid, which the alkali alone cannot effect. But it is evident, that this tedious and expensive operation cannot be made use of except in small, as in our laboratories. If luna cornea were to be reduced in great quantities, either the fixed alkalis, or some of the metallic substances, would be employed, the most part of which has more affinity with the marine acid than silver has. Among the rest of this sort, are regulus of antimony, lead, tin, and iron, &c. If one part of luna cornea, and three of any other substances, are mixed in a crucible, the silver is found reduced at the bottom of the crucible, and the metal employed, united with the marine acid, above it. The ſilver thus ſeparated is very impure. It contains always a portion of the metal uſed for reduction; and as lead, which is directed by Kunckel, is moſt generally employed, the ſilver thus obtained requires cupellation; and by this means never can be brought to that purity of which it is capable by the alkalis or Margraaf's proceſs.

Aqua regia acts very well upon ſilver, which precipitates as the ſolution goes on. This effect is very eaſily conceived: the nitrous acid firſt diſſolves the ſilver which is taken from it by the marine, forming luna cornea, which is depoſited on account of its little ſolubility: in this way may ſilver be ſeparated from gold.

The action of the other acids upon ſilver is not well known; only that a ſolution of borax produces a very copious white precipitate in a ſolution of this metal in nitrous acid, which is owing to a combination of ſedative ſalt and calx of ſilver.

This metal ſeems to be unalterable by the neutral ſalts; at leaſt it is ſo far certain, that it does not detonate with nitre, nor decompoſe ſal ammoniac. This unalterability of ſilver by nitre furniſhes a good way of getting rid of the imperfect metals which may happen to be united with it by means of their calcination; ſuch as copper, lead, &c. The ſilver too much alloyed is fuſed with nitre: this ſalt detonates, and burns the imperfect metals; and the ſilver is found at the bottom of the crucible: it is much purer than it was before.

Almoſt all the combuſtible matters have a more or leſs remarkable action upon ſilver. No metal is more quickly tarniſhed by inflammable matters. Hepatic gas, from whatever ſubſtance it is produced, gives it, as ſoon as it touches it, a blue or violet colour inclining to black, and diminiſhes its ductility. It is a matter of obſervation, that animal hepatic vapours, ſuch as thoſe of privies, of putrrefied urine, warm eggs, produce the ſame effects upon it. The reciprocal action of theſe two bodies has not yet been examined, nor the kind of

 com-

combination formed. Sulphur combines very well with it. For this purpose silver plates are stratified with flowers of sulphur, and quickly fused; a mass of a violet black results, much more fusible than the silver, brittle, and disposed in needles; in short, a true artificial ore. This combination is easily decomposed by the action of fire, on account of the volatility of the sulphur and the fixity of the silver: the sulphur is consumed and dissipated, and the silver remains pure. Liver of sulphur dissolves this metal in the dry way; by fusing one part of silver with three of liver of sulphur, the metal disappears, and may be dissolved in water along with the liver. If an acid is poured into this solution, a sulphureous black precipitate of silver is obtained. Silver-leaf, put into liquid liver of sulphur, very soon assumes a black colour; and it seems that the sulphur leaves the alkali to unite with and mineralise the silver, as we have already seen with regard to mercury.

Silver unites with arsenic, which renders it brittle. The action of the arsenical acid upon it is not known. It difficultly combines with cobalt. It alloys very well with bismuth, with which it forms a brittle mixed metal, whose specific gravity is greater than that of the two metals taken separately.

According to M. Cronstedt, it does not unite with nickel: these metals, fused together, lie at the side of one another, as if their specific gravity was perfectly identical.

It fuses with regulus of antimony, and forms a very brittle alloy. It seems capable of decomposing the antimony, and of uniting with its sulphur, with which it has more affinity than with the regulus. Zinc easily combines with silver by fusion; and a very brittle alloy, granulated at its surface, results from this combination.

In mercury it dissolves completely, even in the cold: silver-leaf is mixed with it; and an amalgam instantly results, varying in consistence, according to the respective quantity of the two substances employed.

This

This amalgam is susceptible of a regular form by fusion and slow cooling: it yields three-sided prismatic crystals, terminated by similar pyramids. The mercury assumes a sort of fixity in this combination; for, in order to separate it from the silver, a more considerable degree of heat is required than that which it requires for volatilisation *per se*. Silver can decompose corrosive sublimate, both by the dry and moist way. It perfectly unites with tin, by which it is totally deprived of all its ductility. It alloys readily with lead, which renders it very fusible, and deprives it of its elasticity and sonorous quality. It alloys with iron; which, although little examined, yet might perhaps become of great use in the arts.

Silver is a metal which is of singular utility on account of its ductility and its indestructibility by the fire and air. Its brilliancy recommends it as a beautiful ornament. It is applied upon the surface of different bodies, and also upon copper. Stuffs are wrought with it, and have their beauty heightened. But its most important use is to form, by its hardness and ductility, vessels of all forms. Table-silver is generally alloyed with a twenty-fourth of copper, which gives it more hardness and adherence; exposing the health to no inconvenience, because the twenty-three parts of silver cover and entirely destroy the deleterious properties of the copper.

Lastly, Silver is employed as the price of all merchandises, and is formed into money; but in this case it is alloyed with a twelfth of copper, and is called *silver of eleven marks*.

LECTURE XLI.

Species 14. Gold.

GOLD, or *Sol* of the chemists, is the most perfect and most unchangeable metal known: it is of a brilliant yellow colour. It is the heaviest body in nature;

ture; it loſes only between a nineteenth and a twentieth part of its weight in water. Its hardneſs is not very conſiderable, as alſo its elaſticity. Its aſtoniſhing ductility, ſo conſpicuous in the art of wire-drawing and of gold-beating, is ſo great, that one ounce of it is able to gild a ſilver thread 444 leagues long; and it may be converted into leaves which the wind may carry away. A grain of gold, according to Lewis's calculation, can be made to cover a ſpace of more than 400 ſquare inches. It is the moſt tenacious of all the metals; a weight of 500 pounds being ſupported by a thread of it only one-tenth of an inch in diameter, before breaking. It ſtiffens eaſily under the hammer; but nealing reſtores all its ductility.

The colour of gold is ſuſceptible of a great number of varieties: there is ſome of it more or leſs yellow or pale; ſome of it which inclines almoſt to white: it is ſuppoſed that theſe differences depend upon ſome foreign ſubſtance. Gold has neither ſmell nor taſte: it is ſuſceptible of cryſtallization, by cooling, into ſhort quadrangular pyramids. In this form it was obtained by Meſſrs Tillet and Mongez.

Gold is found always pure in nature. It is ſometimes found in ſmall iſolated or continued maſſes, and ſcattered in quartz: at other times, it is in ſmall pieces amongſt the ſand at the bottom of waters: laſtly, it is got from ſeveral ores, into whoſe compoſition it enters; ſuch as the galenas, the blendes, red ores of ſilver, and native ſilver. It is united almoſt always with a certain quantity of ſilver, and forms a kind of natural alloy.

Mineralogiſts diſtinguiſh ſeveral varieties of native gold, in plates, in grains, in octaëdral cryſtals, in priſms with four ſtriated ſides, in threads, and in irregular maſſes. M. Sage thinks, that priſmatic native gold is combined with a certain quantity of mercury, which renders it brittle.

The eſſaying of a gold-ore ſhould differ according to its nature: if it is native gold, pulveriſation and waſhing

ſuffice;

uffice; if gold is alloyed with other metals, the ore muſt be roaſted, fuſed, and cupelled with lead, and parted.

The way in which gold is extracted from its ores is very eaſily conceived, from what has been already ſaid upon metallurgy. Native gold requires only ſeparation from its matrix. For this purpoſe it is taken to the ſtamping-mill: it is waſhed, to carry away the matrix, which has been reduced into powder; it is brayed in a veſſel full of water, with ten or twelve parts of mercury; the water is made to run off, and carries away every thing which is earthy. When this amalgam, which is formed in the operation, is freed of all the earth, and it appears very pure, it is ſqueezed through ſhamoy leather; a good part of the mercury paſſes through, and the gold remains united with a certain quantity of it. The amalgam of gold is heated, and the mercury ſeparated by diſtillation; then the pure gold is fuſed which reſults from this operation, and run into bars or ingots. With regard to the gold which is found combined in the ores of other metals, as in thoſe of lead and copper, it is extracted by cupellation and parting from the former; and the lead, which runs off during the liquation of the copper, carries away the ſilver, and gold; then it is cupelled, and the ſilver is ſeparated by parting *.

 Gold

* Beſides parting by quartation, there are two other kinds of parting practiſed in trade, the concentrated parting and the dry parting. The concentrated parting conſiſts in expoſing the metal to very ſtrong acid fumes, by cementation with a mixture of proper materials. The mixture is a compound of calcined vitriol and nitre well rubbed together in powder. Some of this is put into a cementing pot, and the piece of gold to be purged, beat into the ſize of a common coin, is laid over it, and layers of the metal and compoſition, ſtratified alternately, till the veſſel be full. By raiſing the fire, the vitriol acts upon the nitre, and ſets the acid free; and thoſe acid vapours inſtantly penetrate the metallic maſs, and diſſolve the ſilver. It is, however, obſerved, that they do not thoroughly pervade the whole, deſtroying only the ſilver which lies near the ſurface; ſo that this method is only employed when nothing but a ſuperficial purity is wanted.

In

Gold expofed to the fire turns red long before fufion. When it is very red, it appears brilliant and of a clear green colour, refembling the beryl. It does not fufe till it be of a white red: cooled flowly, it cryftallifes. It fuffers no alteration whatever by the ftrength of the fire, or length of expofure to it, fince Kunckel and Boyle kept it feveral months in a glafs-houfe fire without appearance of alteration. However, this kind of unalterability is only relative to the fires which are made by means of combuftible fubftances; fince there appears no doubt, that a much more active heat, as the focus of burning lenfes, is capable of depriving gold of its metallic properties. Homberg obferved, that upon expofing gold to the focus of Tfchirnhaufen's lens, it fmoked, was volatilifed, and even vitrefied. M. Macquer faw gold expofed to the focus of M. de Trudaine's lens fufed, which emitted a fume that gilded filver, and was gold only volatilifed: the globule of gold, when fufed, was

In dry parting, the alloy of gold and filver is granulated, by making the melted metal run from a crucible into water, which is kept in rotatory motion, fo as to divide the metal into grains. Three-fourths of this granulated metal is rubbed in flowers of fulphur, which adheres to the furface while wet; and this is put into a crucible over the fire, and remelted with a gradual heat, that the fulphur may penetrate the filver without flying off; for the procefs proceeds upon the principle of fulphur uniting with filver, while it has no attraction for gold. After the heat has brought the fulphurated filver into thin fufion, the gold-holding granulated filver is added by degrees, which, as it melts, diffufes all over the mafs, and collects the particles of gold which were interfperfed in the fulphurated filver, and precipitates with it to the bottom of the crucible. For although filver has greater affinity to fulphur, and combines intimately with it by a kind of cementation, it does not eafily mix with fulphurated filver which is already in fufion. By this operation, the proportion of filver to gold is much diminifhed, and the purification may afterwards be completed in the ordinary way. This procefs is found to be fo ufeful in great works, that at Rammelfberg, where 100 pounds of ore contain about 40 pounds of tin, and the tin 110 grains of filver, and the filver one-third of a grain of gold, the tin is wrought off, and this minute quantity of gold feparated with profit.

was agitated with a rapid motion, was covered with a dull pellicle, wrinkled, and apparently calciform: that at laſt a violet vitrification was formed in its middle. This vitrefication gradually increaſed, and gave riſe to a ſort of cap of a greater inflection than that of the maſs of gold which was encaſed in this maſs, as the tranſparent cornea is upon the ſclerotic coat. This glaſs increaſes as the gold diminiſhes. Upon the ſupport is always found ſomewhat of a purple ſtain, which ſeemed to be owing to a portion of glaſs which had been abſorbed, M. Macquer has not yet had opportunity to vitrefy a given quantity of gold. This celebrated chemiſt obſerved, that it was neceſſary to reduce this violet glaſs with combuſtible matters, in order to be aſſured of its containing gold, before concluding that it was the calx of this perfect metal. However it be, it is our opinion, that it is a true vitrefied calx of gold; and with the firmer foundation, that in ſeveral proceſſes upon this metal, which we ſhall preſently deſcribe, it aſſumes conſtantly the purple colour; and that ſeveral of its preparations are employed to communicate this colour to enamels and porcelain. Like the other metals, then, gold is calcinable; only, like ſilver, it requires to unite with the air a more intenſe heat, and a longer time, than all the metallic ſubſtances. Theſe circumſtances are, without doubt, relative to its denſity.

Gold is unchangeable by the air; its ſurface is only tarniſhed by foreign ſubſtances which float inceſſantly in it; nor is it altered by water in any degree. From the inveſtigations, however, of M. le Compte de la Garaye, it appears to be capable of dividing it, nearly in the ſame way as it does iron.

Gold, in its metallic ſtate, does not combine with the earths and ſaline earthy ſubſtances; but its calx can enter into the compoſition of glaſs, to which it gives a violet or purple colour.

The moſt concentrated vitriolic acid, aided even by heat, has no degree of power over this perfect metal.

The nitrous acid may take up ſome atoms of it: only, perhaps, in a mechanical way, rather than by a true combination. M. Brandt is one of the firſt chemiſts who has announced its ſolubility in this acid: Meſſrs Scheffer and Bergman have confirmed it. But it is to be obſerved, that, from the experiments of the whole body of chemiſts of the Academy, this acid takes up the gold only under certain circumſtances, which theſe gentlemen have not yet mentioned. M. Deyeaux, member of the College of Pharmacy, has diſcovered, that the nitrous acid diſſolves gold only when it is ſmoking and charged with nitrous gas. The acid in this ſtate is, in his opinion, not pure: he calls it acid charged with gas, and compares it to a kind of aqua regia.

The marine acid *per ſe*, and in a pure ſtate, does not ſenſibly act upon gold. Meſſrs Scheele and Bergman aſſures us, that this acid, dephlogiſticated or diſtilled from the black calx of manganeſe, which, according to theſe chemiſts, ſeizes upon this acid's phlogiſton, diſſolves gold like aqua regia, and with this metal forms the ſame ſalt which is uſually produced by employing this ſame mixed acid. Theſe experiments have not yet been repeated in an accurate manner; and the reſult of theſe experiments has not yet received the ſanction of chemiſts, which ought to eſtabliſh the reliance upon them which they deſerve *.

Aqua regia is ſtill looked upon as the true ſolvent of gold. We ſhall not repeat what we have elſewhere ſaid upon the nature, properties, and differences of this mixed acid, according to the quantity of the two acids combined for its formation. Here we ſhall take notice only of

* Our learned author ſeems here to entertain great doubt of the ſolubility of this perfect metal in the acid in this ſtate. It is a fact of which he might have ſatisfied himſelf with the greateſt facility. There is no metal whatever which marine acid, thus dephlogiſticated, will not diſſolve; and it very readily acts upon gold when conveniently expoſed to its action.

of its action upon gold. As ſoon as aqua regia comes into contact with this metal, it attacks it, raiſing an efferveſcence, which is greater or leſs, according to its concentration, the heat, and the diviſion of the gold. An acceleration of the proceſs may be produced by heat, or at leaſt the heat may favour the commencement of their action. The air bubbles riſe without interruption, till a portion of the metal is diſſolved. This action gradually ſtops till agitation or heat be employed. The gas diſengaged in this ſolution has not been examined. Aqua regia, charged with all the gold which it can take up, is of a more or leſs deep yellow: its cauſticity is conſiderable: it tinges animal matters of a deep purple colour, and corrodes them. When it is cautiouſly evaporated, it gives cryſtals of a fine golden colour, which reſemble topazes, and appear to be truncated octagons, and ſometimes four-ſided priſms. It is difficult to obtain this cryſtalliſation. M. Monnet is of opinion, that it ariſes only from the neutral ſalt formed in aqua regia; and he ſays farther, that in order to obtain it, it is neceſſary to employ an aqua regia made with nitrous acid and ſal ammoniac or marine ſalt: Hence this mixed acid contains either cubic nitre, or nitrous ammoniac. The one or the other of theſe ſalts, according to this chemiſt, is the cauſe of the cryſtalliſation of the gold. It is not, however, entirely demonſtrated, that a ſolution of gold in aqua regia, which conſiſts of uncombined marine and nitrous acids, is incapable of giving cryſtals; ſince, as M. Bergman imagines, this ſalt of gold contains marine acid only which may be extracted by diſtillation. If the cryſtals of gold are heated, they fuſe and take a red colour: this ſalt ſtrongly attracts the air's humidity. If a ſolution of gold is diſtilled in a retort, a fine red liquor is obtained, which has brought over with it a little gold, and ſeems to be marine acid. The alchemiſts, who were very laborious upon gold, gave the name of *red lion* to this liquor. Some cryſtals of a reddiſh yellow are alſo ſublimed. The greateſt part of this metal remains at the bottom of the retort, and requires fuſion only in order to be very pure, and to enjoy all its properties.

The ſolution of gold may be decompoſed by a great number of ſubſtances. Lime and magneſia precipitate its gold in form of a yellow powder. The fixed alkalis preſent the ſame phenomenon: but it muſt be obſerved, that the precipitate forms but very ſlowly, and that the ſolution aſſumes a reddiſh colour, if more of the alkali was put in than was ſufficient; becauſe the exceſs of this ſalt rediſſolves the precipitate. This precipitate may be reduced by heat and in cloſe veſſels: it is a calx which eaſily parts with the air, with which it is combined. However, it is ſuſceptible of fuſion with the vitreſcent ſubſtances, and of giving them a purple colour; as for enamels and porcelain, a precipitate of gold is employed, which is formed by the admixture of a ſolution of gold and of the liquor ſilicum.

Gold, precipitated by the fixed alkalis, preſents a property ſtill very different from thoſe of gold in its metallic ſtate, which is its ſolubility in the vitriolic, nitrous, and marine acids, when uncombined. All theſe acids, heated upon a yellow ſolution of gold, eaſily diſſolve it; but they cannot take up enough of it to give cryſtals. When theſe ſolutions are evaporated, the gold very readily ſubſides, which it does too by ſimple repoſe. M. Monnet has taken notice of a fact which ſhould not be forgotten, in the precipitation of a ſolution of gold by means of the gall-nut*; it is, that this pre-

* As we have made mention only of the precipitation of iron by the gall-nut, we think it proper to give here a brief account of the phenomena which this aſtringent ſubſtance preſents with the moſt part of the metallic ſolutions.

The ſolution of cobalt is precipitated of a clear blue by it; that of zinc, of an aſh-coloured green; that of copper, of a green, which becomes grey and reddiſh; that of ſilver, at firſt in reddiſh ſtriæ, ſoon aſſuming the colour of burnt coffee; that of gold, purple; that of precipitate, which is reddiſh, diſſolves very well in the nitrous acid, which by this means acquires a blue colour.

The volatile alkali throws down from the ſolution of gold a much more copious precipitate. This precipitate, which is of a brown yellow, and ſometimes of an orange, has the peculiar property of detonating with a conſiderable noiſe when it is ſlowly heated: it has been called *aurum fulminans*. The volatile alkali is abſolutely neceſſary for its formation: for this purpoſe, it may be produced either by precipitating with the fixed alkali a ſolution of gold made in aqua regia, which is compoſed of ſal ammoniac; or, what is better, by precipitating with the volatile alkali a ſolution of gold made in aqua regia, conſiſting of the nitrous and marine acids in their pure ſtate. A fourth more of fulminating gold is always obtained than the quantity of gold which has been diſſolved in the aqua regia. There are ſeveral precautions deſerving attention, relatively to the terrible effects of aurum fulminans. It ought to be carefully dried in the open air, without expoſing it to the heat of a fire. As a ſtrong heat is not requiſite for its fulmination, and as friction alone ſuffices for the production of the exploſion, the veſſels in which it is kept ſhould be ſtopped only with cork. Unhappy experience has appriſed us of the dangers incident from the uſe of cryſtal ſtoppers, by the friction which they produce upon the neck of the veſſels when any portion of

of platina, black. Theſe are the facts obſerved and deſcribed by M. Monnet; who has obſerved beſides, that theſe precipitates are ſoluble in the acids, and that the alkalis unite to their ſolutions without any ſign of precipitation.

Meſſrs the academicians of Dijon have added to theſe facts the following obſervations: The ſolution of arſenic is not altered by the gall-nut; that of biſmuth gives a greeniſh precipitate; that of nickel, a white; that of antimony, a blueiſh grey; that of lead, a ſlaty-like depoſition, whoſe ſurface is covered with pellicles mixed of green and red; laſtly, that of tin becomes of a dirty grey by the mixture of the gall-nut; and it gives a precipitate which is copious, and as if mucilaginous.

of the powder ſticks there. A terrible accident happened with M. Baumé; an account of which is found in his Experimental and Rational Chemiſtry, vol. iii. p. 79.

Chemiſts have entertained different notions upon the cauſe of the detonation. M. Baumé thought, that in the detonation a nitrous ſulphur was formed, to which he aſcribed the fulminating property. But M. Bergman has proved the inadmiſſibility of this theory, having diſcovered a method of making fulminating gold without nitrous acid, by diſſolving a precipitate of gold in vitriolic acid, and precipitating it afreſh by the volatile alkali. Nor upon the nitrous ammoniac has the fulmination any greater dependence; for after this compound has been waſhed with plenty of water, to carry off all the ſalt, if it contained any, it does not loſe its fulminating property. Upon a ſtrict examination of what paſſes in the detonation, it is obſerved to inflame the moment it ſhines. If it is heated upon a gentle fire of hot cinders, brilliant ſparks fly off before exploſion, reſembling the electric flaſh: it detonates when it is expoſed to the ſpark produced by the electric bottle; a ſimple ſpark does not kindle it without commotion: laſtly, after fulmination, a part of the gold is left in the ſtate of a purple calx, and another part in its metallic ſtate. Hence it appears, that the fulmination depends upon a combuſtible matter contained in it; and as the alkaline gas, to which we have aſcribed a kind of combuſtibility, is required for the production of fulminating gold, is it not very probable that the exploſion of this ſubſtance depends upon this alkaline gas? This theory is the more ſupported, as M. Berthollet, my colleague, obtained ſome alkaline gas by heating ſome fulminating gold in copper tubes, whoſe extremity was immerſed, by means of a ſyphon, into a pneumato-chemical apparatus with mercury. After this bold experiment, the gold was no longer fulminating, and it was reduced into a calx. M. Bergman made an experiment

riment of the ſame kind, which proves abſolutely the ſame thing. He expoſed gold to a heat which was incapable of making it fulminate; and he deprived it of this property, no doubt, by the gradual volatiliſation of the alkaline gas. However, it muſt be obſerved, that this gas is truly, and even very intimately, combined with the gold; ſince boiling water, M. Bergman ſays, carries none of it away; and the vitriolic acid and fixed alkali are unable to deprive it of its fulminating virtue. However, M. Baumé, who cites this chemiſt's experiments, whoſe work he ſeems to know, ſays, that vitriolic acid and fixed alkali decompoſe fulminating gold. Fulminating gold ſeems ſuſceptible of loſing its property by admixture with ſulphur. Oil very ſoon deprives it of it, as Meſſ. Rouelle and d'Arcet have diſcovered. Æther, and ſeveral other inflammable bodies, whoſe properties ſhall be mentioned in the hiſtory of the vegetable kingdom, are, according to M. Bergman, proper to deprive fulminating gold of this property. A ſingular property of fulminating gold, and which announces its ſtrength, is, that when made to fulminate upon a plate of metal, ſuch as lead, tin, and even ſilver, a mark or hole is made in the plate: farther, it does not ſeem ſuſceptible of kindling in cloſe veſſels, ſince M. Lewis found, that when it was ſhut up in an iron ball, it produced no exploſion. Fulminating gold is very capable of ſolution in the acids: when it is ſeparated from theſe ſalts by the fixed alkali, it is no longer fulminating; and if the volatile alkali is employed, it is found to be in its former ſtate.

The ſolution of gold is precipitated by liver of ſulphur. Whilſt the fixed alkali unites with the aqua regia, the ſulphur which is precipitated combines with the gold; but this combination is not durable: heat alone is required to volatiliſe the ſulphur, and to obtain the perfect metal in its ſtate of purity. We may obſerve here, that gold, precipitated from aqua regia by any intermedium whatever, is perfectly pure; and that it it is even more pure than parted gold; becauſe it is ſeparated from the ſilver which it might have contained by the precipitation of this metal with the marine acid, forming luna cornea, which takes place even during the ſolution of the gold, as has been mentioned more fully above.

Gold is not the metal which has the ſtrongeſt affinity with aqua regia; almoſt all the metallic ſubſtances have the power of ſeparating it from its ſolvent. Biſmuth, zinc, and mercury, precipitate it. A plate of tin put into its ſolution ſeparates this metal in the form of a deep violet powder, called *purple precipitate of Caſſius*. This precipitate is employed in painting enamels and porcelain, and is prepared by diluting a ſolution of tin in aqua regia with a great quantity of diſtilled water, and pouring into it ſome drops of a ſolution of gold. When the ſolutions are pretty copious in metal, a precipitate is preſently formed of a crimſon red, which becomes purple in a few days: the liquor is filtrated, and the precipitate waſhed and dried. This ſubſtance is a calx of tin and of gold: its preparation is one of the moſt ſingular operations in chemiſtry, on account of the variety and inconſtancy of the phenomena which it exhibits. Sometimes it gives a precipitate of a fine red; ſometimes its colour is a deep violet; and, what is more aſtoniſhing, it very often happens that the mixture of theſe two ſolutions occaſions no precipitate. M. Macquer, who has given a full account of theſe varieties, obſerves, that they depend almoſt always upon the ſtate of the ſolution of tin employed. If this ſolution has been made too rapidly, the metal is too much calcined, and it contains too little of it, in order that the aqua regia of the ſolution of gold may act upon it; for it is the action of this laſt upon the tin to which he aſcribes the formation of the purple precipitate of Caſſius. In order to ſucceed, then, in this operation, a ſolution, made very ſlowly, ſhould only be employed, in ſuch a ſtate as to contain the greateſt poſſible quantity

tity of tin without the metal being entirely calined in it. Hence he advises this method of preparing the purple precipitate. Let small pieces of tin be dissolved in aqua regia which has been made with two parts of spirit of nitre and one of spirit of salt, weakened with its equal weight of distilled water: on the other hand, let very pure gold be dissolved by means of heat in an aqua regia composed of three parts of spirit of nitre and one part of spirit of salt: let the solution of tin be diluted with a hundred parts of distilled water, and let it be divided into two portions: to one of the two let a new quantity of water be added: let them both be tried with a drop of solution of gold; then, by observation, distinguish the finest red, in order to treat them both in the same manner; then let the solution of gold be poured in till no more precipitate falls.

Lead, iron, copper, and silver, also have the property of separating gold from its solution. Lead and silver precipitate it in a dirty deep purple; copper and iron separate it in its metallic brilliancy; the nitrous solution of silver and that of martial vitriol, likewise occasion a red or brown precipitate in the solution of gold *.

The

* Gold may also be separated from aqua regia by the addition of certain inflammable substances, which furnish phlogiston to the gold-earth, and so revive it in a metallic state. Thus, when a quantity of any essential oil, or, still better, of vitriolic æther, is poured into a solution of gold in aqua regia, the gold gradually forsakes the acid menstruum to unite with the oil or the æther. This separation will even take place when the æther or essential oil is merely poured upon the surface of the solution, and left at rest; though the process will succeed better, and is sooner finished, when the two fluids are mixed together by agitation. The æther absorbs the whole gold, and keeps it afterwards perfectly suspended; and as it has no effect upon any other metal, this seems the best test of the presence of gold. It is also evident that the gold will be obtained in a state of most perfect purity, from the attraction of the æther being confined to this metal alone. The essential oils acquire a deep yellow colour when allowed to remain some time upon the solution; but do not long retain the gold suspended, as it soon separates,

The neutral ſalts have no remarkable action upon gold; only it is obſerved, that borax fuſed with it alters its colour, and makes it remarkably pale; whereas nitre and marine ſalt heighten it. The ſolution of borax poured into one of gold, occaſions a precipitate of ſedative ſalt charged with the metal.

Sulphur does not unite with gold; and is therefore uſed with advantage in the ſeparation of the metals which are united with gold, but particularly ſilver. This alloy is fuſed in a crucible: when it is fuſed, flowers of ſulphur are projected upon its ſurface: this ſubſtance fuſes and combines with the ſilver, ſwimming in the form of black ſcoriæ above the gold. It muſt be obſerved, that theſe two metals are never completely ſeparated by this operation, which is named *dry parting*, and is employed only to a maſs of ſilver which contains too little gold to be capable of indemnifying the expence ariſing from the uſe of aquafortis.

Liver of ſulphur diſſolves gold completely. It is Stahl's opinion, that by this proceſs Moſes diſſolved the golden calf which the Iſraelites worſhipped. The ſolution is formed by rapidly fuſing a mixture of equal parts of ſulphur and fixed alkali with an eighth of the whole weight of gold-leaf. This matter is run upon a po-

rates, and adheres to the ſides of the veſſel in the form of metallic films: yet after this ſeparation of the gold, the eſſential oils ſtill exhibit a bright yellow colour, which many through miſtake attribute to the particles of gold in the oil, though in reality the tinge ariſes from the action of the acids; as ſtrong acids do at all times produce this change upon the colour of eſſential oils. From this error many have been led to believe themſelves in poſſeſſion of potable gold, by diſſolving thoſe coloured oils in highly rectified ſpirit of wine; and either from fraud, ignorance, or deluſion, endeavoured to impoſe upon the credulity of mankind; for the alchemiſts boaſted of certain mild ſolutions of gold as invaluable ſecrets: they proclaimed them to be efficacious remedies in every diſorder; to cure every human evil without pain; and even to be capable of prolonging life itſelf. How fallacious thoſe hopes were, and how imprudent it is to truſt in the promiſes of pretenders to myſterious diſcoveries, is now ſufficiently known from the detection of many impoſtures of this kind.

poryphyry: it is pounded; ſome diſtilled water is poured upon it: it forms a ſolution of a yellowiſh green, which contains a golden liver of ſulphur. The metal may be precipitated by means of the acids, and ſeparated from the ſulphur, which is depoſited along with it, by the application of heat in open veſſels.

Gold combines with the moſt part of the metallic ſubſtances, and exhibits ſeveral important phenomena in its combinations.

It unites with arſenic. This ſemimetal renders it brittle, and gives it a very pale colour. There is great difficulty in ſeparating by heat the laſt portions of arſenic from this alloy; the gold ſeems to give it fixity: its alloy with cobalt is not known. It unites with biſmuth, which renders it brittle. The ſame is the caſe with nickel and regulus of antimony. As all theſe metals are very calcinable, and the moſt part fuſible, it is very eaſy to ſeparate them from the gold by the action of fire combined with that of the air. Crude antimony was boaſted of by the alchemiſts for the purification of gold. When it is fuſed with this metal, alloyed with any foreign metallic ſubſtances, as copper, iron, or ſilver, the ſulphur unites with theſe ſubſtances, and ſeparates them from the gold, which is found at the bottom of the veſſel. This gold is alloyed with regulus of antimony: it is purified by heating it with a white-red heat. The regulus of antimony is volatiliſed; the laſt portions require a very violent fire: and it is obſerved, that this ſemimetal carries along with it ſome portions of gold in its volatiliſation. This proceſs, ſo celebrated among the alchemiſts has, therefore, no advantage over that in which ſulphur only is employed.

Gold very eaſily alloys with zinc: a mixt metal reſults, whoſe brittleneſs and whiteneſs increaſes according to the greater proportion of this ſemimetal. This alloy, made with equal parts, is of a very fine grain; and it takes ſo fine a poliſh, that it was recommended by Hellot to make mirrors of teleſcopes, which are not ſubject to tarnifh. When the zinc is feparated from the gold by calcination, the flowers of this femimetal are reddifh, and carry along with it a little gold, as Stahl has announced.

Gold has more affinity with mercury than with the other metallic fubftances, and is capable of decompofing their amalgams. It unites with it in all proportions, and forms an amalgam, whofe colour and folidity augment according to the greater or lefs quantity of gold employed. This amalgam liquefies by heat, and cryftallifes by cooling, as all the compounds of this femimetal do. What regular form it affumes is not properly known. M. Sage fays, that its cryftals refemble filver *en plume*, and with a microfcope feem to be quadrangular prifms: he fays alfo that the mercury here acquires fixity. This amalgam is employed to gild with. Although gold be incapable of calcination by the heat of our furnaces, and by the contact of air, it becomes calcinable, however, when heated conjointly with mercury. Some mercury being put along with a forty-eighth of its weight of gold into a matrafs with a flat bottom, whofe neck has been drawn out by means of an enameller's lamp, in order to leave but a very fmall opening; and the mixture being heated in a fand bath, as is done in the preparation of the calx of mercury called *precipitate per fe;* thefe two metallic fubftances calcine at one time: they are changed into a deep red powder; and this double calx is obtained even much more readily than that of mercury heated alone, according to M. Baumé. Here then is a metal which, though very difficult to calcine alone, haftens and facilitates, however, the calcination of another metallic fubftance, which *per fe* is very incapable of fuch an alteration.

Gold alloys very well with tin and lead. Thefe two metals deprive it of all ductility. Its alloy with iron is very hard, and may be ufed to make fharp inftruments, much fuperior to thofe which are made with pure fteel.

This

This mixt metal is grey and attractable by the magnet. M. Lewis proposes the use of gold in soldering properly and solidly small pieces of steel.

Gold combines very well with copper, which gives it a red colour and a great stiffness, and renders it more fusible. This alloy, in different proportions, is used for coin, and in the making of vessels and trinkets. Lastly, gold alloys with silver, which deprives it of its colour, and renders it very pale. This alloy, however, is not made without a certain difficulty, on account of the difference of weight between these two substances, just as Homberg has observed; who has seen them separate during fusion. The alloy of gold and silver forms the green gold of the jewellers.

As gold is of very extensive use, as is the case with silver, the price of all the productions of nature and of art, it is of very great importance to fix the degree of purity of this precious metal, to prevent the excess to which avarice might proceed, and to establish the value of all the masses or pieces of gold vended in commerce. By just and strict laws, the quantity of alloy has been prescribed, which is necessary to communicate hardness and stiffness to the gold destined for making utensils, in which these properties are requisite. Chemistry has furnished means of ascertaining the quantity of imperfect metals combined with gold. The operation is called *essaying the title of gold.* Twenty-four grains of the gold which we want to essay, are cupelled with forty-eight grains of silver and four of pure lead. This last, in its vitrefication, carries off the imperfect metals: the gold remains in combination with the silver after cupellation. These two metals are separated by an operation called *parting.* Parting an alloy of silver and gold, is separating the two metals by means of a solvent, which acts upon the silver without touching the gold. Aquafortis is generally used. The silver is added to the gold, because experience has taught us, that it is necessary that the gold contain at leaſt double its weight of ſilver, in order that the nitrous acid may diſſolve the ſilver entirely. As three parts of ſilver are often added to gold, this operation is called *inquart* or *quartation*, becauſe the gold makes in fact the fourth of the alloy.

After having flattened the button with the hammer, and taken care to heat and agitate it frequently, in order to prevent it from cracking, and the ſeparation of any portions of it by the ſtiffening of the piece; it is rolled upon a feather, and a kind of cornet is formed; it is put into a ſmall matraſs, and above five or ſix gros of precipitated aquafortis, diluted with half its weight of water, are poured in. The veſſel is gently heated till the effervefcence be fairly commenced; then the ſilver diſſolves, and the cornet aſſumes a brown colour. When the action of the acid is over, the acid is poured off and replaced by a freſh portion, which is made to boil above the metal till all the ſilver be diſſolved. The acid is poured off and the cornet waſhed, which has become very thin and full of holes; it is put with ſome water into a crucible, the water is decanted, the crucible made red hot, and the gold then enjoys all its properties. Its weight is found, by which we judge of the alloy which it contained, or of its title. In order to aſcertain exactly the quantity of imperfect metals which it contained, let us ſuppoſe any maſs of gold to be compoſed of 24 parts, called *carats*; and for greater preciſion's ſake, let each carat be ſubdivided into 32 parts, each called $\frac{1}{32}$ of a carat. If the gold which has been eſſayed has loſt one grain upon the 24, it is gold of 23 carats; if it has loſt a grain and a half, gold of $22\frac{16}{32}$ carats; and ſo forth. The weight employed in the eſſay of gold is generally 24 grains, the weight of a mark; it is divided into 24 carats, which are ſubdivided into 32 parts. Half a mark is alſo made uſe of, which weighs 12 grains, but divided into 24 carats, and each carat into 32 thirty-twos *.

There

* A tolerable eſtimate may be formed of the purity of gold from

There are two important observations to be made upon the operation of parting.

1. Some chemists have supposed that the nitrous acid dissolved a little gold along with the silver. M. Baumé,

I 3

pages

inspecting the colour of the mass by persons who are accustomed to judge in this way. The accuracy of their judgment is also assisted by a set of proof-needles, which serve as standards of comparison to point out the quantity of base metal used in alloy. These needles are composed of a certain proportion of the metals supposed to be in the mixture, forming a regular gradation from the purest gold, or gold of 24 carats fine, down to 12 carats or one half, though some extend the series still lower. Dealers in gold are commonly provided with four or five set of touch-needles; the first set, consists of pure gold alloyed with silver alone; the second, of gold, alloyed with a mixture of two parts of silver and one part of copper; the third, equal parts of silver and copper; and the fifth, of pure copper, which is used without any alloy of silver. When a piece of gold is to be valued from inspection, a stroke is to be drawn upon a black stone called *a touch-stone*, and another stroke parallel to this by the needles which approach nearest to the colour of the mass. This touchstone must possess certain properties, in order to answer the purpose well. It must be of a deep colour, nearly black, to show the colour of the metal distinctly. It must not be too smooth, otherwise soft gold will not easily leave a mark upon it; nor too rough, else the mark proves imperfect. When too hard, cleaning the stone from the strokes polishes the surface too much; and when too soft, it is liable to be scratched in cleaning. It must also be insoluble in acids; otherwise the aquafortis, poured upon its surface to try the presence of base metals, will corrode the stone itself. The strokes drawn are in general a quarter of an inch long, and one eighth of an inch broad; and when the first needle does not perfectly resemble the original mass, we make repeated trials with others, until the resemblance be more complete. Then we pour some aquafortis upon the stone, to see whether the metal be only an imitation of the colour of gold. If it contain no gold, the whole stroke is completely dissolved away, and is always eroded in proportion to the quantity of inferior metal with which the gold is debased. In this way the touchstone is employed as a test of the fineness of gold; though, it must be confessed, that it is far from being accurate, unless the nature of the alloying metal be known to the inspector.

But in China, where no alloy is used except pure silver, the least admixture of copper creating suspicion of fraud, and where essays are not commonly permitted in trade, the merchants are said to be very accurate in their judgment, and to estimate the fineness of gold

within

pages 117, 118. of the third volume of his Chemiſtry, has obſerved, that parted ſilver retained a conſiderable quantity of gold. From two pounds of grained ſilver which this chemiſt made uſe of, to make lapis infernalis, he ſays he generally ſeparated more than a half gros of gold in a black powder. However, by performing the parting with an acid, which was not too concentrated, and urging the ſolution not too far, the gold remains pure and untouched, and the ſilver contains none of it. The Profeſſors of Chemiſtry of the Academy having been deſired by Adminiſtration to examine, if, in the proceſs employed for parting, the nitrous acid diſſolved the gold, they made a great number of experiments; from which they concluded, "that, in the parting practiſed according to the regulations and received uſe, a loſs of gold never can take place; and that this operation ought to be looked upon as carried to perfection." This deciſion, extracted from the report publiſhed by the Academy, is well calculated to ſatisfy the public upon this point, and to encourage commerce.

2. Several docimaſtics, and among others Schundlen and Schlutter, have thought that the parted cornet of gold retained a little ſilver. They have given to this the name of *ſurcharge* or *inter-halt*. Meſſrs Macquer, Hellot, and Tillet, intruſted to examine the operation of the eſſayers of money, have proved, that it contained none of it. However, M. Sage aſſures us in his laſt work, entitled *the Art of Eſſaying gold and ſilver*, page 64, that

within $\frac{1}{200}$ part of its fineneſs by means of touching-needles. This conſtant experience gives the Chineſe merchants great advantage in this kind of traffic. But if thoſe who deal with them try the ſpecific gravity of the maſs in a hydroſtatical balance, they will barter upon better terms; becauſe gold, when pure, is the heavieſt of all metals; and when alloyed with ſilver alone, the mixture turns out exactly of the ſpecific gravity of the component metals, ſuppoſing no chemical union to have taken place. This, though a great convenience, is rather ſingular, as almoſt all mixtures of metals prove either heavier or lighter than what they ought to be by computation.

that gold in this ſtate retains always a little ſilver; and that it may be demonſtrated by diſſolving it in a dozen of parts of aqua regia; the ſolution being cooled, depoſites in a little time, often in a dozen of hours after making, a little luna cornea in the form of a white powder.

Gold is employed for a great number of purpoſes. Its ſcarceneſs and price prevent its being made uſe of for utenſils and veſſels as ſilver is: but as its brilliancy and colour are very agreeable to the ſight, art has found means of applying it to the ſurface of a great number of bodies, which it at the ſame time defends from the impreſſions of the air. This art conſtitutes in general gilding; of which there are many varieties. By means of glue, gold-leaf is often applied upon wood. Gold in a calx is prepared by braying with honey ſome gold-leaf, and waſhing it with water, and drying the molecules of gold which are precipitated. Gold in ſhells is the calx of gold intermixed with a mucilaginous water or a gummy ſolution. The following preparation gets the name of *gold en drapeaux*. Cloths are ſteeped in a ſolution of gold; they are dried and burnt. When they are wanted for uſe, a moiſtened cloth is put into the aſhes, and the ſilver is rubbed with it; which being very minutely divided, readily adheres. We have already ſpoken of gilding with painters gold. The copper to be gilt is well cleaned by means of ſand and a weak aquafortis, called *the ſecond water* by the workmen: it is put into a very dilute ſolution of mercury; and the mercury, which is precipitated upon it, makes the amalgam of gold adhere, which is ſpread over the piece after having been waſhed in water to carry off the acid. When the amalgam is uniformly ſpread, the piece is heated upon the coals in order to volatiliſe the mercury; the operation is finiſhed, by ſpreading upon the gold the wax for gilding, which is made up of red bole, verdegris, alum, or martial vitriol, incorporated with yellow wax, and by heating for the laſt time the gilded piece to burn off the wax.

The other uſes of gold for trinkets, laces, are well enough known to ſave the trouble of ſaying any more. As to the medicinal virtues attributed to it, ſkilful phyſicians at preſent refuſe it any; and think that the effects of the different potable golds, held out by the alchymiſts, are entirely dependent upon the ſubſtances in which it is mixed or diſſolved.

Species 15. PLATINA.

PLATINA, which was diſcovered only forty years ago for a paritcular metal, has been as yet found only in the gold mines of America, and in particular in thoſe of Santa-Fe near Carthagena, and of Bailliage de Choco in Peru. The Spaniards gave it this name from Plato, which ſignifies ſilver in their language, comparing it to this metal, whoſe colour it really poſſeſſes. However, the name of white gold ſeems to be more applicable than that of ſilver, approaching in moſt of its properties much nearer to gold than to ſilver.

Several jewels of platina exiſted before the epoch juſt now mentioned; but as this metal cannot be fuſed and wrought alone, it is probable that the ſnuff-boxes, the heads of canes, and other utenſils of this kind, which were ſold under the name of *platina*, were alloys of this metal with ſome metallic ſubſtances which communicated fuſibility to it, as ſhall be taken notice of in the hiſtory of its alloys.

Platina which is depoſited in cabinets, is in the form of ſmall grains or ſpangles of a livid white, whoſe colour inclines always to that of ſilver and iron. Theſe grains are mixed with ſeveral ſtrange ſubſtances: ſpangles of gold are found amongſt them; blackiſh martial ſand-grains, which with the microſcope appear ſcorified like iron droſs; and ſome molecules of mercury.

This

This last metal is separated by heat; washing carries off the sand and grains of iron, which may likewise be separated by the magnet; no more then remains but molecules of gold and grains of platina, which it is easy to pick out separately, as M. Margraaf did. If grains of platina are examined with the microscope, some appear angular, others rounded and flattened like cakes. Beat upon a steel mass, the most part of them is flattened, and appears ductile; some break into several pieces. These last, examined nearly, seem to be hollow, and in their heart pieces of iron and a white powder are found. There is no doubt but that upon these martial atoms, contained in some grains of platina, the property of being attracted by the magnet, found in these grains, depends, though they are separated exactly from the ferruginous sand which they contain. The purity of this metal seems to be very near that of iron. The specific weight of platina, when mixed with all the other strange matters just now mentioned, approaches much to that of gold; it loses in water from a sixteenth to an eighteenth of its weight. Messrs Buffon and Tillet, upon comparing by the weight an equal volume of platina and of gold, reduced into molecules similar to those of platina, found, that the specific weight of platina was about one-twelfth less than that of gold. It comes near in specific gravity to gold when it is purified by fusion.

It is probable that platina is not found in the ores, such as it comes to us; and that its granulous form depends only upon the operations to which the gold ores are subjected for the separation of this last metal. As it is mixed always with gold, it is reduced to powder, and is even half polished by the trituration which the gold suffers in the amalgamation with mercury. It is clearly on this account that it is always mixed with grains of gold and mercurial molecules.

Although platina jewels have been for a long time sold, its par-

particular properties were not known. The workmen in the mines had paid no particular regard to it, and had miſtaken a ſubſtance in whoſe appearance there was nothing flattering, and which, beſides, was of ſo difficult treatment. Don Antonio de Ulloa, a Spaniſh mathematician, who in the famous voyage accompanied the French academicians who were ſent to Peru to determine the figure of the earth, was the firſt perſon to whom we owe the chief knowledge of platina. He ſays a few words about it in the relation of his voyage publiſhed at Madrid *anno* 1748. Mr Charles Wood, an Engliſh metallurgiſt, brought ſome of it from Jamaica, *anno* 1741. He accordingly examined it; and has given an account of his experiments in the Philoſophical Tranſactions for the years 1749-50. At this time, ſeveral of the moſt famous chemiſts in Europe anxiouſly ſet about the examination of this new metal, which promiſed ſo many advantages by its ſingular properties. M. Scheffer, a Swediſh chemiſt, publiſhed his obſervations upon platina, in the Memoirs of the Academy of Stockholm for the year 1751. M. Lewis, an Engliſh chemiſt, gave a ſubſequent, and almoſt complete, deſcription of this metal: it is to be found in the Philoſophical Tranſactions, *anno* 1754. M. Margraaf, in the Memoirs of the Academy of Berlin, has given an account of his experiments upon this new metal. The greateſt part of theſe particular memoirs has been collected by M. Morin, in a work intitled, *Platina, White Gold, or the Eighth Metal*, Paris, 1758. At the ſame time, M. Macquer and Baumé made together a great number of important experiments upon platina, which have been publiſhed in the Memoirs of the Academy for the year 1758. M. de Buffon has related, in the firſt volume of the Supplement to his Natural Hiſtory, a ſet of experiments upon platina, made by himſelf, M. Morveau, and the Count de Milly. And, laſtly, M. le Baron de Sukengen alſo undertook an inveſtigation of the metal. From the accounts given by theſe philoſophers, we ſhall borrow the greateſt part of the hiſtory of the chemical properties of platina.

Platina, expoſed to the moſt violent fire of our furnaces, ſuffers no kind of alteration; only it agglutinates a little. All the chemiſts who have made experiments upon this metal agree in this point. M. Macquer and Baumé kept ſome of it expoſed ſeveral days to the continued fire of a glaſs-houſe furnace, without its grains ſuffering any alteration except that of joining ſlightly to one another; this agglutination was even ſo weak, that a ſmall force eaſily ſeparated them. In theſe experiments they obſerved the colour of platina become brilliant when it was in a white heat; that it aſſumed a dull grey colour when it was long heated; and, laſtly, that it increaſed conſtantly in weight, as M. Margraaf had ſaid, which could ariſe only from the calcination of ſome ſtrange ſubſtances which are mixed with it. Theſe chemiſts expoſed ſome platina to the focus of a great burning mirror: it began to ſume; it gave vivid and very ardent ſparks; laſtly, the portions of this metal, expoſed to the centre of the focus, were fuſed in a minute. The fuſed portions were of a white brilliant colour, and preſented the form of a button: they could be cut into laminæ with a knife. Struck upon a piece of ſteel, one of them was flattened, and reduced into a thin plate, without cracking or ſplitting; it was ſtiffened with the hammer. This elegant experiment ſhows us, that platina is fuſible by a fire of the moſt extreme force; that it is as malleable as gold and ſilver; and that it is unalterable by the action of the fire; for in all theſe experiments, the moſt part of which were made in the open air, platina exhibits no ſigns of calcination. M. de Morveau alſo fuſed platina by making uſe of the air-furnace deſcribed by M. Macquer, with his reductive flux, compoſed of eight parts of pounded glaſs, one part of calcined borax, and a half of pounded charcoal.

Platina expoſed to the air is nowiſe changed. However,

ever, it is not known what might be the effect by a long-continued red heat with access of air: perhaps it would be calcined, as Juncker asserts gold and silver are when treated in the same way.

This metal suffers no alteration from water, the earthy, saline-earthy, nor alkaline substances.

The most concentrated vitriolic acid, the strongest and most fuming nitrous and marine acids, even when they are assisted by a boiling heat, have not the least action upon platina; nor does distillation, so efficacious a means with all the chemists for favouring the action of the acids upon the metallic substances, promote their action; only, according to Messrs Lewis and Baumé, the vitriolic acid tarnishes its grains. The nitrous acid, on the contrary, renders it brilliant. M. Margraaf says, that, at the end of the distillation of this acid with platina, he obtained some arsenic; a phenomenon which other chemists have not taken notice of. The marine acid does not change it in the least degree. M. Margraaf likewise obtained from this acid, distilled from this metal, a white sublimate, which appeared to him to be arsenic; and a reddish sublimate, the properties of which he could not examine on account of its too small quantity. All these substances evidently appear foreign to platina. This metal then resembles gold in the small action which the simple acids have upon it; but this analogy is still more remarkable in its solubility in aqua regia.

The aqua regia which dissolves platina the best, is a composition of equal parts of nitrous and marine acids. To make this solution, which is in general more difficultly made than that of gold, an ounce of platina must be put into a retort with a pound of the aqua regia described; the retort is set in a sand-bath, and a recipient is adapted. As soon as the acid is hot, several bubbles arise from the mixed acid, which is beginning to act upon the platina: the gas disengaged has not been examined. It seems not to be copious: the action of the menstruum goes on without violence or rapidity;

mean

mean time, it aſſumes at firſt a yellow, which paſſes to an orange colour, and deepens gradually, becoming of a very obſcure brown red. When the ſolution is finiſhed, ſome reddiſh and black ſand is found at the bottom of the retort, which is ſeparated by decantation: the liquor, when ſaturated, depoſites gradually ſmall ſhapeleſs cryſtals, of a fallow colour, which are a combination of acid and platina. The ſolution of platina is one of the moſt coloured metallic ſolutions; though it appears of a deep brown, if it is diluted with water it aſſumes at firſt an orange colour, which very ſoon becomes yellow, and very ſimilar to the ſolution of gold: it ſtains animal-matters of a blackiſh brown, but not at all purple. M. Baumé ſays, that platina, fuſed with the focus of a burning-glaſs, when diſſolved in aqua regia, never aſſumes a brown colour like that of platina grains, and that this ſolution is of a deep orange yellow.

M. Macquer aſſerts, that in evaporating the ſolution of platina, and allowing it to cool, much larger and much more beautiful cryſtals are obtained than thoſe it depoſites when it is ſaturated. Mr Lewis having allowed this ſolution to evaporate in the open air, obtained cryſtals of a deep red, tolerably large, of an irregular figure, reſembling benzoin flowers, though they were more thick. This ſalt is ſharp, and not cauſtic; it partly melts, allows its acid to diſſipate, and for a reſiduum gives an obſcure grey calx. Concentrated vitriolic acid occaſions a precipitate of a deep colour, which is undoubtedly a vitriol of platina. The marine acid in ſome time produces a yellowiſh depoſition.

The alkalis and ſaline-earthy ſubſtances decompoſe the ſolution of platina, and precipitate this metal in the ſtate of a calx. Mild fixed alkali occaſions an orange precipitate. This precipitate is not the pure calx of platina. Meſſrs Macquer and Baumé have obſerved, that it owes its colour to a certain quantity of acid which it contains. Its colour is then to be conſidered as depending upon a portion of calx and a portion of ſalt of platina

 mixed

mixed together. This opinion is proved; becauſe, by waſhing the precipitate with hot water, this fluid is coloured by diſſolving the ſalt of platina; and the reſiduum, which is a pure calx of this metal, is grey. The fixed alkali, boiled upon this precipitate, readily carries off its colour by decompoſing the ſalts, and leaves a calx of platina, which is of a pearl-white colour, according to the experiments of M. Baumé. This chemiſt convinced himſelf of the ſolubility of the precipitate of platina in the alkali, by pouring, drop by drop, the ſolution of this metal into hot fixed alkali, no precipitate falling down. It is on this account that this ſolution, precipitated by the fixed alkali, retains always a deep colour; and the platina is eaſily obtained by evaporation to dryneſs. M. Margraaf diſcovered, that the mineral alkali did not precipitate the ſolution of platina: this phenomenon is ſurely owing to the immediate rediſſolution of the platina in the alkali. The alkali, ſaturated with the colouring part of Pruſſian blue, forms a copious blue precipitate, which, according to M. Baumé, is owing to the iron contained in the alkali; ſince, if Pruſſian alkali is employed, which is deprived of the iron which it contains by means of the proceſs mentioned by this chemiſt, it gives with the ſolution of platina only a few atoms of blue, owing to the ſmall portion of iron which this metal always contains. The Pruſſian volatile alkali forms two precipitates in this ſolution, as Mr Lewis has obſerved; one, which is yellow and very copious, occupies the inferior part of the mixture; the other is blue, and above it. This laſt is ſo ſcanty, that it diſappears upon agitation.

The volatile cauſtic alkali precipitates platina of an orange yellow. This precipitate is almoſt entirely ſaline; the greateſt part diſſolving in the water, and becoming coloured, like a ſolution of gold. After the action of the water upon this precipitate, a blackiſh ſubſtance remains, which appears to be ferruginous. There is this great difference between the precipitate of platina

tina and that of gold by the volatile alkali, that the firſt does not fulminate as the ſecond does.

The gall-nut precipitates platina of a deep green, which turns pale gradually by repoſe.

All the precipitates obtained from the ſolution of platina by means of the alkaline ſubſtances, are not ſuſceptible of vitrefication and of colouring glaſs. In the trials which Meſſrs Lewis and Baumé made to this purpoſe, the platina was always reduced into ſmall grains, into ramifications, or reſemblances of lace. A kind of button of platina may be obtained, by ſubjecting theſe precipitates to the action of ſome reductive fluxes, as borax, cream of tartar, glaſs. Meſſrs Macquer and Baumé, in thirty-five minutes, fuſed a precipitate of platina, mixed with fluxes, by a fire of a forge blown with two bellows. Under a hard blackiſh glaſs, reſembling that of bottles, they obtained a brilliant button of platina, which appeared to have been well fuſed. This button was not ductile; it was broken into two pieces; the internal part of which was hollow. This metal preſented a granulated texture and coarſe fracture: it was nearly as hard as forged iron, and it ſcratched deeply gold, copper, and even iron. Although we have ſaid that the precipitates of platina did not appear ſuſceptible of vitrefication or mixture with glaſs, M. Baumé, however, fuſed them into a vitreform matter by two different proceſſes. The precipitate of platina, mixed with borax and a very fuſible white glaſs, and expoſed for thirty-ſix hours to the hotteſt place of a potter's furnace, produced a greeniſh glaſs, inclining to yellow, without the appearance of metallic globules. This glaſs, heated with cream of tartar in a violent fire, exhibits the platina reduced into grains, which it has ſeparated from the blackiſh glaſs in which they were diſperſed, by pulveriſation and waſhing. This chemiſt, along with M. Macquer, then expoſed the precipitate of platina to the focus of the ſame burning mirror with which they had fuſed this metal.

metal. This precipitate exhaled a very thick and very luminous ſmoke, which ſmelled ſtrongly of aqua regia: it loſt its red colour, and reſumed that of platina, and was fuſed into a ſmooth brilliant button, which was only an opaque vitreſcent matter, of a hyacinth colour at its ſurface, and blackiſh internally; which may be conſidered a true glaſs of platina. It is neceſſary, however, to obſerve, that the ſaline matters with which it was impregnated, contributed without doubt to its vitrefication.

The precipitate of platina does not appear to be ſoluble in the ſimple acids: but it diſſolves very well in aqua regia; to which it gives only an orange colour, never coming up to the brown which platina in grains exhibits.

The ſolution of platina is not precipitable by the alkaline or perfect neutral ſalts; but ſal ammoniac occaſions a copious depoſition. It is not yet quite certain what happens here. It appears that the orange precipitate, which is obtained by pouring a ſolution of ſal ammoniac into a ſolution of platina, is a true ſaline ſubſtance, entirely ſoluble in water. A very important property was diſcovered in this precipitate by M. de Liſle; which is its fuſibility *per ſe*, and without addition, with a ſtrong fire of a furnace, or of that of an ordinary forge. Platina fuſed by this proceſs is in the form of a very denſe and compact brilliant button; but it loſes malleability, and does not become ductile, till it is expoſed to a very intenſe heat. M. Macquer thinks, that the ſame thing happens in this fuſion as in the fuſion of platina in grains when expoſed to the action of a violent fire; that it is only an agglutination of the ſoftened molecules; which, being infinitely more divided and more ſubtile than the grains of platina, approach nearer, and touch one another in many more points, than they do: which circumſtance renders the texture of this metal much more compact, although it has undergone no true fuſion. However,

ſerve, that if platina in grains fuſes with
, and acquires a very conſiderable duc-
ipitate of this metal by ſal ammoniac
fuſed on account of its extreme divi-
if it is not as ductile as the button of
y the rays of the ſun, it depends per-
retention of ſome matter which it has
with it in its precipitation, of which it
ed by the action of fire.
 diſſolved platina in an aqua regia com-
n parts of nitrous acid and one part of
Upon diſtilling this ſolution to dryneſs,
ttom of the retort was red, a ſalt of a
ublimed; and the reſiduum was in the
iſh powder. It is not known if, from
platina in a ſimple aqua regia, that is,
e nitrous and marine acids, the ſame
d be obtained by diſtillation.
raaf, Baumé, and Lewis, mixed the ſo-
a with ſolutions of all the other metallic
he reſult of their experiments is, that
netals precipitate platina in the form of
brown powder; and that each of theſe
joys metallic properties, as is the caſe
he other metals. This is an analogy
lſo between gold and platina, although
ot with tin give a purple colour as gold
er a brown precipitate, inclining to red.
o the effect of the different metallic ſolu-
tina, it will be ſufficient to obſerve, that
ad in the nitrous acid, iron and copper
t acids, and gold in aqua regia, produce
in the ſolution of platina, according to
and that, on the contrary, the ſolutions
l neutral ſalt, zinc, and ſilver, by the ni-
cipitate it; the firſt, in a cryſtalliſed ſub-
pious, of a fine golden colour; the ſe-
orange red matter; and the third, in a

yellow-coloured matter: these different precipitates have not yet been properly examined, nor is the decomposition occasioned understood.

The most part of the neutral salts have no action upon platina. M. Margraaf, in a violent fire, heated platina with vitriolated tartar and Glauber's salt: the salts were fused, and the platina remained in grains without alteration: it had communicated only a small reddish colour to the saline matters, no doubt from the iron with which it was mixed.

Nitre alters platina in a singular manner, according to the experiments of Messrs Lewis and Margraaf. Although it does not detonate with platina when a mixture of these substances is projected into a red hot crucible; however, upon the application of a strong and long-continued heat, such as Mr Lewis used for three days and nights successively to a mixture of one part of platina and two of nitre, the metal acquires the colour of rust. If the mixture is boiled in water, the alkali is dissolved, and carries along with it a brownish powder; and the platina separated from this ley is found dissolved more than a third. The brown powder is separated by filtration. This powder seems to be a kind of calx of platina mixed with a little crocus martis. Mr Lewis at last gave this calx a whitish-grey colour by repeated distillations with sal ammoniac. M. Margraaf, who repeated this excellent experiment, has added two important facts; the one is, that the platina combined with the alkali of the nitre, and diluted with a certain proportion of water, forms a gelly; and the other, that upon calcining the portion of this metal separated from this gelly, diluted with water, and filtrated, it took a black colour like pitch. This experiment announces a great alteration in the platina; and it would be of great consequence to continue the experiment, to see if, by the force of repeated calcinations with nitre, it would be possible to reduce all the metal into a brown powder, like that of which we have

been

been ſpeaking; and in particular to determine the ſtate of the platina thus calcined.

Marine, febrifugal, boracic, and earthy ſalts, produce no alteration upon platina, nor facilitate its fuſion. Sal ammoniac, diſtilled with this metal, gives a few flores martiales from the iron which the platina contains.

Chemiſts are not agreed upon the reciprocal action of arſenic and platina. M. Scheffer ſays, that arſenic fuſes this metal: but the experiment did only in part ſucceed with Mr Lewis, and not at all with Meſſrs Margraaf, Macquer, and Baumé.

The combination of platina with cobalt, nickel, and mangancſe, never has been tried.

This perfect metal alloys very well with biſmuth; which renders it the more fuſible the greater the quantity of this laſt.

This alloy is brittle: it becomes yellow, purple, and blackiſh by the air: it cannot be cupelled without the greateſt difficulty: it forms a maſs always of little ductility.

Platina fuſes eaſily with regulus of antimony: a brittle metal in facets reſults; from which the regulus cannot be ſeparated by fire, but it retains always enough of it to render the platina more light and brittle.

Zinc renders platina very fuſible, and combines eaſily with it. This metal is brittle and hard to the file: it inclines to a blue when the platina is more abundant than the zinc. They are ſeparated by heat, which volatiliſes the zinc: platina, however, always retains a little of it.

Platina does not unite nor form an amalgam with mercury, notwithſtanding ſeveral hours trituration. The ſeparation of gold from platina, beſides, is known to be practiſed in America. Several intermedia, ſuch as water, employed by Meſſrs Lewis and Baumé, and aqua regia by M. Scheffer, do not facilitate in any degree their union : this property ſeems to bring it near to iron, whoſe colour and hardneſs it already poſſeſſes.

Platina alloys very well with tin. This alloy is very fuſible, and runs eaſily. It is brittle and fragile even by a ſtroke, when equal parts are employed. When the tin is in the doſe of a dozen, and even more parts, to one of platina, this mixt metal is very ductile ; but it has an unpoliſhed and coarſe grain, and it turns yellow with the air. Platina remarkably diminiſhes the ductility of tin, and appears incapable of ſeparation from this alloy : however, when it is well poliſhed, it may remain a long time without alteration by the air. Mr Lewis, to whom we are indebted for almoſt all our knowledge of the alloys of platina, ſeems to have ſucceeded in its calcination, and in diſſolving it in marine acid by means of tin.

Platina and lead unite very well by fuſion ; for which a ſtronger fire is neceſſary than the preceding alloy requires. Platina deprives lead of ductility. From their combination a mixt metal reſults, inclining to purple, more or leſs brittle according to the proportion of platina, ſtriated and granulated in its texture, and readily altered by the air. Cupellation was one of the moſt important experiments to be made upon platina ; indeed this operation was alone capable of purifying it from the ſtrange metals which it might contain. Mr Lewis, and ſeveral other chemiſts, have in vain attempted to cupel platina in the ordinary cupelling furnaces, however intenſe the heat might be which they produce; The vitrefication and abſorption of the lead took place in the commencement of the operation, on account of the exceſs of the fuſible metal : but in proportion as this laſt diminiſhes, the platina reſumes its infuſibility which it communicates to the laſt portions of the lead, ſo that it fixes upon the cupel, and weighs more after the operation than before it, on account of the lead which it retains. Meſſrs Macquer and

Bau-

Baumé, however, have ſucceeded in the complete ſeparation of the lead. They expoſed upon a cupel an ounce of platina and two ounces of lead in the hotteſt place of a furnace for baking porcelain of Sèves. The fire of wood applied continues fifty hours. At the expiration of this time, the platina was flattened upon the cupel; the ſuperior ſurface was dull and wrinkled; it was eaſily taken off; its inferior ſurface was brilliant; and, what is of more conſequence, it was capable of extenſion by the hammer. This chemiſt was convinced, that this platina contained no lead, and that it was very pure. M. de Morveau likewiſe ſucceeded in cupelling a mixture of one gros of platina and two of lead, by employing the furnace with M. Macquer's vent. This operation continued from eleven to twelve hours, and was repeated four times. M. de Morveau obtained a button of platina, not adhering, uniform, of a colour reſembling that of tin, a little rough, which weighed a full gros, and ſeemed nowiſe affected by the magnet. This then is a proceſs calculated to obtain platina fuſed into plates, which may be forged, and be conſequently employed for different precious utenſils, on account of their hardneſs and unalterability. M. Baumé has alſo taken notice of a very uſeful property, that of being ſoldered and forged like iron, without the aſſiſtance of any other metal. After the application of a white heat to two pieces of platina, which had been cupelled in a furnace of ſèves, he placed them on one another, and ſmartly ſtruck with a hammer; they were as completely and ſolidly ſoldered as two pieces of iron would have been. We need not ſay any more upon this experiment, to ſhow the advantages which will hence accrue to the arts.

Mr Lewis could not obtain an alloy of forged iron and platina. This mixt metal might have the great advantage of uniting the hardneſs of ſteel with a great ductility; at leaſt it would not be fragile and brittle like ſteel. This Engliſh chemiſt fuſed a mixture of caſt iron and platina. This alloy was ſo hard, that a file could not affect it: it had ſome ductility, but broke clean when it was red.

Platina gives hardneſs to copper, with which it very eaſily fuſes. This alloy is ductile when the proportion of the copper is three or four parts to one of platina. It is ſuſceptible of a fine poliſh, and does not tarniſh when expoſed to the air in ten years.

Platina in part deſtroys the ductility of ſilver, augments its hardneſs, and tarniſhes its colour. This mixture is very difficultly fuſed; the two metals ſeparate by fuſion, and repoſe. Mr Lewis obſerved, that ſilver, fuſed with platina, is thruſt to the ſides of the crucible with a kind of exploſion. This phenomenon ſeems to belong to the ſilver alone, as M. d'Arcet ſaw this metal break little porcelain veſſels in which it was contained, and driven above theſe veſſels by the action of heat.

Platina does not readily combine with gold without an intenſe heat. It greatly alters its colour, unleſs it be in very ſmall quantity; for inſtance, a forty-ſeventh of platina, and all inferior proportions, make no great alteration upon gold. Platina changes the ductility of gold only in a ſmall degree: it is indeed one of the metals which change it in the leaſt degree. The weight of platina, which is almoſt equal to that of gold, might be favourable to deceit; and on that account the Spaniſh premier ſhut up the ores of platina. However, ſince the diſcoveries in chemiſtry of the teſts for detecting ſuch fraud, of gold alloyed with platina, and even platina alloyed with gold, all apprehenſions may be given up; and it is ſtrongly to be wiſhed that the mines of platina were opened, that ſociety might reap, by commerce, thoſe advantages which this new metal promiſes.

The ſolution of ſal ammoniac, as we have ſeen, has the property of precipitating platina. If, then, gold is ſuſpected of being alloyed with platina, we may try its ſolution

lution in aqua regia with a ſolution of ſal ammoniac; the little platina which it contains, will occaſion an orange or reddiſh precipitate: if there is no precipitate, it is a proof that the gold contains no platina. If it happens that the excellent properties of platina ſhould render it every day more rare and more deſirable than gold, avarice could not deceive us more eaſily by alloying gold with this metal, as a ſolution of martial vitriol, which has the property of precipitating gold without changing that of platina in any way, would inſtantly diſcover the fraud. A lamina of tin, immerſed into a ſolution of platina alloyed with gold, would alſo diſcover this laſt, by being covered with a purple precipitate; whilſt platina gives it only a brown colour inclining to red; beſides, this laſt precipitate does not colour glaſs, whereas the precipitate of gold gives it a purple colour.

All the properties of platina which we have pointed out, ſeem to prove, that this ſubſtance is a particular metal, the moſt perfect and moſt unalterable of them all. Its ſmall ductility and infuſibility, by ſome conſidered as two ſtrong objections againſt this opinion, cannot, however, refute it; ſince there is perhaps leſs difference between the fuſibility of platina and that of forged iron, than there is between the fuſibility of forged iron and that of lead; and ſince its little ductility, hitherto known, depends only upon the incomplete fuſion which it has been made to undergo. As to the opinion of the learned philoſophers, who look upon platina as an alloy of gold and iron, however ingenious and ſatisfactory it may appear, it cannot be admitted; for this metal will not ſeparate into two by an accurate analyſis; and at preſent platana cannot be imitated by artificially alloying gold and iron. Laſtly, M. Macquer has made a very ſtrong objection to this laſt opinion, obſerving, that the more completely platina is freed of the iron which it contains, the more it differs from the external characters and properties of of gold

It is sufficiently understood of what important use this precious metal might be, if introduced into commerce, since it unites the indestructibility of gold with a hardness almost equal to that of iron; baffles the action of the most intense heat, and of the most concentrated acids. The arts and chemistry would doubtless reap vast advantages from it *.

* All metals owe their fluidity to heat, and are found in a solid or in a fluid state, according as the intensity of the heat is augmented or diminished. Even platina and mercury have proved subject to this general law; from which they were long supposed to be exempted. More accurate and recent experiments have shown, that platina may be fused without addition by the action of heat alone, and that mercury may be congealed into a solid mass by exposure to a sufficient degree of cold. The fusion of platina, and the congelation of mercury, require the most intense degrees of heat and of cold which it is in the power of art to produce. Metals suffer no change from the mere act of fusion, but are found, after they have cooled, to be quite unaltered, and to be the same in every respect as they were before melting. Many metals, in passing from a fluid to a solid state, lose their heat so gradually as to pass into an intermediate degree of cohesion, where some part is fluid and some part solid, and where the whole mass bears a distant resemblance to a parcel of wet sand. Lead is the most remarkable of metals for this property: other metals instantly congeal upon losing their heat, and pass from a fluid to a solid state with a very sudden transition, as water into ice. In some cases, the external surface of the metal congeals, while the more internal parts continue fluid. This happens in the cupellation of silver with lead: the moment the whole lead is scorified and absorbed by the cupel, the same degree of heat is no longer able able to keep the silver in fusion; so it congeals on the surface; and as the hardened shell of silver contracts during congelation, the fluid metal within is squeezed out through different crevices, and forms what is called *the vegetation of the button.*

The contraction of silver during congealation is a property common to it with most other metals. Gold is very remarkable in this respect. It seems, when in fusion, to rise in the crucible with a convex surface as it melts, and to contract very greatly as it fixes in the mold. But though this be the general case, there are some exceptions to be remarked, in which metals expand upon cooling. Bismuth and iron expand in fixing; so that a piece of cold iron will float on the surface of melted iron: and on account of this property bis-

bismuth is added to compositions of type-metal, to give a sharp impression from the mould.

All metals, when in fusion, become solvents, and are capable of dissolving some other metal. But they are not able to dissolve any other kind of matter, neither earths nor metallic calces; nor is there any calx which is soluble in the metal from which it was made. Those cases in which a solution, or even a mixture, was supposed to take place between a calx and a metal, are entirely founded upon a misconception of the fact. When iron is said to unite with the white calx of arsenic by fusion, we entirely overlook the change which happens before the union. There is in this combination a regular series of chemical processes performed. The arsenic first calcines a portion of the iron, becomes a regulus itself, and in this state combines with the metallic iron, while the martial calx is rejected from the mixture: so that the instance of arsenic and iron, instead of being an exception to the general rule, proves the strongest confirmation of its truth. But when every obstacle is removed, and two metals are combined by the assistance of heat, we obtain a mixture, possessed of properties which we would not expect to result from the union of the component metals. The mixture is termed an *alloy;* and we find universally, that all alloys are more fusible than either of the metals of which they are composed. By the addition of gold to iron, an alloy is formed of very easy fusion: and Mr Rosen mentions a mixed metal, consisting of lead, zinc, and bismuth, which will melt with a heat not exceeding that of boiling water. The fusible mixture of Sir Isaac Newton has long been known to chemists, tho' it requires a greater degree of heat than the composition mentioned by M. Rosen. When we compare the fusibility of alloys with the fusible quality of compound stones, and consider that all simple earths are quite unalterable by heat, but that they acquire a fusibility by mixture with each other; we perceive an analogy, which may at last lead to a discovery of the cause upon which the whole rests.

The specific gravity of alloys is but rarely the intermediate gravity of the two metals. In most cases it proves heavier than computation, in others lighter. But what is most worthy of notice, we often find the alloy heavier than the heaviest of the two metals of which it is composed; as proves to be the case in mixture of tin with copper, and in many other alloys. On the contrary, other alloys turn out specifically lighter than the lightest of the two metals, as we are told happens in many mixtures of platina with other metals. All the phenomena are very striking, and deserve the most serious consideration to investigate their cause. They clearly demonstrate a most intimate penetration of parts, and a complete chemical solution and union of the integrant parts of the metals with each other; they show the futility of trusting to the specific gravity of an alloy to judge of the proportional quantities of known component metals. If we should even attempt to form tables of specific gravity by com-

 putation,

putation, we muſt beware leſt we fall into an error in the mode of our computing. It is a capital miſtake, and makes an eſſential difference in the reſult, to ſubſtitute equal weights inſtead of equal bulks. To underſtand the proof of this, let us take an example of two metals; the ſpecific gravity of one of which we ſhall ſuppoſe 9, and of the other 18. The ſpecific gravity of an alloy of equal bulks, conceived to be mechanically mixed, will prove an arithmetical mean of $13\frac{1}{2}$; becauſe $9+18=27$; the weight in air being divided by 2, the loſs which the component metallic ſubſtances ſuſtain by weighing in water, is $=13\frac{1}{2}\ \frac{27}{2}=13\frac{1}{2}$. But if, inſtead of equal bulks, we aſſume equal weights, then as 18 parts of the heavier metal loſe 1, and 18 of the lighter loſe 2, the mixture of both will loſe exactly 3 parts, and the ſpecific gravity of the mixed metal will be $\frac{36}{3}=12$, greatly leſs than the other mode of computation. Errors of this kind are ſo frequently committed by chemical calculators, that it is proper every perſon beappriſed of them, before he undertakes a ſet of experiments upon the ſubject.

The effects of metals in an alloy are not reciprocal upon each other. For although the addition of gold or of ſilver to lead or tin, renders both theſe baſe metals vaſtly more fuſible, yet the ſmalleſt impregnation of lead or tin will impair the ductility of gold and ſilver in a wonderful degree. So remarkable is the effect of tin in debaſing the ductile quality of gold, that even the vapours of a ſingle grain of tin will render a whole pot-full of gold quite brittle and uſeleſs, as workmen often experience to their coſt; who are frequently compelled, on this account, to pull down a brick furnace and rebuild it, in order to remedy the defect.

Metals ſeem only to be miſcible with each other in certain proportions, and to have a fixed point of ſaturation, in the ſame way as the mixture of other ſubſtances; though it is a difficult matter to determine the fact by a regular ſet of experiments. They ſhould alſo ſeem to admit of different degrees of intimacy in their union; as ſome of them, though to all appearance completely blended together, may be ſeparated by the mere effect of heat, without the aid of chemical attraction. This is the caſe in many compoſitions with copper. When mixed with lead, it may be ſeparated by eliquation; that is, by expoſing the alloy to a degree of heat ſufficient to fuſe the lead, but not to melt the copper; ſo that the liquid lead may run off from the pores. In the ſame way the zinc may be entirely conſumed from braſs, and the copper reſtored to its original purity. Had the incorporation of thoſe metals been very complete, they ſhould have come into fuſion at one time, as gold and iron, though poſſeſſed of very different degrees of fuſibility. It is indeed very difficult to mark all the gradations in the ſcale of union among metals, as their incorporation may be more or leſs perfect, and they may be either chemically diſſolved by a thorough penetration of parts, or only mechanically mixed together. When two

metals

metals possess very nearly the same degrees of fusibility, their parts may be so thoroughly blended by heat, as to assume the appearance of a perfect alloy. Thus the alloys of gold and silver would seem to be nothing more; for if an essay be made of different parts of the mass, it is found to consist of an unequal proportion of ingredients: and when kept fused for some time in a crucible, they separate from each other, the gold falls to the bottom, and the silver floats a-top, from their difference of specific gravity. In order to form some judgment, whether two metals be chemically dissolved, or only mechanically mixed, we should attend to the following circumstances. If the mixed mass possess an intermediate specific gravity, be of an intermediate colour and ductility, and be separable by heat alone; we may then safely conclude, that they are only mechanically mixed: but if the colour, ductility, and specific gravity, be very different from that of the component metals, and they are totally inseparable by heat, we have then just reason to believe in a most complete chemical solution.

Alloys are useful to obtain the exact combination of properties which we wish a metal to have. They are used in brass, type-metal, bell metal, pewter, optical specula, and many other purposes in life.

In the conversion of metals into calces, by exposure to the influence of heat and air, we meet with a diversity of character in the different metals. Some of them endure a great heat before they begin to calcine; some calcine easily, but far from completely; and some calcine both easily and most perfectly. Copper and iron sustain a violent heat before they throw off scorified scales. Bismuth and lead soon lose their metallic appearance, but are not afterwards reduced to the state of perfect calces: whereas tin and antimony are both of them most perfectly calcined in a short time, and with no great degree of heat. Those varieties suggest some radical differences in the constitution of the metals, which our knowledge of the subject has not hitherto been able to explain. The calces of most of them are convertible into glass; and this disposition to vitrefy seems to be intimately connected with the degree of calcination which they have sustained. When the calcination is very complete, the calx cannot be made to vitrefy by any degree of heat. Upon this account we find the calx of tin, and the diaphoretic calx of antimony, quite infusible: but when the calx of antimony is mixed with any of the reguline metal, it is then brought into a state capable of vitrefication. The calx of tin cannot be made to vitrefy by any chemical treatment, unless it be mixed with some very powerful flux. Even with the most proper additions it is not easily fused, and never flows thin, or forms a very perfect glass. Litharge, on the contrary, which is the glass produced from lead, is not capable of being converted into an earthy calx. It never loses the appearance of a glass, and is very easily restored into the form of a metal. In the state of calces, we formerly observed, that the metallic earths were not miscible with metallic reguluses;

luses: but in this state they readily unite with all earths, promote their vitrefication, and are the most powerful fluxes. Glass of lead is on this account commonly employed to scorify ores, and to separate the metallic matters from the matrix and mineralizing earths. Other calces, which are not so fusible, and which do not easily vitrefy, form the bases of enamels or imperfect glasses. Putty, the calx of tin, is used in the formation of white enamel: and all colours which must withstand a violent heat, burnt into vessels of porcelain, enamel, or glass, are prepared from the calces of some metal or other.

All metals are not subject to calcination by the effects of heat and air, as we find that gold, silver, and platina, are exempted from their influence; though those three perfect metals, and every other metallic substance, may readily be calcined by solution in an acid menstruum. Some species of an acid in a liquid form, either simple or mixed, is to be found capable of dissolving any given metal; and the metal is afterwards to be obtained in form of a calx, either by evaporation to dryness, and expulsion of the acid by heat; or, if the end be not attainable in this way, it will be by precipitating the metallic earth by means of an alkaline salt. Metallic earths, in this degree of calcination, are easily dissolved in the acids from which they fell, and even become soluble in weaker acids which had no power over them before, and could not be made to dissolve them in a reguline form. What change they have undergone to prepare them for more ready solution, will appear by considering the action of acids upon manganese. It is sufficiently known, that the addition of some inflammable substance is requisite to promote the solution of black manganese in acid menstrua. The manganese soon becomes soluble, falls from the acid in form of a white precipitate, and may thereafter be restored to its former blackness by sufficient calcination: so that this white precipitate is less calcined, and approaches more nearly to the state of a regulus, than the black manganese. From these appearances, we are led to conclude, that tho' all metals must be somewhat calcined before acids can dissolve them, they must not be too much so; and that there is some intermediate degree which answers the purpose of solution best. Before the confirmation of these facts, we knew that metals might be too highly calcined to be readily soluble in acids; but we had no direct proof that this power was recoverable by a nearer approximation to a metallic state. There are indeed many facts which serve to illustrate this principle; as, for example, a solution of iron, which becomes turbid when exposed to the air, from the precipitation of an ochry matter. This is no more than the metal becoming too much calcined to be longer capable of suspension in the acid. Other proofs might be given, if the doctrine were now not thoroughly established, and if it were not certain, that neither reguline metals, nor perfect calces, are soluble in their proper menstrua. The degree of calcination

nation which metals muſt undergo, varies in different caſes; and is affected by circumſtances of the ſolution of the ſame metal in the ſame acid, and may be more or leſs calcined and yet perfectly diſſolved. We know that nitrous acid will diſſolve mercury in the cold; and after the action has ceaſed, the aſſiſtance of heat will enable the acid to diſſolve more. Now, what is laſt taken up is in a metallic form; and the proof of the opinion is, that calomel may be precipitated by the addition of common ſalt (See article *Mercury*), which cannot be done from a ſolution made by gold; and calomel, prepared in the common way, is nothing but a neutral marine ſalt, incorporated by trituration with ſome metallic mercury. The effect of heat to promote the ſolution of a metal in its entire form, is the more remarkable, when we conſider the ſolution of tin in aqua regia is effected in the oppoſite manner, where the metal is calcined by the ſmalleſt degree of heat, and rendered inſoluble.

Once a metal has been diſſolved in an acid, it may be thrown down from the ſolvent by the addition either of an alkaline ſalt, or of ſome other metal which poſſeſſes a ſtronger attraction to the menſtruum. But in the caſe of metallic precipitations, we meet with many anomalous facts, which it is not eaſy to reduce under any general law. A ſolution of green vitriol will precipitate gold from aqua regia; and, what is ſtill more extraordinary, a ſolution of tin in aqua regia will precipitate gold from aqua regia in form of a purple powder, provided the ſolutions be ſufficiently diluted with water. Both theſe facts, eſpecially the laſt, are moſt ſingularly remarkable; and, in a caſe ſo ſtrange, it ſeems more prudent to ſuſpend our judgment of the cauſe, than to adopt an opinion which may be inconſiderate and erroneous.

Although one metal is precipitated by the addition of another metal, the precipitation is not always occaſioned by the calx of the precipitant. The reaſon of this difference is obvious. One metal precipitates another by means of a double elective attraction, and gives a metallic precipitate; ſo that the conjoined action of the two attractions will operate more powerfully than when but one is employed, as in the caſe of one calx diſplacing another without any ſuch aſſiſtance. And it is for the ſame reaſon that the order of attraction of acids to metals, and to their calces, is not always the ſame; and whoever means to conſtitute a table of elective attractions with much accuracy, will do well to attend to the variation, as M. Bergman has done. M. Bergman has likewiſe endeavoured to employ the precipitation of one metal by another, as a means of eſtimating the proportional quantity of phlogiſton which each of them may contain. He diſſolves 100 parts of ſilver in nitrous acid, and employs different metals as precipitants. The 100 parts of ſilver were found to require, in order to precipitate it in a metallic ſtate,

Parts.

Parts.		
135	of	mercury,
31	—	copper,
234	—	lead,
55	—	zinc,
74	—	bifmuth.

From thefe experiments he infers, that the phlogifton contained in 100 parts of filver is equivalent to that contained in the abovementioned proportion of precipitants; and thus he computes the proportional quantity of phlogifton in the different metals. If the mode of computation be juft, we fhould find all metals tally with each other, whatever ftandard of comparifon be ufed. Thus, as the calculation by precipitating filver gives the proportion of phlogifton in zinc to the proportion of the fame principle in mercury, as 55 to 135, 55 parts of zinc fhould precipitate 135 of mercury; but it requires a larger proportion, nearly 58. One hundred and feventy-four parts of bifmuth fhould fall, according to M. Bergman's computation, by 55 parts of zinc, though they confume about 85; a prodigious irregularity, from which we may fufpect that the method is by no means accurate. Indeed, if we confider, that in order to make the fuccefs complete, the acid menftruum muft be faturated with quantities of different metals, which contain equal portions of phlogifton, we perceive that there is little chance this teft will hold. As 135 parts of mercury are requifite to furnifh phlogifton to 100 parts of filver, the fame quantity of acid muft be faturated with 135 parts of mercury and 100 of filver, elfe many inconveniences will enfue. If we fuppofe that the precipitant metallic fubftance is richer in phlogifton than the one to be precipitated, but that its calx has a lefs powerful attraction for the folvent, the confequence neceffarily muft be, that when a fufficient quantity is immerfed to furnifh phlogifton to the whole of the diffolved metal, only a portion of it will fall, correfpondent to that portion of the acid now faturated with the precipitant. The whole cannot fall, elfe part of the acid would be left unfaturated, which will not probably happen, and phlogifton will efcape in the folution, and not be accounted for in the computation. If a fufficient quantity of the precipitant metallic fubftance be not foluble in the menftruum, the inconvenience is ftill more obvious, as the whole of the diffolved metallic fubftance will not be faturated with phlogifton; fo that no juft eftimate can be obtained. This difficulty actually occurs in precipitations of filver by tin, where a portion of tin is only partially calcined, and difturbs the accuracy of the refult. There are fome other objections which might be ftated to this propofal, if it were proper to dwell at greater length upon a fubject of fpeculation fo little inveftigated.

Metallic calces have all been fuppofed to refemble the nature of an earth, and were commonly fpoken of as fuch until very lately, that M. Scheele difcovered that the calx of arfenic might be refolved into

a

a pure acid. Since that time, several gentlemen have endeavoured to prosecute the subject upon the same principles; and some of them have had the good fortune to meet with radical acids, which were convertible into the regulus of a metal by proper treatment. The number of those newly-discovered metallic acids is three; the acid of the siderum, which was mentioned as the cause of cold shortness in iron; the acid of the lapis ponderosus; and the acid of molybdæna. But so little is yet known of the distinctive properties possessed by the metals which those acids produce, that it will be sufficient to announce the discovery of them here, and to observe, that they may all be formed into perfect metals, by means of the same treatment and additions, as the calx of arsenic.

The doctrine which believes in the simplicity of the metallic calces, whether they be of an acid or of an earthy nature, is exactly the converse of M. Lavoisier's theory, which supposes the metals to be simple substances, and their calces to be combinations of them with pure air. Whether the calces assume a saline or earthy appearance, this system is alike applicable, as M. Lavoisier also asserts, that all acids are chiefly composed of the same kind of air. According to this system, a metallic regulus is not soluble in an acid from a deficiency of air in the acid, or from a greater attraction for the air by the acid, and not from the redundancy of phlogiston; and is rendered soluble by impregnating the acid with more air, as happens in digesting marine acid upon manganese. In the precipitation of one metal by another, the calx furnishes the air to the metal, and is left destitute itself under a reguline form. When the solution of tin in aqua regia precipitates gold from aqua regia, the reason is obvious. The tin possesses a strong attraction for pure air, which it receives from the gold, that received it from the acid; and thus by robbing the gold and enriching itself, both metals become no longer soluble in their common menstruum. So far is consonant to common experience, that gold not sufficiently calcined, or tin too much calcined, are neither of them capable of suspension in aqua regia.

LECTURE XLII.

Genus 5. BITUMENS *.

THE bitumens are combuſtible, ſolid, ſoft, or fluid ſubſtances, of a ſtrong acrid aromatic ſmell, ſeeming to be much more compounded than the bodies of the mineral kingdom juſt now examined. They are found either forming layers in the heart of the earth, or dropping through rocks, or ſwimming upon the ſurface of waters. Their character is, they burn for the moſt part with flame, when they are heated with the contact of air, as the ſubſtances do which are formed by the organs of animals and vegetables, and to which we give the name of *oils*. Their analyſis is much leſs accurate than the analyſis of the earthy, ſaline, or metallic bodies; becauſe the action of fire alters them remarkably, and principles are extracted from them which react upon one another in proportion as they are volatiliſed. The bitumens have an analogy with vegetable and animal ſubſtances. By diſtillation they give over an odorous water or phlegm, more or leſs coloured and ſaline; an acid, frequently concrete; ſometimes ſome volatile alkali and oils; which, from being light at firſt, become thicker and more coloured as the diſtillation advances, and according to the force of the heat applied. After this analyſis, there remains a greater or leſs quantity of groſs, light, rare, brilliant, or compact charry matter, according to the different kinds of bitumens. This analyſis ſhows, that inflammable bodies have a vegetable or animal origin, as will be more clearly ſeen in the enumeration of their properties.

The bitumens ſuffer ſome alteration from light: when they are fluid, their colour deepens, and their ſmell is modified in tranſparent veſſels. The air thickens them by

* It ought to be recollected, that we divided the mineral combuſtible ſubſtances into five genera; which are, the diamond, inflammable gas, ſulpur, the metallic ſubſtances, and the bitumens.

by the successive evaporation of their humidity, which flies off the more quickly the drier the air is. Their spiritus rector, or smell, dissipates in the same proportion; and they are gradually converted from the state of fluidity to tenacity and solidity; but a great number of years is requisite for completing this alteration.

Water in which the bitumens are boiled does not dissolve them; but it takes up their spiritus rector, and emits the smell which is peculiar to them: hence it seems, that water has more affinity with their smelling principle than with the oily matter of the bitumen, and might perhaps thus deprive these bodies of all their odour.

The action of the saline-earthy matters upon the bitumens has not been examined. However, lime, and also the pure alkalis, seem capable of uniting with them, and of forming compounds soluble in water, which have got the name of *soaps*. Nor do we know the way in which the mineral acids act upon them; it is probable that they would dissolve and put them into a soapy state, as they do to the oils. Nor has the action of the neutral salts, inflammable gas, sulphur, and the metallic bodies, been examined: in general, indeed, the chemical properties of these bodies are but very little known. This subject is entirely new, and experiments might produce perhaps useful results.

The naturalists have been much more busied about the origin and formation of the bitumens, than the chemists about their analysis. There are several opinions upon this point. Some think, that these combustible bodies belong properly to the mineral kingdom, and that they are to the minerals what the oils and resins are to the organic substances. The analogy, which is somewhat agreeable to the imagination, does not agree with the facts; for nothing is known in the mineral kingdom which possesses the character of oils. For this reason, the opinion of those who ascribe the bitumens to vegetable substances buried in the heart of the earth, and altered by the action of the mineral acids, has been more generally adopted than the former. In fact, every thing proves that the bitumens originate from organic substances. In their neighbourhood a great deal of these substances is constantly found, whose form is distinguishable; besides, they have themselves the chemical characters of substances formed by life; and, in a certain degree, they may be imitated, by combining some oils with concentrated vitriolic acid. In the chemical history of the vegetable substances, we shall take notice, that the oil of vitriol, put in contact with essential oils, blackens, hardens, and gives them a strong pungent odour, resembling that of the bitumens. But are these bodies formed only by vegetables buried in the earth, as most naturalists have advanced, and do animals contribute nothing? The great quantity of bitumens which exists in the heart of the earth, compared with the little wood found in their neighbourhood, and in particular the small quantity of oily substances which these vegetables contain, seem to oppose the opinion that the origin of bitumens is entirely owing to the individuals of the vegetable kingdom: on the contrary, the abundance of these combustible substances, in places where only traces of vegetables are found, and the almost constant appearance of the carcases of animals heaped above the bitumens, should lead us to suppose, that the organic substances have greatly, and perhaps even more than vegetables, contributed to their formation. Let us observe also, that the successive layers of some bitumens which are found in continued masses in the heart of the globe, announce that these bodies have been slowly deposited by the waters, and that their formation corresponds with the epoch when the immense heaps of shells and other marine substances were formed by the sea. They have therefore been in a fluid state, and blackened through time by the action of saline bodies, or of other agents, which the heart of the earth contains in great quantity. Such is the opinion

of

of M. Parmentier, member of the College of Pharmacy, which he has given upon the origin of the charcoal of earth, in a memoir which he read at the opening of the course of this company. The oils and the fats of marine animals appear, then, to be materials which nature uses for the formation of certain bitumens, whilst the origin of others is manifestly vegetable, and is owing to the resins or essential oils buried and altered in the earth.

The number of bitumens is very great. The naturalists have made several genera of them. Chemically considered, we look upon them as species; because in fact they have all the same characters relatively to their chemical properties. Some are liquid, others are of a soft consistence: some are solid; and among these last some are hard, and susceptible of a polish; others are friable. We distinguish six very distinct species of them, which comprehend a great quantity of varieties, which we shall point out. These six species, whose history we are going to give, are, amber, asphaltes, bitumen of Judea, jet, charcoal of earth, ambergris, and petroleum.

Species I. *Amber*.

Amber, called *yellow amber* or *karabé*, is the finest of all the bitumens in its external characters: it is in irregular pieces, of a yellow or brown colour, transparent or opaque, formed of layers or scales. It is susceptible of a very pretty polish. When it is rubbed some time, it becomes electrical, and capable of attracting straws. The ancients, who knew this property, gave succinum the name of *electrum*, from whence came *electricity*.

This bitumen is of a very hard consistence, approaching to that of certain stones; which circumstance induced some authors, and in particular Hartman a naturalist, who lived about the end of the last century, to rank them among the precious stones. However, it is friable and brittle. When it is pounded, it emits a very agreeable odour. In its heart insects are often found very well preserved and very distinguishable; which proves that it has been liquid, and that in this state it enveloped the substances found in it.

Amber is for the most part buried more or less deeply: it is found under coloured sands, in small incoherent pieces, and dispersed in beds of pyritous earth; above it, wood charged with a blackish bituminous matter is found: hence it is believed, that it is formed of a resinous substance, which has been altered by the vitriolic acid of the pyrites. It swims also at the sides of the sea. It is collected on the sides of the Baltic sea in Ducal Prussia. The mountains of Provence near the city of Sisteron, the marquisate of Ancona, and the duchy of Spoletto in Italy, Sicily, Poland, Sweden, and several other countries, produce it. The colour, the texture, the transparency, and opacity of this bitumen, have distinguished a very great number of varieties. According to Wallerius, they may be reduced to the following:

1. White transparent succinum.
2. Transparent succinum, of a pale yellow.
3. Transparent succinum, a citron yellow.
4. Transparent succinum, of a golden yellow; chryselectrum of the ancients.
5. Transparent succinum, of a deep red.
6. White opaque succinum; leucelectrum.
7. Yellow opaque succinum.
8. Brown opaque succinum.
9. Succinum, coloured green, or blue by foreign matters.
10. Veined succinum.

There might still be a greater number of varieties from its internal appearances. But we should be cautious with regard to the value which we put upon specimens of it, which are remarkable for their size, transparency, and the well-preserved insects which they contain internally;

internally; ſince there is a poſſibility of deception, ſeveral perſons poſſeſſing the art of giving it tranſparency, of colouring it at pleaſure, and of ſoftening it ſo as to introduce foreign ſubſtances.

Wallerius takes notice, that the golden-coloured amber always owes its tranſparency to nature, and that artificial tranſparency is always mixed with a pale colour.

Although it be very probable that this bitumen owes its origin to vegetable reſinous matters, ſeveral naturaliſts have had different opinions upon its formation. Some have conſidered it as the hardened urine of certain quadrupeds; others as an earthy juice which the ſea has carried away, and which, carried by the waters upon the ſhore, is dried and hardened by the rays of the ſun. This claſs of naturaliſts conſiders it as a particular mineral juice. Such was the opinion of an ancient naturaliſt called *Philemon*, and mentioned by Pliny. George Agricola afterwards revived it. Frederic Hoffman imagined it to be formed of a light oil, ſeparated from the bituminous woods by heat, and made thick by the vitriolic acid. This opinion of Hoffman cannot be adopted; for we do not underſtand how an oil, ſeparated in the interior parts of the earth, could contain animals which live only at its ſurface. It is more than probable, that amber is owing to a reſinous juice which firſt flows fluid from ſome tree: this juice, buried more or leſs deep in the earth, by the revolutions which the globe has ſuffered, is hardened, and impregnated with mineral and ſaline vapours, which circulate through it. There is not even an appearance that it has been altered by concentrated acids; for experiment ſhows us, that the action of the acids would have blackened and converted it into a charry matter. Pliny thought, that amber was nothing but the reſin of the pine, hardened by the coolneſs of autumn. Some authors have thought, that it was produced from a kind of reſin, improperly called *gum-copal*, which had for ſome time continued in the earth. But whatever analogy there may ſeem to be betwixt these two substances, we cannot absolutely specify the kind of vegetable matters to which the amber belonged.

Amber, exposed to the fire, does not liquefy without a very strong heat; it softens, and bubbles up greatly. When it is heated with the contact of air, it kindles, and emits a very thick and very fragrant fume. Its flame is yellowish, variegated green and blue. A black shining charcoal is left after combustion, which by incineration yields a very small quantity of brown earth. M. Bourdelin, in his Memoir upon Amber, Acad. 1742, obtained only eighteen grains of this earth by burning two pounds of amber in a matrass. A half pound of the same bitumen, burnt and calcined in a crucible, yielded him, in a second operation, twelve grains of an earthy residuum, from which he procured iron by means of a magnetic bar.

If amber is distilled in a retort and with a graduated fire, a red phlegm at first comes over, which is manifestly acid. This acid spirit retains the strong odour of the succinum; then a volatile acid salt, which crystallises in small white or yellowish needles in the neck of the retort; next, a white and light oil of a very vivid colour. This oil gradually turns coloured as the fire turns stronger; and ends in a brown blackish, thick, viscous, matter, like empyreumatic oils. During the passage of these two oils, a certain quantity of volatile salt sublimes, more or less coloured. After this operation, there remains in the retort a black mass, moulded upon the bottom of the retort, brittle, and resembling the bitumen of Judea: George Agricola, near three centuries ago, made this observation upon distilled amber. If the operation is conducted with a gentle and well-regulated fire, and if the amber is in great quantity, all these products may be obtained separately by changing the receiver. Generally they are received in the same vessel, and then rectified with a gentle heat. The spirit is somewhat discoloured in this rectification. The oil,

which

which comes over black at the end of the operation, on account of its carrying along with it a charry matter, and of the reaction of the acid upon its principles, may be rendered very white and light by several successive distillations. M. Rouelle has given a very good process for obtaining it in this state by one operation. For this purpose, this oil is to be put along with some water into a glass alembic, and distilled with the heat of boiling water; the most pure and the only portion which is volatilised with this degree of heat, on account of its lightness, passes over with the water above which it is collected. If we want to preserve it in this state, it must be kept in vessels of free-stone; for in glass vessels, the rays of light penetrating through them, in no long time give it a yellow and even brown colour.

This analysis demonstrates, that amber is formed of a great quantity of oil, rendered concrete by an acid. It also contains a very small quantity of earth, whose nature has not been examined, and some atoms of iron.

Oil of amber seems to approach to essential oils; it possesses their volatility and smell: it is very inflammable: it seems capable of forming soaps with the alkalis.

The volatile salt of amber was for some time considered as an alkaline salt. Glaser, Lefevre, Charas, and Jean Maurice Hoffman professor at Altdorf, were of this opinion. Barchusen, and Boulduc the father, are the two first chemists who, in the last century, discovered the acid nature of this salt. Since their time, all the chemists have adopted that opinion, but have not agreed about the nature of this acid. Freferic Hoffman, because amber is found in Prussia in layers filled with pyrites, imagined that its salt is formed of vitriolic acid. Neuman seems to be of the same opinion. M. Bourdelin, in the above-mentioned Memoir, mentions several experiments which he made, with a view to determine the nature of this salt. He observes first, that the sal succini, obtained by distillation from this bitumen, however white and pure it is, contains always an oily matter: it is without doubt to this ſubſtance that it owes its ſmell and the kind of combuſtibility which it poſſeſſes and exhibits when it is thrown upon burning coals. Several methods have been tried to rid it of this ſubſtance. When we come to examine the nature and properties of ardent ſpirits, we ſhall ſee that this fluid is unfit for the purpoſe.

The fixed alkali, digeſted upon amber with a view to carry off the fat and oily matter, and to obtain the ſalt in a ſimple ſtate, has no better ſucceſs; it diſſolved only a little bitumen, and acquired a lixivial ſaline taſte like marine ſalt.

M. Bourdelin could not find a better proceſs for the union of the acid of amber in its pure ſtate, deprived of all oily matter, with the fixed alkali, than to detonate a mixture of two parts of nitre with one of this bitumen. He waſhed the reſiduum of this operation with diſtilled water. This ley conſiſted of a ſalt of amber: it precipitated the ſolution of ſilver in white ſcales; and that of mercury in the ſame colour. Several other metallic ſolutions were likewiſe decompoſed; but M. Bourdelin conſidered theſe two facts as inconcluſive. They ſeemed to him to ſhow, that the acid of amber was the ſame with marine acid, the ſame phenomenon being preſented by this laſt with the nitrous ſolutions of mercury and ſilver. The ley of the reſiduum, evaporated with the air, yielded a mucilaginous matter; in the middle of which elongated ſquare cryſtals were gradually depoſited, whoſe form, ſaline taſte, decrepitation upon burning coals, and in particular the conſiderable efferveſcence and ſmell of marine acid which they raiſed by the affuſion of oil of vitriol, ſhowed him, that the ſpirit of ſalt was here united with the baſe of the nitre. In ſpite of this analyſis, which is very accurate, conſidering the time in which M. Bourdelin wrought, the chemiſts, who ſince him have examined the ſal ſuccini, have found it not analagous to the marine acid, and

have

have discovered in it all the characters of an oily vegetable acid.

The investigation of the chemical properties of this bitumen has not been carried farther. Even the way in which the acids act upon it is not known. Frederic Hoffman asserts, that it may be entirely dissolved in a caustic fixed alkaline ley, and in the acid of vitriol. It is known also, that the essential oil of amber can unite with the volatile caustic alkali; and, by simple mixture and agitation, form a kind of liquid soap of a milky whiteness and penetrating smell, which is known in pharmacy by the name of *eau de luce:* lastly, that this same oil dissolves sulphur with the heat of a sand-bath, and constitutes a medicine called *balsamum sulphuris succinatum.*

Amber is employed in medicine as an antispasmodic: it is recommended in hysterical and hypochondriacal affections, amenorrhœa, gonorrhœa, fluor albus, &c. It is employed in substance, after washing with hot water and porphyrization. It is used in strengthening and resolving fumigations, by putting it powdered upon a hot-brick, and directing the fume which it emits upon the part to be subjected to its action.

The volatile spirit and the salt of amber are considered as expectorants, cordials, and antiseptics: they are likewise used as powerful diuretics. The oil of amber is employed externally and internally for the same purposes as amber itself: it is prescribed in less doses, on account of its greater activity. Balsamum sulphuris succinatum, which is given to the quantity of some few drops in proper drinks, or mixed with other substances to form pills, has been successfully used in humoural and pituitous affections of the breast, kidneys, &c. With the spirit of amber and opium a syrup is made, called *spirit of karabé*, which is used successfully as a calming anodyne and antispasmodic medicine. L'eau de luce, which is prepared by pouring some drops of oil of amber into a vessel of volatile caustic alkali, and agi-

agitating this mixture till it has acquired a white milky colour, has been long in uſe as a very active ſtimulant in aſphyxies: it is held near the noſtrils, and ſtimulates the nerves; and by the agitation which it excites, it reproduces the motion of the fluids, and recovers the patient.

The fineſt pieces of amber are cut and turned, to make vaſes, heads of canes, collars, bracelets, ſnuff-boxes. Theſe kinds of trinkets were no longer to be found after diamonds and ſtones came into uſe; but they are ſent to Perſia, China, and ſeveral other nations, which eſtimate them ſtill as great rareties. Wallerius ſays, that the moſt tranſparent bits may be employed for microſcopes, burning-glaſſes, priſms, &c. It is ſaid that the King of Pruſſia has a burning-glaſs of amber a foot in diameter; and that there is, in the cabinets of the Duke of Florence, a column of amber ſix feet high, and a very beautiful luſtre. Two pieces of this bitumen may be united, by applying them to one another, after being wet with oil of tartar and heated.

Species 2. *Aſphaltum.*

ASPHALTUM, or Bitumen of Judea, called alſo *gum of funerals*, *karabé of Sodom*, *mountain pitch*, *mummy's balſam*, is a black, heavy, ſolid, very brilliant, bitumen. It is eaſily broken, and its fracture is vitreous. A ſmall lamina of it appears red when placed between the eye and the light. Aſphaltum has no ſmell when it is cold: when it is rubbed it acquires a lightneſs. It is found upon the waters of the lake Aſphaltides or Dead Sea in Judea, near which the ancient cities of Sodom and Gomorrha were ſituated. The inhabitants, diſturbed with the ſmell of this bitumen amaſſed upon the waters, and encouraged by the profit which is got from it, collect it with care. Lemery, in his Dictionary of Drugs, ſays, that the aſphaltum is diſgorged like a liquid pitch from the earth covered by the Dead Sea, and that, raiſed to the ſurface of the water, it is

there

there condenſed by the heat of the ſun, and by the action of the ſalt, which theſe waters contain in great quantity. It is found alſo in ſeveral lakes of China.

Aſphaltum of commerce is extracted, according to M. Valmont de Bomare, from mines of Daunemore, and particularly in the principality of Neufchatel and of Wallengin. It is of two colours, according to this naturaliſt; blackiſh, greyiſh or fallow: but this aſphaltum is not nearly pure; and it ſeems to be only an earth hardened and penetrated by the bitumen.

Naturaliſts are divided upon the origin of aſphaltum, as they are upon that of all the bitumens. Some ſuppoſe it a mineral production, formed by the union of an acid with a greaſy ſubſtance in the heart of the earth: others conſider it as a vegetable reſinous ſubſtance, buried under earth, and altered by the mineral acids. The moſt general and moſt probable opinion is, that it is of the ſame origin with amber, and that it is formed by this laſt bitumen, which has undergone the action of a ſubterraneous fire.

Aſphaltum, expoſed to the fire, liquefies, bubbles up, and burns with a flame and thick ſmoke of a ſtrong acrid diſagreeable ſmell. By diſtillation, it gives over a coloured oil like petroleum. This natural product has not been chemically examined.

Aſphaltum is employed as a covering to ſhips by the Arabs and Indians. It enters into the compoſition of black China varniſhes, and into the artificial fires which burn upon water. The Egyptians uſe to embalm their bodies with it; but for this purpoſe it was uſed only by the poorer ſort, who could not procure more precious antiſeptics. Wallerius ſays, that merchants prepare a kind of aſphaltes with thickened pitch, or by mixing and fuſing this with a certain quantity of true balſam of Judea: but this fraud may be detected by means of ſpirit of wine, which entirely diſſolves the pitch, and aſſumes only a yellow colour with aſphaltum.

Spe-

Species 3. *Jet.*

JET, called by the Latins *gagas*, *black amber* by Pliny, *pangitis* by Strabo, is a black bitumen, compact, and hard like some stones; brilliant and vitreous in its fracture, and susceptible of a fine polish. When it is rubbed for some time, it attracts light bodies, and appears electrical like succinum. It has no smell: when heated, it acquires one nearly resembling that of bitumen of Judea.

Jet is found in France in Provence, in the county of Foix: there is also a quarry of it which is wrought at Belestat in the Pyrenees. It is found also in Sweden, in Germany, and Ireland. The quarries of jet are disposed in beds; they contain pyrites, just like pit-coal, and the most part of the bitumens.

This bitumen softens and fuses when strongly heated: it emits a fetid odour upon burning. Some oil is extracted by distillation.

Among the different opinions upon the formation of jet, the most probable is, that it is asphaltum made hard by the lapse of time. This was adopted by the learned Wallerius.

Jet is employed to make mourning trinkets. Wirtemberg is the place where it is wrought upon. It makes buttons, bracelets, boxes.

Species 4. *Fossile Coal.*

THE name of *fossile coal*, *pit-coal*, *lithantrax*, *earth-coal*, *stone-coal*, is given to a black, laminous, shining or dull substance, which breaks easily, but has not properly the consistence and purity of the bitumens described.

This bitumen has received its name from its combustible property, and the use for which it is employed in several countries. It is found in the heart of the earth, below stones more or less hard, and aluminous or pyritous masses. These last constantly bear the mark of several

ve-

vegetables, which for the moſt part are exotic, accor- ding to the obſervation of M. Bernard de Jeſſieu. Pit-coal is at a greater or leſs depth in the earth. It is diſpoſed always in horizontal or inclined layers: the latter diſpoſition is the moſt frequent. The layers which compoſe it differ in thickneſs, conſiſtence, colour, weight. More or leſs extenſive layers of ſhells and foſſile madrepores are often found above this bitumen; which have made ſome moderns believe, and particularly M. Parmentier, that pit-coal had been formed in the ſea by the depoſition and alteration of the oily or fat matters of marine animals. Quarries of pit-coal are wrought, like mines, by digging pits, and forming galleries; and the coal is detached by means of pick-axes. The workmen are often in the danger of loſing their lives by the choak-damp. This ſubſtance is called *touſſe* or *pouſſe* by the workmen. It extinguiſhes candles, and ſeems to be fixed air. In theſe pits a kind of inflammable gas is alſo produced, very deleterious, which produces ſometimes dangerous exploſions.

Pit-coal ſeems to be very copious in the earth. It is found in Scotland, England, Ireland, Hainault, the countries of Liege, Sweden, Bohemia, Saxony, &c. Several countries of France furniſh a great quantity of it; and particularly Bourgogne, Lyonnois, Forez, Auvergne, Normandy, &c.

Pit-coal is diſtingniſhed into earth-coal and ſtone-coal, according to its hardneſs or friability; but the manner in which it burns, and the phenomena which it preſents in combuſtion, furniſh characters of very great importance to diſtinguiſh the different ſorts. Upon this view Wallerius diſtinguiſhes three ſorts.

1. Scaly earth-coal, which remains black after combuſtion.

2. Compact and luminous earth-coal, which after combuſtion yields a ſpongy matter reſembling ſcoriæ.

3. Earth-coal, fibrous like wood, which is reduced into aſhes by combuſtion.

This bitumen, heated in contact with a body in combustion and with the air, kindles the more slowly and difficultly as it is more weighty and compact: once it is kindled, it emits a brisk and very durable heat, and is a long time burning before consumption: it may also be extinguished, and serve several times for a new combustion. Its inflammable matter appears very dense, and as if fixed to another incombustible substance, which retards its destruction. Upon burning, it exhales a particular strong smell, but it is nowise sulphureous when the earth-coal is very pure and contains no pyrites. The combustion of this bitumen appears to be very analogous to that of organic substances, in this, that it is susceptible of being stopped and of being divided into two distinct periods. The combustible, oily, and most volatile part which the earth-coal contains, is dissipated and set on fire by the first application of heat; and if, when all this principle is dissipated, the combustion is stopped, the bitumen retains only the most fixed and least inflammable part of its oil, reduced to a true charry state, and combined with an earthy base. This is the process the English employ for the preparation of their coaks, which is only earth-coal deprived of its oily fluid part by the action of the fire. It is very easily seen what passes in this experiment, by heating it in close vessels in a distilling apparatus. There is obtained an alkaline phlegm, some concrete volatile alkali, an oil which deepens in colour, and becomes more weighty as the distillation advances. At the same time, a great quantity of an elastic inflammable fluid comes over, which is considered as an oil in vapour, but which might be rather a particular inflammable gas. A scorified matter remains in the retort, charry, and still capable of being kindled: this is the coaks of the English. If the action of the fire upon the pure earth-coal is carefully attended to, we observe, that it undergoes an evident softening, and seemingly a semifusion: now it is known, since this state injures the fusion of ores, that it

is

is essential to deprive earth-coal of this property. This has been done by carrying off this softening principle, that is, the oil which it contains in great abundance, and reducing it to a state analogous to that of the coal of vegetables. Let us not forget to observe, that the volatile alkali, which is formed in very great quantity by the earth-coal, favours the opinion which we have given upon its animal origin: since, as we shall take notice elsewhere, the bodies which belong to the animal kingdom, yield always this salt in their distillation.

Earth-coal is singularly useful in countries where there is no wood. It is used as fire materials, and without the fear of any danger, which some persons have ascribed to its use: the sulphureous vapour which it has been said to emit, ought not to be regarded as any objection to it, since the most accurate analysis has convinced all chemists, that when earth-coal is pure, it contains not an atom of sulphur. Hence we see how false and deceiving the pretensions of some illiterate men are, who boast of processes to deprive this bitumen of sulphur. Another consideration which ought particularly to induce the working of earth-coal, particularly in France, is, that the working of ores consumes enormous quantities of charcoal of wood, and it is to be feared that the wood may in time fail: it is particularly in these kinds of works that industry should endeavour to find out the earth-coal, as the English have done long ago *.

* Earth-coal being become an object of great utility, deserving all the attention of philosophers, on account of the different uses for which it may be successfully employed; it is proper to consult what authors have said about it, who have published treatises or particular memoirs upon the subject. Such are M. de Genssane, M. Venel, Messrs Jars, and particularly M. Morand, who has undertaken and executed a complete and very extensive work upon earth-coal.

The great abundance of pit-coal found in the island of Great Britain constitutes a chief part of the riches of the country; as coal affords a very powerful species of fuel in all the arts in which heat is employed, and as it can, by the improved modes of working, be applied

Species 5. *Ambergris.*

AMBERGRIS is a concrete ſubſtance, of a ſoft and tenacious conſiſtence like wax, grey-coloured, ſtained yellow or black, of a ſtrong agreeable ſmell when it is heated or rubbed. It is in irregular, ſometimes round maſſes,

plied to the manufacture of iron, and to other purpoſes for which it was formerly ſuppoſed to be unfit. When coal is deprived of the more fluid matters by uſtulation, a ſpecies of charcoal or cinders remains, which is known by the name of *coaks*; and the coal in this ſtate is capable of exciting the moſt intenſe heat. In the former method of converting coal into coaks, the volatile matters were diſſipated in pure loſs; but more recently the workmen have learned to collect what is expelled by the heat, as it is found to yield a moſt excellent kind of pitch, admirably calculated to ſerve all the uſes of common pitch, and to be in ſome inſtances preferable to what is obtained from wood. The pitch comes over in the form of a volatile vapour, which are made to condenſe in the cold by means of a large quantity of cold water and of cold air; ſo that the reſervoir or refrigeratory muſt be of a capacious ſize, and placed at a diſtance from the diſtillation oven, beyond the reach of the fire. I have uſed the term *diſtillation*, though it is not perhaps proper in this caſe, ſince no external heat is employed to ſeparate the component parts of the ſubject from one another. The ovens are ſo contrived to admit an under ſupply of air, ſo that the coals, after they are kindled, ſpontaneouſly decompoſe themſelves by a ſlow incomplete ſpecies of combuſtion, which does not deſtroy the ingredients. The reſiduum, left in the oven proves moſt excellent cinders or coaks. The product collected in the refrigeratory is in part a volatile alkaline ſpirit, but chiefly a viſcid bituminous ſubſtance, capable of being inſpiſſated into a more ſolid conſiſtence, and of affording pitch with ſomewhat of an offenſive ſmell; which is rather of an advantage, as it may tend to prevent the worms of warm climates from penetrating into the bottoms of ſhips; a calamity much dreaded by all ſeamen.

The legiſlature of Great Britain has thought proper to eſtabliſh a difference in the rate of duty payable by culm and other ſpecies of ſmall coal, carried coaſtwiſe, without ſpecifically marking the circumſtance in which their diſtinction conſiſts. In order to underſtand what this diſtinction ought to be, it is neceſſary to remark, that ſome kinds of coals in burning begin to melt, and ſo far come into a liquid ſtate, that the ſeparate pieces run together and unite by the adheſion of their ſurfaces. Such coals are called *caking coals*; and in them maſſes, formed by layers of different natures, and more or leſs thick, according as a greater number of them is united. Pieces of it have been ſeen weighing more than 200 lb. This ſubſtance has been manifeſtly liquid; and it has enveloped ſeveral ſtrange ſubſtances which are found in it; ſuch as heads of birds, feathers, bones of fiſhes, and other marine bodies. It is found floating upon the waters of the ſea about the Molucca iſlands, Madagaſcar, Sumatra, the coaſts of Coromandel.

Naturaliſts make ſeveral varieties of amber. Wallerius ſtates the ſix following.

1. Ambergris, ſtained yellow,

2. ——— ſtained black.

These two varieties are the moſt deſired and moſt precious.

 3. White

them it ſeems immaterial of what ſize the pieces be, ſince the ſmalleſt atoms will by fuſion coaleſce into one conſolidated maſs; ſo that the refuſe or droſs will furniſh a fuel fit for all the œconomical purpoſes of life. Other coals do not fuſe, unite, and cake by the application of heat, but retain this original form, and keep in detached pieces in the midſt of the fire. The ſmall of this coal cannot therefore be applied to domeſtic uſe; becauſe when in powder the coal-duſt inſinuates itſelf between the crevices, prevents the circulation of air, and choaks up and extinguiſhes the fire as completely as a parcel of incombuſtible ſand. It is the fragments and duſt of this coal which conſtitues culm, and which is only applicable to burn limeſtone, and make bricks; two very valuable articles in rural œconomy. I think that, after what has been ſaid upon the nature of coal, it will be an eaſy matter for any perſon to diſtinguiſh culm from the ſmall of caking coal. He has only to try whether or not he can make a fire of it in a common grate, without the addition of other fuel: if he be able, it is good coal; if not, it is culm. Or ſhould this trial be thought unſatisfactory, though it is in reality the beſt, let him put ſome upon an ignited iron ſhovel, to try whether it melts and runs together; or elſe let him ſend it to a cinder-baker, and if he can make cinders from it, he may ſafely pronounce the ſpecimen excellent coal. There never was any difficulty on the ſubject, and there would be no trouble in collecting the tax, were it not for the inſuperable ignorance and love of oppreſſion, which ſo often pervades the underling officers of the revenue.

3. White amber, of one colour only.
4. Yellow ——
5. Brown ——
6. Black ——

These two last are got from whales, and are distinguished by a disagreeable animal smell. It is to be observed, that these varieties depend only upon the mixture of some strange substances.

The origin of amber is much disputed. Some look upon it as a kind of petrole which flows from rocks, and thickens by the sun and the action of salted water: others consider it as an animal product; and among these, some ascribe it to the excrements of birds who live upon odoriferous herbs; others to the filth voided by sea-calves, crocodiles, &c. Several have imagined that ambergris is formed by the whales, in the stomach of which it is often found. Pomet and Lemery have supposed that it was a mixture of wax and honey baked in the sun, and altered by the sea-water. M. Formey, who has adopted this opinion, has supported it with an experiment; which consists in digesting a mixture of wax and honey. He says that a product is yielded of a sweet smell, very analogous to that of amber. Lastly, some English authors look upon ambergris to be an animal juice, deposited in bags placed near the origin of the genital organ of the male whale; and several others imagine that it is formed in the vesica urinaria of this fish. If this opinion were demonstrated, ambergris would be very far from being a bitumen, and it should be ranked in the class of resinous animal juices, such as musk and civet. Notwithstanding, this substance, analysed by Messrs Geoffroy and Neuman, yielded them the same principles as the bitumens; that is to say, an acid spirit, a concrete acid salt, some oil, and a charry residuum; which induced them to rank it among these bodies.

Ambergris is an antispasmodic, a stomachic, and cor-

 dial.

dial. It is employed in proper drinks to the quantity of some grains; or it is mixed with other substances to form pills. Physicians make no extensive use of it, because they have observed, that the odorous principle, in which alone the virtues of this medicine consist, is often too active, too penetrating, and injurious. Several persons cannot indure its smell without suffering all the affections peculiar to irritations of the nerves: it ought therefore to be administered with great moderation. It is considered also as a powerful aphrodisiac.

The greatest use of ambergris is to furnish a perfume to the toilette; it is generally mixed with musk, which so attenuates its smell, as to render it sweeter and more supportable; still it is not agreeable to every body.

As ambergris is very dear, it is counterfeited and mixed with different substances. The true is distinguishable by the following characters: It is scaly, of a sweet smell, insipid; it fuses without giving bubbles or scum when it is exposed to the flame of a candle in a silver spoon; it swims above water; it does not stick to hot iron. That which does not exhibit all these properties, is alloyed and impure.

Species 6. *Petroleum.*

THE name of *petroleum* has been given to a liquid bituminous substance which runs between the stones upon rocks. This sort of oil differs in its lightness, smell, consistence, and inflammability. Authors have distinguished a great number of varieties. They have given the name of *naphtha* to the lightest, most transparent, and most inflammable petroleum; that of *petroleum*, to a liquid somewhat thick bitumen of a deep brown colour; and lastly, that of *mineral pitch* to a black bitumen, thick, moderately liquid, tenacious and adhesive to the fingers.

The varieties described by Wallerius and several other naturalists are these.

 1. Naph-

1. Naphtha, white.
2. ——— red.
3. ——— green or deep.
4. Petroleum, mixed with earth.
5. ——— dropping through ſtones.
6. ——— ſwimming upon waters.
7. Mineral pitch or maltha,
8. Piſſaphaltes. It is of a middle conſiſtence betwixt that of ordinary petroleum and aſphaltes or Jews pitch.

Different naphthas are formed in Italy in the duchy of Modena, on mount Ciaro, twelve leagues from Plaiſance. Kempfer relates, in his Amœnitates Exoticæ, that it is collected in great quantity in ſeveral places of Perſia. Petroleum runs in Sicily and in ſeveral places of Italy: in France, in the village of Gabian in Languedoc; in Alſace; at Neufchatel in Switzerland; in Scotland, &c. The piſſaphaltes and mineral pitch were formerly got from Babylon, where it was uſed inſtead of mortar to the walls; from Ragufa in Greece; and from the pond of Samoſate, capital of Comagene in Syria. To day they are got from the principality of Neufchatel and Wallengen; from the wells of Pege, one league from Clermont-Ferrand in Auvergne; and from ſeveral other places *.

With

* The vapour of naphtha, which iſſues through crevices of the earth, is ſuppoſed to be the cauſe of the flame which is ſometimes ſeen on the waters of fountains.

There are many fountains where the vapour on the ſurface will take fire at the approach of a lighted candle. But there is a very noted fountain of this kind at Chittagon in the Eaſt Indies, which burſts out into flame of its own accord, whenever it has been extinguiſhed by any accident. The credulous inhabitants of the country conſecrated the ſpot to a favourite deity, and appointed a ſet of prieſts to watch over the waters of the fountain: But ſome European travellers, ſuſpecting the whole to be a pious fraud contrived to impoſe upon the credulity of the people, pulled down the wall to ſatisfy their curioſity, and found that the vapour actually kindled ſpontaneouſly whenever it was put out. I had this anecdote from a gentleman who was preſent at the time.

With regard to the different varieties mentioned, it muſt be obſerved, that they all ſeem to have the ſame origin, and that they differ from one another only in ſome particular modification. Moſt naturaliſts and chemiſts aſcribe their formation to the decompoſition of the ſolid bitumens by the operation of ſubterraneous fire. They obſerve, that the naphtha appeared to be the lighteſt oil, which the fire diſengages firſt; and that what follows, acquiring colour and conſiſtence, forms the different ſorts of petrolea; and that theſe laſt, united with ſome earthy ſubſtances, or altered by the acids, aſſume the characters of *mineral pitch* or of *piſſaphaltes*. For the ſupport of their opinion, they have a very exact compariſon with the phenomena which the diſtillation of amber preſents; and which in fact furniſhes a kind of naphtha, and a petroleum more or leſs brown, according to the degree of heat and the length of the operation. Laſtly, they obſerve, that nature preſents frequently all kinds of petrolea in the ſame place, from the lighteſt naphtha to mineral pitch. Such are the fluid bitumens got from mount Feſtin in the duchy of Modena. Although this opinion be very probable, ſome authors think that petroleum is a mineral oily combination, formed of the vitriolic acid and ſome fat matter.

The chemical properties of petroleum have not yet been examined: it is known only that naphtha is very volatile, and ſo combuſtible, as to take fire when in the neighbourhood of any ſubſtance in combuſtion: its volatility ſeems to be the cauſe of attracting the flame. An acid phlegm is procured from brown petroleum, and an oil, which at firſt reſembles naphtha, and becomes more coloured as the diſtillation advances. A thick matter like piſſaphaltum remains, which may be rendered dry and brittle like aſphaltum, and reduced entirely to a charry ſtate by a more active fire.

The different kinds of petroleum are employed for different uſes in the countries where they abound.

Kempfer tells us, that in Perſia it is uſed for giving light in lamps furniſhed with wicks. It may alſo be made uſe of as fuel. For this purpoſe, M. Lehman tells us, that ſome naphtha is poured upon ſome handfuls of earth, which are ſet fire to with a piece of paper; it inſtantly takes fire with activity, but it emits a very copious thick fume, which attaches itſelf to all bodies, and has a very diſagreeable odour. Petroleum is alſo ſuppoſed to enter into the compoſition of wild-fire. Thick petroleum is alſo employed to make a very hard and durable mortar. By the decoction of piſſaphaltum in water, an oil is got from this bitumen which is uſed to ſecure veſſels.

Laſtly, ſome phyſicians have ſucceſsfully uſed it in diſeaſes of the muſcles, the palſy, lameneſs, &c. by rubbing the ſkin with it, or expoſing it to its fume. Van Helmont conſidered friction with it as a very good remedy againſt colded members, and directed it as an excellent preſervative againſt the impreſſions of cold.

LECTURE XLIII.

Of MINERAL WATERS.

AFTER the hiſtory of all the bodies which compoſe the mineral kindom, and the examination of their phyſical properties, we conclude their account by that of the mineral waters; becauſe, ſince theſe fluids hold diſſolved, frequently, earthy, ſaline, and metallic matters together or ſeparately, it was impoſſible to comprehend their exiſtence, without being previouſly acquainted with the principles which mineralize them. Beſides, we place the inveſtigation of the mineral waters with more advantage here, as it will ſerve as a recapitulation of what has been already delivered upon the minerals, the methods of analyſing them recalling to mind the moſt part of the principles.

§ 1. *De-*

§ 1. *Definition and History of the Mineral Waters.*

THE name of *mineral waters* has been given to those fluids which contain any minerals in solution. However, as there is not a water, even among the purest which nature produces, which is not impregnated with some of these substances, this name ought to be restricted to those which hold a sufficiency of matter dissolved, to have a sensible effect upon the animal-œconomy, and to have the property of curing or preventing the diseases to which our bodies are liable *: on this account the name of *medicinal waters* would appear better suited to these fluids than that by which they are commonly known, and which use forbids us to change.

The first acquaintance we had with mineral waters was owing to chance, as is the case with all the knowledge which man possesses. The good effects which they produce upon those who made use of them, were undoubtedly the cause of their distinction from common waters. Ancient philosophers, who considered their properties, paid little attention, except to their sensible qualities; such as the colour, the weight or lightness, the smell, and taste.

Pliny, however, distinguished a great number of waters, either from their physical properties, or from the utility resulting from their use. But not until the seventeenth century were the methods begun to be found out of investigating the different principles, which they held dissolved, by making them the subject of experiments, which chemistry alone was able to conduct. Boyle, in his beautiful Experiments upon Colours, which he published at Oxford *anno* 1663, among the first chemists made known several tests, capable of indicating by the

 al-

* It should be observed, that waters which contain no sensible principles upon analysis, may, however, produce remarkable effects upon the animal-œconomy; to do this, they only require to be very light, very brisk, and their temperature exceeds that of common waters. Such is the cause of the action of the waters of Plombières and Luxeuil, which differ from pure waters only in their warmness.

alteration of the colour those substances which were dissolved in water.

The Academy of Sciences, at its institution, was sensible of the great importance of the analysis of waters; and Duclos, in the year 1667, undertook the examination of those in France. This chemist's investigation of the subject is found in the ancient memoirs of this Company. Boyle studied the subject towards the end of the seventeenth century, and published a work upon this subject, *anno* 1685. Boulduc, 1729, published a method of analysing waters, much more perfect than any that had appeared before him: it consists in evaporating their fluids; stopping at different times to separate by filtration the substances which are deposited as the evaporation advances.

Several celebrated chemists after that had great success in their investigation of waters. Every one of them made valuable discoveries relating to the different principles contained in these fluids. Thus Boulduc found natrum in them, of which he has determined the nature. M. le Roi, a physician at Montpelier, found some calcareous marine salt; M. Margraaf some marine salt with base of magnesia; Dr Priestley some aërial acid; Messrs Monnet and Bergman hepatic gas. These two last-mentioned chemists, besides the discoveries with which they have enriched the analysis of waters, have given also complete treatises upon the manner of proceeding in this analysis, and have carried this part of chemistry to a much higher degree of precision than had been done before them. Besides, particular analyses of several mineral waters are extant, which have been made by very able chemists, and throw great light upon this matter, that has been justly considered as the most difficult of all chemical investigations. The limits which we must necessarily prescribe to ourselves, hinder us from entering into the details to be found in every work: however, we shall carefully mention the authors of the discoveries as opportunity shall offer.

§ 2. *Prin-*

§ 2. *Principles contained in Mineral Waters.*

It is a few years ago only that all the fubftances which may be diffolved in mineral waters were found out. We eafily fee the reafon of this; Chemiftry had not then acquired all the knowledge requifite for the determination of the nature of thefe fubftances; and it is in proportion to the difcovery folely of certain tefts, that their exiftence has been afcertained. Another reafon, which has alfo retarded the progrefs of the fcience on this head, is, that the mineral fubftances diffolved in waters are almoft always only in very fmall quantity; and befides, they are always mixed feveral together, fo that they cover reciprocally their diftinct properties. However it be, the multiplied refearches of the chemifts juft now mentioned, and of a great number of others to be mentioned afterwards, have difcovered, that there are feveral mineral fubftances to be very frequently found in waters; that fome are found but rarely, and others never. Let us take a view of each clafs of thefe fubftances, according to the order in which we have examined them.

The quartzy earth is fometimes fufpended in waters; and as it is in a very great ftate of divifion, it remains fufpended in them without precipitating; but its quantity is always exceedingly fmall.

Clay alfo feems to be found in them: the extreme finenefs of this earth, which caufes it to be found diftributed through all its parts, is at the fame time the caufe of their tranfparency being affected. Indeed, the argillaceous waters are whitifh, and of a pearl or opal colour: they are alfo greafy to the touch; and are called *foapy*.

Lime, magnefia, terra ponderofa, are never pure in waters; they are always combined with acids.

Nor are the fixed alkalis to be found in their pure ftate; but they are frequently found in the ftate of neutral falts.

The

The ſame is the caſe with the volatile alkalis, and with the moſt part of the acids. Nevertheleſs, the aërial acid is often uncombined, poſſeſſing all it properties in waters. It conſtitutes even a particular claſs of mineral waters, known by the appellation of *gaſeous*, *ſpirituous*, or *acidulated waters*.

Among the perfect neutral ſalts, there are few, except Glauber's ſalt, marine ſalt, febrifugal ſalt, and mild fixed mineral alkali, which are often held diſſolved in mineral waters. Nitre and mild vegetable alkali are found very rarely.

Selenite, calcareous marine ſalt, chalk, Epſom ſalt, marine ſalt with baſe of magneſia, and magneſia united with the aërial acid, are the earthy neutral ſalts moſt commonly to be found in waters. As for calcareous nitre and nitrous magneſia, of which ſome chemiſts have told us, theſe ſalts are generally found only in ſaline waters, and almoſt never in mineral waters properly ſo called.

The argillaceous neutral ſalts, and thoſe with baſe of terra ponderoſa, are almoſt never diſſolved in waters: alum ſeems to exiſt in ſome *.

Pure inflammable gas has not yet been found to be diſſolved in mineral waters.

Pure ſulphur has not been found in theſe fluids. Sometimes, though very rarely, it exiſts in ſmall quantity in the ſtate of liver of ſulphur; but for the moſt part it is the hepatic gas, or the vapour of liver of ſulphur, which mineraliſes and conſtitutes the ſulphureous waters.

Laſtly, among the metals, iron is the moſt frequently diſſolved in waters: and it may be found in two ſtates; either combined with the aërial or vitriolic acid.

Several

* We do not ſpeak of the opinion of Givre and other chemiſts, who conſider alum as one of the moſt conſtant principles of mineral water; but of the accurate analyſis from which M. Mitouart found that alum exiſted in the waters of Dominique de Vals, and M. Opoix the ſame ſalt in thoſe of Provins.

Several chemiſts have thought, that it might be diſſolved in its metallic ſtate, and without the intermedium of an acid. But as this metal almoſt never exiſts in nature, but in that of ruſt or in the ſtate of a vitriol, the opinion of theſe learned men could not be adopted, unleſs when the aërial acid was not known, and when people were at a loſs to conceive the ſolubility of iron in water without the help of the vitriolic acid.

With reſpect to the bitumen, which ſeveral authors have admitted in waters, down to the time of M. le Roy, the moſt part of chemiſts at preſent deny its exiſtence. In fact, as it was particularly from the taſte that this oily ſubſtance was ſuſpected in waters, it is known, that this taſte, which does not belong to the bitumen, depends entirely upon calcareous marine ſalt. It is not difficult to conceive how water, which runs in the heart of the globe, and particularly of mountains, may become impregnated with the different ſubſtances juſt now enumerated. It is alſo conceivable, from the nature of the layers of earth over which the waters run, why they are more or leſs impregnated with principles, why the quantity and nature of theſe principles ſometimes vary; particularly if we conſider the change of direction which theſe fluids may ſuffer from the multiplied alterations of which the globe is ſuſceptible, eſpecially at its ſurface, and in the more elevated grounds.

§ 3. *Different Claſſes of Mineral Waters.*

From what we have juſt now related concerning the different ſubſtances generally contained in mineral waters, it is evident, that we might form as many claſſes of theſe fluids as there are earthy, ſaline, and metallic bodies, which may be held diſſolved; and thus the number of claſſes might be very conſiderable. But it muſt be here obſerved, that none of theſe ſubſtances mentioned is found alone and uncombined in the waters; on the contrary, they are often found diſſolved in the number three, four, five, or more. This, then, is

a

a difficulty which prevents us from making a methodical division of the waters with respect to the principles which they contain. However, by attending to that substance which is most copious, and possesses most powerful properties, we shall have a distinction which, though not extremely exact, will suffice to characterise and facilitate our judgment of each of these fluids. This is the plan of most of the chemists who have investigated the properties of mineral waters in general. M. Monnet has established three classes of mineral waters; alkaline, sulphureous, and ferrugineous. The discoveries that have been made since his time require their distribution into a greater number of classes. M. Duchanoy, who has published an estimable work upon the art of imitating mineral waters, distinguishes ten; to wit, the gaseous, the alkaline, the earthy, the ferrugineous, the simple warm waters, the gaseous warm waters, the saponaceous, the sulphureous, the bituminous, and saline. Although this author may be blamed for having multiplied the classes of waters, since no pure gaseous and bituminous waters are known, his division is without contradiction the most complete, giving a most accurate idea of the nature of the different mineral waters, and best adapted to its subject. For the sake of an orderly arrangement of the waters relatively to the principles which they contain, and in order to complete what we have already said upon the subject, we shall propose a less extensive division, and one which seems more methodical than that of M. Duchanoy; observing, that we do not consider the simple hot waters as mineral ones, they being taken by the best chemists only for hot water; nor shall we say any more of the bituminous waters, being unacquainted with any such real ones in nature.

Class I. *Acidulous Waters.*

The gaseous waters, which we would rather call *acidulous*, are those in which the aërial acid prevails. They are

are diſtinguiſhable by their ſharpneſs and facility in boiling and forming bubbles by ſimple agitation. They redden the tincture of turnſol, precipitate lime-water and liver of ſulphur. As no waters are as yet known which contain only this acid pure and uncombined, we are of opinion, that we ſhould ſubdivide this claſs into ſeveral orders, according to the other principles which are contained in them, or according to the modifications which they exhibit. All appear to contain more or leſs alkali and calcareous earth; but their different degrees of heat furniſh a very good method of dividing them into two orders. The firſt comprehends the acidulous and alkaline cold waters; ſuch as thoſe of Seltz, Sainte Myon, Bard, Langeac, Chateldon, Vals. In the ſecond order we would place the acidulated and alkaline hot waters; as thoſe of Mount d'Or, Vichy, Chatelguyon, &c.

Claſs 2. *Saline Waters.*

We underſtand, as M. Duchanoy does, by the term *ſaline waters*, thoſe which contain a very great quantity of neutral ſalts in ſolution, ſo as to affect in a very remarkable manner, and for the moſt part as purgatives do, the animal-œconomy. The theory and nature of theſe waters are eaſily diſcovered: they are entirely ſimilar to ſolutions of ſalts made in our laboratories; only they contain almoſt always two or three different kinds of ſalts. Glauber's ſalt is very rare; Epſom, marine, and calcareous marine ſalts, are the ſaline principles by which, together or ſeparately, they are mineraliſed. The waters of Sedlitz, Seydſchutz, Egra, are charged with Epſom ſalt, frequently mixed with calcareous marine ſalt. Thoſe of Balaruc contain marine ſalt, chalk, and ſome marine ſalt with earthy baſe; thoſe of Bourbonne, marine ſalt, ſelenite, and chalk; thoſe of Mothe are more compound than the preceding, and hold diſſolved marine ſalt, ſelenite, chalk, Epſom ſalt, marine ſalt with baſe of magneſia, and an extractive matter. It is to be here

here obſerved, that the ſalts with baſe of magneſia are much more common in waters than has been hitherto thought, and that there are ſtill few analyſes in which they have been properly known, and in a particular manner well diſtinguiſhed from calcareous marine ſalt.

Claſs 3. *Sulphureous Waters.*

THE name of *ſulphureous* has been given to thoſe mineral waters which appear to poſſeſs ſome of the properties of ſulphur; as the ſmell, and the power of colouring ſilver. Chemiſts were long ignorant of the true mineraliſing principle of theſe waters. Moſt of them believed it to be ſulphur; but they never were able to demonſtrate its exiſtence, or at leaſt they never found any atoms of it. Thoſe who ſet about the ſtudy of ſome of theſe waters, admitted in them either a ſulphureous ſpirit or liver. Meſſrs Venel and Monnet firſt oppoſed this opinion. The latter has very nearly come to the truth; he conſiders the ſulphureous waters to be impregnated ſolely with the vapour of liver of ſulphur. M. Rouelle the younger has told us, that theſe fluids may be alſo imitated, by agitating water with air diſengaged from liver of ſulphur by means of an acid. M. Bergman has greatly enlarged this doctrine, having examined the properties of hepatic gas, of which we have taken notice in the article *Sulphur:* he has proved, that it is this gas which mineraliſes the ſulphureous waters, which he has on this account called *hepatic;* and he gives teſts for diſcovering the preſence of ſulphur. Notwithſtanding theſe diſcoveries, M. Duchanoy, in ſpeaking of the ſulphureous waters, admits in them liver of ſulphur, ſometimes alkaline, calcareous, or argillaceous; and in this opinion he follows M. le Roy of Montpelier, who, as we have ſaid in the hiſtory of Sulphur, propoſed to imitate theſe waters, by making a liver of ſulphur with baſe of magneſia. Perhaps there are waters which contain really a little liver of ſulphur, whilſt others are mineraliſed only by hepatic gas. In this caſe, it would be neceſſary

neceſſary to diſtinguiſh two orders of ſulphureous waters. Thoſe might be perhaps called *hepatic* which contain a little native liver of ſulphur; and *hepatiſed*, thoſe which are impregnated with hepatic gas. The waters of Bareges, Cauterets, and Bonn, ſeem to belong to the firſt order; and thoſe of Saint Amant, Aix-la-Chapelle, and Montmorency, to the ſecond. All theſe are hot waters.

Claſs 4. *Ferrugineous Waters.*

IRON being the moſt copious and moſt alterable metal, it is no wonder that water is eaſily impregnated with it. On this account, ferrugineous waters are the moſt abundant and common of the mineral waters. Modern chemiſts have thrown much light upon this claſs of waters. Anciently they were all believed to be vitriolic. M. Monnet has aſſerted, that the moſt of them contain no vitriol; and he has thought that the iron was diſſolved without the intervention of an acid. At preſent, it is known that iron, which is not in the ſtate of vitriol, is diſſolved by means of the aërial acid, and forms the ſalt which we have called *creta ferrea.* Meſſrs Lane, Rouelle, and ſeveral chemiſts, have put this truth out of doubt. The greater or leſs quantity of aërial acid, and the ſtate of the iron in the waters which owe to it their virtues, lead us to diſtinguiſh this fourth claſs into three orders. The firſt comprehends the martial acidulous waters in which the iron is held diſſolved by the aërial acid; the ſuperabundance of which renders them ſharp and ſouriſh. The waters of Buſſang, Spa, Pyrmont, Pougue, and Dominique of Vals, are of this firſt order. The ſecond comprehends the ſimple martial waters, in which the iron is diſſolved by the aërial acid not to exceſs; and conſequently theſe waters are not acidulous. Thoſe of Forge, Aumale, Condé, as alſo the greateſt number of ferrugineous waters, are of this order. This diſtinction was made by M. Duchanoy: but we add a third order from

M.

M. Monnet; the vitriolic waters. Although theſe waters be extremely rare, ſuch however exiſt. Into this order M. Monnet put the waters of Paſſy. M. Opoix admits the vitriol of mars, and even in very great quantity, to be in thoſe of Provins: it is true that M. de Fourcy has denied its exiſtence, and conſiders the iron of theſe waters to be diſſolved by ſixed air: but we cannot be deciſive on this point; for the reſults of theſe chemiſts are entirely oppoſite, and require a new examination. It muſt be added, that iron is not found alone in waters; it is mixed with chalk, ſelenite, and marine ſalts, with magneſia, lime, or the alkalis as baſes. However, as the metal which they contain is the principle baſe of their properties, they ought to be called *martial*, from the principles which we have eſtabliſhed *.

As to the ſaponaceous waters admitted by M. Duchanoy, we muſt wait, in order to give our opinion upon this head, till chemical and medical knowledge determine the cauſe of their ſaponaceous property which this phyſician attributes to clay, and of the effects which they, as medicines, may produce upon the animal-œconomy on account of this property.

§ 4. *Examination of Mineral Waters from their Phyſical Properties.*

After having given a view of the different ſubſtances which may be collected in the waters, and after a ſlight ſketch of the manner in which they may be divided into claſſes and orders, from the principles they contain; it becomes neceſſary to mention the methods of analyſis, and of aſcertaining, with the utmoſt preciſion,

* In the enumeration of the waters divided into claſſes, we do not ſpeak of thoſe which contain arſenic and copper, becauſe they are conſidered as poiſons. We likewiſe paſs over the waters which contain volatile alkali or ſal ammoniac, which is the production of the putrefaction of organic matters upon which they have been ſtagnant: Theſe kinds of waters do not belong to the medicinal waters.

ſion, the ſubſtances which they hold in ſolution. This analyſis has been looked upon as the moſt difficult part of chemiſtry; and with the greater reaſon, on this account, that a complete knowledge of all the chemical phenomena, joined to the habit of making experiments, is required: and upon that account we have placed this article at the end of the mineral kingdom. In order to obtain a precise knowledge of any water which we wiſh to examine, 1ſt, The ſituation of the ſource muſt be obſerved, the neighbourhood deſcribed with exactneſs, and particularly the layers of minerals of which the ſoil is compoſed: parcels of the ſoil at different depths muſt be taken; and from inſpection, we endeavour to find the ſubſtances which may impregnate the water. 2dly, Then we examine the phyſical properties of the water; ſuch as its taſte, ſmell, colour, tranſparency, weight, temperature. For this purpoſe, we ought to be ſupplied with two thermometers which nearly tally with each other, and which are made of a weighty liquor: and theſe preliminary experiments ought to be made in different ſeaſons, at different hours of the day, and at different times, according to the ſtate of the atmoſphere. A long-continued drought, or copious rains, have ſingular influence upon the waters. Theſe previous eſſays generally point out the claſs to which the water ought to be referred, and direct the reſt of the analyſis. 3dly, The depoſitions at the bottom of the channels, the ſubſtances which ſwim upon the water, and the ſublimed matters, are alſo of ſuch importance as not to be overlooked. Then we proceed to the analyſis; which is made in three ways, by means of teſts, diſtillation, and evaporation.

§ 5. *Examination of the Mineral Waters by Teſts.*

The name *teſts* has been given to the ſubſtances which are mixed with the waters, in order to know, from the phenomena which preſent themſelves, the nature of the ſubſtances which the waters have diſſolved.

 The

The best chemists have always considered the use of tests as a very uncertain means to discover the principles of mineral waters; upon this account, that their action did not indicate, in an exact manner, the nature of the substances dissolved; because we were often ignorant of the cause of the changes produced in these fluids by their admixture: and in fact, the saline substances which are generally employed in analysis, are capable of producing a great number of phenomena, of which it is very difficult to give our opinion. On this consideration, the most part of those who have studied this subject, have had little confidence to put in tests; they have believed, that evaporation was a much more certain method of ascertaining the nature and quantity of the principles of mineral waters: and in the best accounts of the analysis of these fluids, it is constantly recommended to make use only of these substances as auxiliaries, which are at most capable of showing or causing the suspicion of the nature of the principles which constitute those waters. On this account, modern analysers have admitted only a certain number of tests, and have greatly diminished the list of those which the first chemists had employed.

However, we cannot at present doubt, but that the heat which is necessary to evaporate the waters, however feeble it may be, may produce sensible alterations in their principles; and so strip them of their nature, that their residuum, examined by the different means which chemistry furnishes, yields compounds quite different from those which were held in solution in the waters. The loss of the gaseous substances which often make one of the active principles of mineral waters, changes their nature in a singular manner; and, besides the precipitation of several bodies which owe their solubility only to the presence of these volatile bodies, produces a reaction between the fixed matters which alter their properties. The phenomena of the double decompositions which heat is capable of producing, be-

between compounds which have no mutual action in the cold, are not be comprehended without, a numerous set of experiments, about which we can have as yet only conjectures. Without entering into long details, it will be sufficient for us to know, that this assertion is clearly demonstrated to all chemists, so as to convince us, that we must not depend entirely upon evaporation. Does no method, then, remain of ascertaining the particular nature of the substances dissolved in mineral waters, without having recourse to heat? and does the complete knowledge of modern compositions, which have extended chemistry so far, contain no process, with a view to correct the errors into which evaporation may lead us? The account which I am going to give, and which I extract from a memoir that I read to the Royal Society of Medicine, will prove, that the tests which are in a very pure state, and which are employed in a particular manner, may be much more useful in analysing the mineral waters than has hitherto been imagined.

Among the considerable number of tests which have been proposed for the analyses of mineral waters, those from which we should expect the greatest light are, the tincture of turnsol, syrup of violets, lime-water, caustic fixed alkali, volatile caustic alkali, oil of vitriol, nitrous acid, the ley, saturated with the colouring part of Prussian blue, the spiritous tincture of the gall-nut, and the nitrous solutions of mercury and silver. To these M. Bergman has added paper, coloured by the aqueous tincture of Brasil wood, which becomes blue by the alkalis; the aqueous tincture of terra merita, which these same salts turn to a brown red; the acid of sugar, to indicate the presence of the smallest possible quantity of lime; and several others, which have been proposed by most of the chemists: but those which we have mentioned, suffice to ascertain all the substances contained in mineral waters.

The effects and use of these principal tests have been

plained by all the chemiſts; but enough has not been ſaid about their ſtate. Before they are uſed, it is of great importance to know exactly their nature, that we may not be deceived in their effects. M. Bergman has explained at great length the alterations which they are capable of producing. This celebrated chemiſt ſays, that a paper coloured with the tincture of turnſol, aſſumes a deeper blue by the alkalis, but is not altered by fixed air or the aërial acid. As it is chiefly to diſcover the preſence of this acid that this colouring-part is uſeful, he directs uſing only its tincture in water with proper dilution, that it may take a blue colour. He rejects entirely the ſyrup of violets, becauſe it is ſubject to fermentation, and becauſe it is never got genuine in Sweden. M. de Morveau, in a note, adds, that it is eaſy to diſtinguiſh a ſyrup, coloured by bluebottle or tournſol, by means of corroſive ſublimate, which gives it a red colour; whereas it renders the ſyrup of violets green.

Lime-water is one of the moſt uſeful teſts in the analyſes of mineral waters, though few chemiſts have made expreſs mention of it in their works. This fluid decompoſes the metallic ſalts, particularly martial vitriol, from which it precipitates the iron. It ſeparates clay or magneſia from the vitriolic and marine acids, to which theſe ſubſtances are frequently joined in mineral waters. It can alſo ſhow, by precipitation, the preſence of aërial acid. M. Gioanetti, phyſician at Turin, has made a very ingenious uſe of it, to diſcover the quantity of aërial acid in the waters of St Vincent. This chemiſt, after having obſerved that the bulk of this acid, from which its quantity is always judged, might vary according to the temperature of the atmoſphere, mixed nine parts of lime-water with two parts of water of St Vincent. He found the exact weight of the calcareous earth, formed by the junction of the aërial acid of the mineral water with the lime of the lime-water; and he found, according to the calculation of M. Jacquin, which demonſtrates the exiſtence of thirteen ounces of this

this acid in thirty-two ounces of chalk, that the water of St Vincent contained a little more than fifteen grains: but as the lime-water might ſeize upon the aërial acid joined to the fixed alkali, as well as that which is free, M. Gioanetti, to diſcover the quantity of this laſt, made the ſame experiment with water deprived of its free acid by means of boiling. This proceſs might then be employed, to determine exactly, and in an eaſy manner, the weight of free aërial acid in any gaſeous mineral water.

One of the principal reaſons which have led chemiſts to conſider the action of teſts in the analyſis of mineral waters as very uncertain, is, that they may ſhow ſeveral ſubſtances diſſolved in the mineral waters; and it is on this account very difficult to know exactly the effect which they produce. This truth reſpects particularly the fixed alkali conſidered as a teſt, ſince it decompoſes all the ſalts formed by the union of the acids with clay, magneſia, lime, and the metallic ſubſtances. When the alkali precipitates a mineral water, we cannot know by inſpection alone the nature of the earthy ſalt decompoſed in this experiment. Its effect is ſtill more uncertain when it is employed at the point of ſaturation with aërial acid, as it is generally done, ſince the acid which is united with it may increaſe the confuſion. On this account I propoſe the uſe of the cauſtic alkali in its very pure ſtate. It has alſo an advantage that the mild alkali does not poſſeſs; which is that of ſhowing the preſence of chalk, diſſolved in a gaſeous water by means of the ſuperabundant aërial acid. As it ſeizes upon the ſuperabundant part, the chalk, which ceaſes to be longer ſoluble in the water, precipitates. I have aſſured myſelf of this fact, by pouring ſome of the ſoap-boilers ley, freſh made, into an artificial gaſeous water, which kept chalk in ſolution: the chalk precipitated in proportion as the cauſtic fixed alkali ſeized upon the aërial acid which kept it diſſolved. By filtration, and evaporation to dryneſs, I obtained ſome ſal ſodæ, making a very briſk effervefcence with ſpirit of vitriol. The cauſtic fixed alkali may alſo occaſion a precipitates in the mineral waters, without their containing earthy ſalts: it is enough that they hold diſſolved a leſs ſoluble alkaline neutral ſalt, and then the alkali precipitates it by uniting with the water, juſt as ſpirit of wine does. M. Gioanetti obſerved this phenomenon in the waters of St Vincent; and it is alſo eaſy to convince one's ſelf of this, by pouring ſome cauſtic alkali into a ſolution of vitriolated tartar or marine ſalt, and theſe two ſalts are ſoon precipitated.

The cauſtic volatile alkali is in general leſs ſuſceptible of error when it is mixed with mineral waters; becauſe it decompoſes only the ſalts with earthy baſes, as clay and magneſia, without affecting calcareous ſalts. But it is of importance to make two obſervations on this head: the firſt is, that we muſt have the alkali in its moſt cauſtic ſtate, containing not an atom of aërial acid: without this precaution, it decompoſes the ſalts with baſe of lime by means of double affinity; the ſecond is, that we muſt not leave this mixture expoſed to the air, when we wiſh to know its action ſeveral hours after it has been made; becauſe, as M. Gioanetti has very well obſerved, this ſalt ſeizes in a little time upon the aërial acid of the atmoſphere, and becomes capable of decompoſing the calcareous ſalts.

To leave no doubt upon this important ſubject, I made three deciſive experiments. After having diſſolved in diſtilled water ſome grains of ſelenite, made with tranſparent calcareous ſpar and ſpirit of vitriol, an indiſpenſable precaution, becauſe the chalk or Spaniſh white contains magneſia as well as river-water does, I divided the ſolution into two parts. Into the firſt I poured a few drops of recently prepared and very cauſtic volatile alkali: I put this mixture into a well-ſhut phial: in twenty-four, even forty-eight hours after, it was clear and tranſparent, without any depoſition: no decompoſition, then, had taken place. The ſecond portion

tion was heated in the ſame way; but was put into a veſſel whoſe large opening communicated with the air: in a few hours, on the ſurface, a cloud was formed, which augmented in thickneſs, and at laſt precipitated. This depoſition made a briſk effervеſcence with ſpirit of vitriol, and formed ſelenite. The aërial acid which this precipitate contained had therefore been ſupplied by the volatile alkali, which had attracted it from the atmoſphere. This combination of the aërial acid and alkaline gas forms aërated ſal ammoniac, capable of decompoſing the calcareous ſalts by means of a double affinity, as has been demonſtrated by Meſſrs Black, Jacquin, and ſeveral other chemiſts; of which we may make ourſelves certain, by pouring a ſolution of concrete volatile alkali into a ſolution of ſelenite, which the volatile cauſtic alkali does not affect. Laſtly, to be confirmed in the etiology of this ſecond experiment, I took the firſt portion, which was united with the cauſtic volatile alkali, and was preſerved in a cloſe veſſel, and had loſt nothing of its tranſparency. I put the veſſel which contained it upon the ſyphon of a very ſmall pneumato-chemical apparatus, and directed into this mixture the aërial acid which was iſſuing from an effervеſcing mixture of fixed alkali and ſpirit of vitriol. As the bubbles of this acid paſſed through the mixture, it turned turbid, as lime-water does. After filtration, chalk was found upon the filtre; and the water, when evaporated, yielded vitriolic ammoniac. The aërial acid ſpirit produced the ſame decompoſition in another mixture of pure ſelenite and cauſtic volatile alkali. This deciſive experiment completely proves, that it is ſolely by means of the double affinity, which inſtantly takes place by the addition of the aërial acid, that the volatile alkali becomes capable of decompoſing ſelenite. Hence it is evident, that when we are to keep the mixture of a mineral water with the volatile alkali during ſeveral hours, as is neceſſary, becauſe it decompoſes certain earthy ſalts but very ſlowly, this experiment ought to be made in a veſſel which can be completely ſhut up, to prevent the contact of the air from producing a falſe reſult. This precaution is in general of very great importance in the uſe of all the teſts. It is beſides recommended by Meſſrs Bergman and Gioanetti.

I ſhall add one obſervation upon the uſe of the volatile alkali. As it is very difficult to have volatile alkali perfectly cauſtic, in which ſtate it muſt be when employed to analiſe mineral waters, a very ſimple means may be employed, which I have practiſed with ſucceſs. It conſiſts in pouring a little volatile alkaline ſpirit into a retort whoſe neck is immerſed into the mineral water; by the application of a gentle heat to the retort, the alkaline gas is diſengaged, and paſſes over very cauſtic into the water. If a precipitate appears, the mineral water contains ſome martial vitriol, which is proved conſtantly by the colour of the precipitate, or ſalts with baſe of earth, of alum, or of magneſia. It is very difficult to judge by the phyſical properties of the earthy precipitate, formed in a water by the cauſtic volatile alkali, to which of theſe two laſt baſes the precipitate ought to be referred. However, the manner in which it is formed may tell what its character is. By diſſolving ſix grains of Epſom ſalt in four ounces of diſtilled water, and ſix grains of alum in an equal quantity; and by directing into each of theſe ſolutions a little alkaline gas, that of the Epſom ſalt was inſtantly turned turbid, whilſt that of the alum began to precipitate only in twenty minutes after. Care had been taken to put the mixture into a veſſel cloſely ſhut. The ſame phenomenon happens with the nitrous and marine ſalts with baſe of magneſia and earth of alum, when diſſolved in equal quantity in diſtilled water, and managed with the ſame precautions. The celerity or ſlowneſs of precipitation of a mineral water by the addition of the alkaline gas, thus ſerves to ſhow what is the earthy ſalt decompoſed. In general, the ſalts with baſe of magne-

ſia

ſia are infinitely more common in the mineral waters than thoſe with baſe of aluminous earth. I ſhould not forget mentioning a fact obſerved by M. Bergman; which is, the volatile alkali is ſuſceptible of forming with Epſom ſalt a compound, in which an undecompoſed portion of this neutral ſalt is combined with a portion of Glauber's ſecret ammoniacal ſalt. Perhaps this undecompoſed portion of Epſom ſalt forms with the vitriolic ammoniac a mixed neutral ſalt, analagous to ſal alembroth. The volatile alkali then precipitates only a part of the magneſia, and cannot exactly determine the quantity of Epſom ſalt, of which it is the baſe. For this reaſon, lime-water appears to me preferable for aſcertaining the nature and quantity of the ſalts with baſe of magneſia contained in mineral waters. It has alſo the property of precipitating the ſalts with baſe of aluminous earth, much more copiouſly and readily than alkaline gas does *.

The concentrated vitriolic acid precipitates of a dirty white a water containing terra ponderoſa, according to M. Bergman: but as this earth, according to him, is very rarely found in mineral waters, I muſt proceed to the other effects of this teſt. When it produces bubbles in a water, it ſhows the preſence of chalk, mild fixed alkali, or pure aërial acid. Each of theſe ſubſtances may be diſtinguiſhed by particular phenomena. If a water charged with chalk is heated, into which ſome vitriolic acid has been poured, a pellicle and a ſelenitical depoſition is readily formed; which does not happen in ſimple alkaline waters. At firſt view, it would appear, that the ſelenite ought to precipitate as ſoon as the vitriolic acid is poured into a water containing chalk.

* It will be eaſily perceived that I repeat ſeveral facts which have been already explained through the work. I do this deſignedly, in order to render this ſmall account of the analyſis of waters more clear and complete; and to collect, about the means of their analyſis, all the knowledge which I imagine indiſpenſably neceſſary for the ſtudy of this ſubject.

chalk. However, this very rarely happens without the aid of heat; becaufe thefe waters contain for the moft part a fuperabundance of aërial acid which promotes the folution of the felenite; and it is requifite to deprive them of this acid before this falt can be feparated. We may be convinced of this by throwing a few drops of concentrated vitriolic acid into a certain quantity of precipitated lime-water, and afterwards made clear by aërial acid. If the lime-water abounds greatly with regenerated calcareous earth, a felenitical precipitate falls in a few minutes, or more flowly and in proportion to the diffipation of the free aërial acid. If it does not fall by fimple repofe, as it happens when the water contains but little felenite and a great quantity of fuperabundant aërial acid, a flight heat is enough to form a felenitical pellicle and precipitate.

Concentrated fpirit of nitre is recommended by M. Bergman to precipitate fulphur from hepatic waters, formerly called *fulphureous*. Without adopting the theory of this celebrated chemift, which confiders the hepatic gas as fulphur converted into a gafeous ftate by heat and phlogifton; without thinking, with him, that heat precipitates the fulphur from them by feizing upon the phlogifton and difengaging the heat, I have thought proper to mention this method, which is very ufeful to difcover the prefence of fulphur in fulphureous or hepatic waters. To be affured of this fact, it is fufficient to pour a few drops of fuming fpirit of nitre into diftilled water, which has been, by means of a pneumato-chemical apparatus, impregnated with the gas difengaged from cauftic liver of fulphur heated in a retort. This artificial hepatic water, which flightly differs from the natural fulphureous waters in this, that it is difficult of filtration, and appears always a little whitifh, gives in a few moments a precipitate with nitrous acid. This precipitate is of a yellowifh white; collected upon a filtre and dried, it burns with the flame and fmell peculiar to fulphur, of which it poffeffes all the characters.

ms that ſpirit of nitre alters hepatic gas, as
ıflammable matters by means of the great
pure air which it contains. If it is the only
ıg this property, it is owing to the very in-
union of the air, which makes a conſti-
iple in this acid. For the ſame reaſon, it
l combuſtible matters, and reduces them to
burnt bodies much more rapidly than the

is as yet ſo little known, with regard to
of its acting, as the phlogiſticated alkali.
time ſince chemiſts knew that this liquor,
th ox-blood, contained Pruſſian blue already
has been imagined, that this blue might
l by means of an acid; and in this ſtate it
ropoſed as a ſubſtance capable of demon-
exiſtence of iron in mineral waters. The
quet having obſerved, that a phlogiſticated
ipitated by an acid, ſtill contained Pruſſian
ı was gradually precipitating, took care to
s blue by filtration: the alkali depoſited it
l it was filtered more than twenty times in
f two years, without being totally deprived of
ue. I have ſome of this liquor, prepared more
ears ago; and it ſtill has a pretty blue colour.
colouring part of Pruſſian blue be contained
ogiſticated ley, as M. Bucquet thought, and
mach ſaid long ago? However it be, we
ive up this ley in the employment of teſts.
ıer, from his famous diſcovery of the de-
n of Pruſſian blue by the alkalis, propoſed
, ſaturated with the colouring part of this
letect iron in mineral waters: however, as
contains a little Pruſſian blue, which may be
from it by an acid, as M. Macquer has ſhown,
directs adding to this Pruſſian alkali, two
unces of diſtilled vinegar in the pound, di-
with a gentle heat till all the blue be precipitated;

tated; then pure fixed alkali is poured upon it, to ſaturate the vinegar. Notwithſtanding this very ingenious proceſs, we have had occaſion to obſerve, I mean M. Bucquet and myſelf, that this Pruſſian alkali, when puriſied by vinegar, depoſited Pruſſian blue a long time, and particularly by evaporation with heat. M. Gioanetti, whom I have had occaſion to mention ſeveral times with approbation, has made the ſame obſervation, by evaporating to dryneſs the Pruſſian alkali, purified by M. Baumé's proceſs. He has propoſed two methods for obtaining this liquor more pure, and totally exempt from iron. In the one, he directs ſurcharging the Pruſſian alkali with diſtilled vinegar, evaporating it to dryneſs with a gentle heat, diſſolving the remaining maſs in diſtilled water, and filtrating the ſolution. All the Pruſſian blue remains upon the filtre, and the liquor contains no more. The other proceſs conſiſts in neutraliſing this alkali with a ſolution of alum; the vitriolated tartar is filtrated, and ſeparated by evaporation. Theſe two liquors give not an atom of Pruſſian blue with pure acids, nor by evaporation to dryneſs. The volatile alkali, ſaturated with the colouring matter of the Pruſſian blue, has alſo been propoſed, which has the ſame inconvenience, and may be purified in the ſame way. Lime-water ſaturated with the colouring matter of the Pruſſian blue, of which I have ſpoken in the article Iron, does not appear exempt from theſe inconveniences. Poured upon a ſolution of martial vitriol, it forms inſtantly a Pruſſian blue, quite pure, and deſtitute of a green colour. The acids precipitate not an atom of blue from it. It contains, therefore, no iron; and it is preferable to the Pruſſian alkalis for eſſaying mineral waters. This phenomenon undoubtedly depends upon this, that lime, diſſolved in water, has not near the ſame action upon iron as the alkalis have. This Pruſſian lime-water appeared to me well adapted to detect iron in waters, whether gaſeous or vitriolic. For indeed this aërial gas, which holds the iron diſſolved in waters, being of an acid nature, decomposes as well the Prussian leys, by means of the double affinities, as it does the martial vitriol. I tried the Prussian lime-water upon the waters of Spa and Passy: I instantly obtained a blue precipitate, not very remarkable, though sensible, in the former, and very apparent in the latter. This, then, is a liquor easily prepared, which contains not an atom of Prussian blue, and which is very proper to show the presence of the smallest quantity of iron in waters. It is a kind of neutral salt, formed by the colouring part of the Prussian blue and lime.

The gall-nut, as well as all bitter and astringent vegetable substances, as oak-bark, the fruit of the cypress tree, the bark of walnuts, have the property of precipitating solutions of iron, and giving to this metal different colours, according to its quantity and its state, and according to the state of the water which kept it in solution. This colour exhibits a great number of varieties, from a pale rose to the deepest black. It has been discovered, that the purple colour which the waters assume with the tincture of the gall-nut, is not an indication that iron is contained in a metallic state, as M. Monnet had believed; since martial vitriol and iron, united with aëriel acid, which I call *creta martialis*, are likewise coloured purple by the infusion of the gall-nut. It is rather the quantity of the iron, its greater or less degree of adherence to the water, and the more forward or backward state of this solution's decomposition, which occasion the differences of colour, which are observed in these precipitations, as M. Duchanoy has very well observed in his Essay upon the Art of imitating Mineral Waaters. Yet, notwithstanding this test has been known and employed with much success in the analysis of mineral waters, since the time M. Duclos recommended it in the year 1667, and although Messrs Macquer, Monnet, and the Dijon Academicians, have made some pretty experiments upon the gall-nut, the nature of the astringent principle is still unknown. We can only suspect

pect that it is a kind of particular acid, as it unites with the alkalis, reddens the blue colour of vegetables, decomposes liver of sulphur, and combines with the metals. The gall-nut is employed to discover the presence of iron in a mineral water, either in the form of a powder, or its infusion made in the cold and the tincture with spirit of wine. This last form is to be preferred, because it is much less subject to alteration than the solution in water, which is very apt to grow mouldy. What is most singular is, that the products by distillation of the gall-nut likewise colour martial solutions. The solution in the acids, alkalis, oils, æther, all present the same phenomenon. The iron which this matter precipitates from the acids, is in a condition little understood, and forms a kind of neutral salt, which is not attractible by the magnet, although very black: it dissolves slowly and without a sensible effervescence in the acids; it loses these properties by the action of the fire, and becomes attractible. The gall-nut is a test of so vast a sensibility, that a single drop of its tincture gives, in the space of five minutes, a purple colour to a water containing only a twenty-fourth of a grain of martial vitriol in near three pints of water.

The two last tests I shall propose in the examination of waters are, the solutions of silver and mercury in nitrous acid. They have been usually employed to detect the viriolic or marine acids; but several other substances may likewise precipitate themselves without the need of the presence of the smallest quantity of these acids. The white and heavy striæ, which a lunar solution occasions in a water containing only half a grain of a marine salt in the pint, very easily and very surely announce the acid of this salt. But they do not discover in the same degree the presence of the vitriolic acid, as it requires, according to M. Bergman's calculation, at least thirty grains of Glauber's salt in a pint of water, to produce an immediate sensible effect; add to this, that the fixed alkali, chalk, and magnesia, may, in a much more remarkable

able manner, precipitate the nitrous ſolution: therefore the precipitation of a mineral water by means of this ſolution, cannot be depended upon as an accurate teſt of the preciſe ſaline or earthy ſubſtance to which it is owing.

The nitrous mercurial ſolution is ſtill more apt to miſlead. It not only indicates the preſence of the vitriolic and marine acids in a water, but it is precipitated by the mild fixed alkali, of a yellowiſh powder, which we might miſtake for the effect of the vitriolic acid. A very ſimilar depoſition is cauſed by lime and magneſia. It is commonly believed, that the very copious white precipitate which it forms in a water, is owing to the preſence of a marine ſalt; but the mucilages and extractive ſubſtances exhibit the ſame phenomenon, as all chemiſts now know. Beſides theſe ſources of error and uncertainty, from the property which ſeveral ſubſtances have of producing a ſimilar effect with the nitrous ſolution, we muſt have regard to other properties which depend upon the ſtate of the ſolution itſelf, and of which it is very requiſite we ſhould be warned, ſo as not to commit groſs miſtakes in the analyſis of waters. M. Bergman has taken notice of a few of the ſingular differences which are obſerved in this ſolution, according to the manner in which it has been made in the cold or with heat, chiefly relating to the colour of the precipitates which it gives by different intermedia: but he has not ſaid a word of the property which this ſolution has of being precipitated by lime-water when it is very copious in calx of mercury, although M. Monnet has taken notice of the fact in his Treatiſe upon the Solution of the Metals. As this is of great importance in the analyſis of waters, I ſet about aſcertaining the fact completely; and I ſucceeded in a very ſimple way. I made a great number of ſolutions of mercury in very pure nitrous acid of different quantities of theſe two ſubſtances, without and with heat, and with acids of different degrees

 of

of ſtrength. Theſe experiments yielded different reſults.

1. The ſolutions made in the cold are more or leſs readily charged with a greater or leſs quantity of mercury according to the concentrated ſtate of the acid; but whatever quantity of mercury is taken up in the cold by a concentrated acid, this ſolution never is precipitated by water. I diſſolved in the cold two gros and a half of mercury in two gros of fuming ſpirit of nitre, weighing an ounce four gros and five grains, in a bottle which contained an ounce of diſtilled water: the combination took place with ſingular rapidity; more than the fourth of the acid was loſt in a very copious thick nitrous gas and aqueous vapours, which were diſcharged by the heat of the mixture: the ſolution was of a deep green, very tranſparent. I poured a few drops of it into an ounce of diſtilled water; ſome whitiſh ſtriæ formed, which were diſſolved by agitation, and gave no precipitate. This is, however, the moſt copious ſolution which I was able to make in the cold, which excites the greateſt motion and efferveſcence, and moſt deep red vapours. As it had depoſited cryſtals, I added two gros of diſtilled water, which diſſolved the whole without appearance of precipitation. Farther, thoſe made in the cold with ordinary nitrous acid, and half their weight of mercury, never will be precipitated by water, and might be employed ſucceſsfully in the analyſis of mineral waters.

2. However ſmall be the degree of concentration of the nitrous acid, if it is ſtrongly heated upon mercury, it will diſſolve a greater quantity than the ſtrongeſt acid will do in the cold; and the ſolution, ſlightly coloured yellow, will appear thick; it will let fall by repoſe an irregular yellow maſs, which may be changed into a beautiful turbith by means of boiling water. This ſolution, poured into diſtilled water, forms a very copious precipitate of a yellow colour reſembling turbith. A ſolution made in the cold will exhibit the ſame phe-

phenomenon, if it is ſtrongly heated, and if much nitrous gas is diſengaged. Theſe ſolutions with heat ſhould never be uſed in the analyſis of mineral waters, ſince they are decompoſed by diſtilled water.

3. It appears that theſe two kinds of ſolutions differ from one another only in the quantity of calx of mercury, which is much greater in that precipitated by water than that which is undecompoſable by this fluid. I demonſtrated the truth of this preſumption, by evaporating equal quantities of the one and of the other of theſe ſolutions. I obtained a fourth more of this precipitate from the ſolution precipitable by water, than from that which was not precipitable. The ſpecific weight ſeemed ſtill a better means of ſhowing the reſpective quantity of mercurial calx contained in theſe different liquors. I compared the relative weight of an equal volume of three mercurial nitrous ſolutions, which differed from one another. The one not at all precipitable by diſtilled water; and the reſult of the firſt experiment, mentioned above, weighed one ounce one gros and ſixty-ſeven grains in a bottle, which contained juſt an ounce of diſtilled water. The ſecond ſolution had been made by a very gentle heat, and gave a ſlight opal colour to diſtilled water, without producing a very remarkable precipitate; in the ſame bottle it weighed one ounce ſix gros and twenty-four grains. A third mercurial ſolution, ſtrongly heated, which precipitated a true turbith mineral, of a dirty yellow, by means of diſtilled water, weighed in the ſame bulk, one ounce, ſeven gros, and twenty-five grains. To confirm this opinion the more, a deciſive experiment remained to be made. If the ſolution which water precipitated, owed this property to too great a quanty of mercurial calx relatively to the quantity of acid, it ought to loſe this property by the addition of acid neceſſary to ſupport the mercury: it ſo turned out. By pouring ſome aquafortis upon a ſolution which water decompoſed, it became undecompoſable by water; and it was abſolutely in the

ſame ſtate as that which was made ſlowly and by the heat of the atmoſphere alone. M. Monnet has already taken notice of this proceſs, as hindering the cryſtals of mercurial nitre from being reduced into a turbith by the contact of air. By an inverſe proceſs, evaporating a portion of the acid of a proper ſolution which is not precipitated by water, this ſolution is converted into one, which abounds in the calx, and conſequently is precipitable by water. It may regain its firſt quality by reſtoring to it the acid which it loſt during evaporation.

Such are the different conſiderations upon which I have thought proper to enter, to render leſs uncertain the effect of the teſts upon the waters. But however preciſe we may be in theſe inveſtigations; however extenſive our knowledge of the degree of purity, and of the different ſtates of the different ſubſtances which we mix with mineral waters in order to diſcover their principles; if it is only to be alleged, that each teſt is ſuſceptible of indicating two or three different matters diſſolved in theſe waters, a doubt will always remain of the reſult of their action. Lime, for example, attracts aërial acid; it precipitates the ſalts with baſe of clay and magneſia, and alſo the metallic ſalts: the volatile alkali has the ſame effect; the fixed alkali precipitates, beſides theſe ſalts, thoſe with baſe of lime; lime-water, charged with the colouring part of Pruſſian blue, the Pruſſian alkali, and the ſpirituous tincture of the gall-nut, precipitate vitriolum ferri and creta martialis; the nitrous ſolutions of ſilver and of mercury, decompoſe all the vitriolic and marine ſalts, which may vary or be found ſeveral in the ſame water; they themſelves are decompoſed by the alkalis, chalk, and magneſia. Amongſt this great number of complicated effects, how are we to diſtinguiſh what happens in the water we are examining? how are we to know, whether it is ſimple or whether it is compound?

Theſe queſtions, though very difficult at the time when chemiſtry was not furniſhed with all its reſources, may, conſidering our preſent knowledge, be now agitated;

tated; and we are hopeful of anſwering them in a ſatiſfactory manner. I obſerve, in the firſt place, that ſince the nature of the teſts is much better known than it was a few years ago, and their reaction upon the principles of the waters is much more fully appreciated, this is already a ground of ſtrong preſumption, that their uſe may be of much more utility than has hitherto been imagined. Among the great number of excellent chemiſts who have applied themſelves to the analyſis of waters, there have appeared none, except Meſſrs Baumé, Bergman, and Gioanetti, who have perceived that a greater part of them might be extracted than has as yet been done. It has been a long time the cuſtom, in examining mineral waters by teſts, to uſe very ſmall doſes, and to perform the experiment in glaſſes; the obſerved phenomena of precipitation are noted down, and the experiment is puſhed no farther. M. Baumé in his Chemiſtry has adviſed to ſaturate a certain quantity of a mineral water with the fixed alkali and the acids, to collect the precipitates, and to examine their nature. M. Bergman thought that we might judge by the weight of the precipitates of the quantity of the principles contained in the waters. Some other chemiſts have alſo employed this method, but always in ſome particular view; and no perſon has propoſed analyſing mineral waters by this means. To inſure ſucceſs, I am of opinion, that ſeveral pounds of a mineral water ſhould be mixed with each teſt, till the teſt ceaſe to precipitate the water. The precipitate is then to be allowed to collect for twenty-four hours in a veſſel cloſely ſhut; the mixture is to be filtrated: and by the known means, the precipitate remaining upon the filtre is to be examined after it has been weighed and dried. By this means, we ſhall diſcover completely the nature of the ſubſtance upon which the teſt has acted, and determine the cauſe of the decompoſition. We might follow a certain or- in theſe operations, by firſt mixing the waters with the ſubſtances which are leaſt ſuſceptible of altering them; and thus advancing from these substances to those which are capable of producing the most varied changes, and changes of the most difficult estimation. This is what I used to do in such species of analysis. After having examined the taste, the colour, the weight, and all the other properties of a mineral water, I pour upon four pounds of this fluid as much lime-water; if no precipitate falls in twenty-four hours, I think myself certain that this water contained no disengaged aërial acid, nor mild fixed alkali, nor earthy salts with base of aluminous earth or magnesia, nor metallic salts; but if a precipitate instantly or gradually forms, I filter the mixture, and examine the chemical properties of the deposition. If it has no taste, if it is insoluble in water, if it effervesces with acids, and if it forms with spirit of vitriol an insipid salt almost insoluble in water, I hence conclude that it is chalk, and that the lime-water attracted only the aërial acid dissolved in the water. If, on the contrary, it is not copious, if it collects difficultly, if it excites no effervescence, if, with the vitriolic acid, it produces a styptic salt, or a bitter one and very soluble, it is formed of magnesia or aluminous earth, and often by both. There is no need of being more explicit upon the means used to distinguish these two substances, as they ought to be very well known. I only add, that they might be repeated often enough, that no doubt of their nature may remain.

After their examination by means of lime-water, I pour upon four pounds of the same mineral water a gros or two of volatile caustic alkali; or I pass through it some alkaline gas disengaged by means of heat. When the water is saturated with it, I leave the mixture in a close vessel for twenty-four hours; then, if a precipitate is formed, which can arise only from the salts with base of iron, magnesia, or aluminous earth, I investigate its nature by the different methods spoken of in treating of the action of lime. But the action of the alkaline gas being less to be depended upon than that of lime-

wa-

water, which affects the same decompositions which it does, it is proper to observe, that we should employ it only as an auxiliary; the results from which cannot be expected to be so accurate as those given by the former test.

When the salts, with base of aluminous earth or of magnesia, have been discovered by means of lime-water or alkaline gas, the caustic mineral alkali serves to detect those with base of lime; such as selenite and calcareous marine salt. Therefore I precipitate some pounds of the water I am examining by means of this alkali dissolved, till no farther turbidness is occasioned, as it decomposes the salts with base of aluminous earth as well as those formed by lime; if the precipitate resembles in form, colour, and quantity, that which is yielded by lime-water, it is to be presumed that the water contains no calcareous salt; and the chemical examination of this precipitate generally confirms the suspicion. But if the mixture is much more turbid than that with the lime-water; if the deposition is more weighty, more copious, and collects more quickly; then it contains lime mixed with magnesia or aluminous earth. I satisfy myself by treating the deposition with the different means already mentioned. It is known, that iron precipitated by the tests at the same time with the saline earthy substances, is easy to be detected by its colour and taste; and that the small quantity of this metal, separated by these processes, is not capable of affecting the results.

It were useless to dwell upon the substances which the oil of vitriol, spirit of nitre, the gall-nut, alkalis, or lime, saturated with the colouring part of Prussian blue, when employed as tests, might discover in the mineral waters. What has been delivered already about the general effects of these matters ought to suffice: give me leave only to say, that upon mixing them in great quantity with the waters, we may, by collecting the precipitates, discover more completely the nature and the quantity of their principles, as M. Bergman and M. Gioanetti have done.

I ſhall dwell longer upon the products given by the nitrous ſolutions of ſilver and of mercury when mixed with mineral waters. It is particularly with theſe teſts that it is of advantage to operate upon large quantities, to be able to determine the nature of the acids which theſe waters contain. The analyſis of theſe fluids will become complete by the knowledge of their acids, ſince theſe laſt are in them combined frequently with the baſes which the preceding teſts have diſcovered. The colour, the form, and the abundance of the precipitates formed by the nitrous ſolutions of mercury and ſilver, have completely indicated to chemiſts the nature of the acids to which they are owing. A copious weighty depoſition, which is inſtantly formed by theſe ſolutions, diſcovers the marine acid. If it is not ſo copious, white, and cryſtalliſed, with the lunar nitre; yellowiſh and irregular with mercurial nitre; if it collects but ſlowly, it is aſcribed to the vitriolic acid. However, as theſe two acids are found frequently in the ſame water, as an alkali and chalk alſo decompoſe theſe ſolutions, our reſults are uncertain when we refer only to the phyſical properties of the precipitates. We muſt then ſubmit them to a more thorough examination. For this purpoſe, the lunar and mercurial ſolutions ſhould be mixed with five or ſix pounds of the water which we wiſh to analyſe; the mixtures filtrated twenty-four hours after; the depoſitions dried, and treated *ſecundum artem*. Upon heating in a retort the precipitate made by the nitrous ſolution of mercury, the portion of this metal, united with the marine acid of the waters, volatiliſes in form of corroſive ſublimate or ſweet mercury: that which is combined with the vitriolic acid remains at the bottom of the veſſel, and is of a reddiſh colour. Theſe two ſalts may alſo be diſtinguiſhed by putting them upon a burning coal: the vitriol of mercury exhales ſulphureous vapours, and

and is coloured red; the marine ſalt remains white, and is volatiliſed without ſmell of ſulphur. Theſe phenomena ſerve alſo to diſtinguiſh the precipitates which might be formed by the alkaline ſubſtances contained in the waters, ſince theſe laſt exhale no ſulphureous odour, and are not volatile without decompoſition.

The precipitates produced by the combination of the mineral waters with the nitrous ſolution of ſilver, may be examined as eaſily as the preceding. The vitriol of ſilver being more ſoluble than the luna cornea, diſtilled water may be ſucceſsfully employed to ſeparate theſe two ſalts. Luna cornea is diſtinguiſhed by its fixity, fuſibility, and particularly its leſs decompoſibility than that of the vitriol of ſilver. This laſt, put upon coals, emits a ſulphureous odour, and leaves a calx of ſilver fuſible without addition. I do not ſpeak of all the proceſſes which chemiſtry might furniſh for diſcovering and ſeparating the two lunar ſalts juſt now mentioned; I think it ſufficient for me to mention ſome of them.

§ 6. *Examination of the Mineral Waters by diſtillation.*

In the analyſis of mineral waters, diſtillation is employed to find out the gaſeous ſubſtances united to them. Theſe ſubſtances are, either air, or aërial acid, or hepatic gas. To know their nature and quantity, we muſt take a few pounds of the mineral water, and put them into a retort, which they fill only to a half or two thirds; we muſt adapt to this veſſel a bent tube, which is to be ſunk in a veſſel full of mercury. The apparatus thus fixed, the retort is heated till the water be fully boiling, or till no more elaſtic fluid paſſes into the veſſels. When the operation is finiſhed, we ſubtract from the volume of gas obtained the quantity of air contained in the retort; the remainder is the aëriform fluid which was contained in the mineral water, whoſe nature we ſoon find by means of a lighted candle, tincture of turnſol, and lime-water. If it burns, and has a fetid odour, it is hepatic gas; if it extinguiſhes the candle, reddens turnſol, and precipitates lime-water, it is aërial acid: laſtly, if it ſupports combuſtion without inflaming, if it has no ſmell, nor alters turnſol nor lime-water, it is atmoſpheric air. It may happen that this laſt fluid is purer than atmoſpheric air; then we judge of its degree of purity by the manner in which it ſupports combuſtion. The proceſs, which is followed to obtain the gaſeous matters contained in the waters, is derived quite from modern chemiſtry. Formerly a moiſtened bladder was uſed, which was tied to the neck of a bottle containing a mineral water; the fluid was agitated; and by the ſwell of the bladder we judged of the quantity of gas contained in the water. Now we know that this method is not to be relied on; becauſe the water can part with all its gas only by ebullition, and becauſe the ſides of the moiſtened bladder alter the elaſtic fluid obtained. There is no need of obſerving, that by this proceſs only the diſengaged aërial acid is obtained which is contained in the mineral water we are examining; that we muſt carefully mark the phenomena which the water exhibits, in proportion as the gas is ſeparated: laſtly, that we ſhould diſtil a quantity of water, ſo much the leſs as it is ſhown to contain more gas, by its taſte, ſparkling appearance, and lightneſs.

§ 7. *Examination of Mineral Waters by Evaporation.*

EVAPORATION is generally conſidered as the ſureſt means of obtaining all the principles of mineral waters. It has been taken notice of already, and we repeat it again from the works of Meſſrs Venel and Cornette, that a long ebullition may decompoſe the ſaline matters diſſolved in the water; and for that reaſon we have directed the examination of them by teſts in large quantity. However, evaporation may be ſo far uſeful, when joined to analyſis by teſts, that it may be always conſidered as one of the principal means of analyſing waters; and that therefore the moſt proper method of conducting this proceſs ſhould be ſought after. The aim of this operation

ration being to collect the fixed principles contained in a mineral water, it is evident, that to know the nature and proportion of these principles, so much the more water must be evaporated as the water appears to be less charged. We ought to work upon twenty pounds at once when the water seems to contain much saline matter: if there is but little dissolved, we must indispensably evaporate a much larger quantity, sometimes a hundred pounds. The nature and form of the vessels for the evaporation is not at all indifferent. Those of metal, except those of silver, are alterable by water; those of glass, of a certain thickness, are very subject to ruption; those of glazed and well-baked earth the most proper, though the splitting of their covering coat sometimes occasions the absorption of the saline matters. Those of unglazed porcelain would be without contradiction the best; but their dearness is a considerable obstacle to their use *. Different methods have been proposed by chemists for evaporating mineral waters. Some have proposed evaporating them to dryness in close vessels, to prevent the admixture of foreign substances, which may be floating in the air, with the residuum; but this operation is tedious. Others have directed a heat to be employed in their evaporation somewhat below the boiling point; imagining, that this degree of heat alters the fixed principles, and always carries off a part of them. This is the opinion of Venel and Bergman. M. Monnet, on the contrary, chooses to boil the water; because its motion prevents the admission of foreign substances. M. Bergman gets rid

* It were greatly to be wished that a manufacture of common porcelain was raised, which might supply all the necessary vessels to the kitchen, pharmacy, and chemistry. There is no need of a white, fine, precious porcelain; but of a very fine earth, with its surface polished, of any colour, hard enough to resist the heat and a sudden heat and cold. We have long known how to make easily a good porcelain, which has at the same time not the beauty and whiteness so much valued, which raises the price considerably; now it is a porcelain of this kind which is wanted.

rid of this inconvenience, by covering the evaporating veſſel with a lid with a hole in the middle to give paſſage to the vapours. This laſt method greatly retards the evaporation, the ſurface of the fluid being conſiderably diminiſhed. It ought to be employed in the beginning till the vapours be ſtrong enough to diſperſe the powder. But the greateſt difference in the conduct of this experiment conſiſts in this, that ſome, copying Boulduc, ſeparate the ſubſtances which are depoſited as the operation goes on, in order to obtain each of the principles of the waters pure and detached. Some, on the contrary, urge the evaporation to dryneſs. We agree with M. Bergman, that this laſt method is more expeditious and more ſure; becauſe, whatever precaution we take in the firſt way, in order to ſeparate the different ſubſtances which are depoſited or which cryſtalliſe, they never are got pure; and we muſt always analyſe them farther: beſides, this method is never accurate, on account of the frequent filtrations, which occaſion great loſs; and in ſhort, it is very embarraſſing, and renders the evaporation very tedious. Waters, then, ſhould be evaporated to dryneſs; in which operation different phenomena are obſerved. If the water is charged with gas, bubbles appear in it upon the firſt impreſſion of the heat: in proportion as the aëial acid is diſengaged, a pellicle and depoſition is formed, ariſing from calcareous earth and ferrum aëratum. The cryſtalliſation of the ſelenite ſucceeds theſe firſt pellicles; and at laſt the marine and febrifugal ſalts cryſtalliſe in cubes at the ſurface, and the deliqueſcent ſalts are obtained only by conducting the evaporation to dryneſs: next, the reſiduum is weighed; it is put into a ſmall phial with three or four parts of ſpirit of wine: the whole is ſhaken; and after ſome hours reſt it is filtrated; the ſpirit of wine kept apart; the reſiduum which the ſpirit has not touched is dried with a gentle heat: after it is well dried, it is accurately weighed; and by the loſs the reſiduum has ſuffered, we know the quantity which it contained of calcareous marine ſalt and marine magneſia, which are very ſoluble in ſpirit of wine. We ſhall ſay more afterwards of the manner of aſcertaining the preſence of theſe two ſalts in the ſpirit.

Then the reſiduum, treated with the ſpirit of wine, and well dried, is waſhed with eight times its weight of cold diſtilled water; and after repoſe of ſome hours it is filtrated: the reſiduum is dried a ſecond time, and then boiled half an hour in four or five hundred times its weight of diſtilled water: it is again filtrated; and nothing remains but what is inſoluble in cold and boiling water. The former diſſolves the neutral ſalts, ſuch as Glauber's, marine, febrifugal, and Epſom ſalts: if nitre or alum were preſent, which rarely happens, they are likewiſe ſoluble in cold water. The boiling water, in conſiderable quantity, diſſolves but little of the ſelenite. There are, then, four ſubſtances to be examined after theſe different proceſſes are over: 1. The reſiduum inſoluble in ſpirit of wine and in water of different temperatures; 2. the ſalts diſſolved in the ſpirit; 3. thoſe ſeized upon by the cold water; and, laſtly, thoſe diſſolved by the boiling water. Let us proceed to the requiſite experiments for their examination.

1. The reſiduum, which has reſiſted the action of the ſpirit of wine and of water, perhaps compounded of calcareous earth, aërated magneſia, aërated iron, clay, and quartz; theſe two laſt ſubſtances are very rare, but the three firſt are very common: the deeper or lighter brown or yellow colour indicates the preſence of iron. If the reſiduum is of a whitiſh grey, it contains none of this metal. When iron is preſent, M. Bergman directs to moiſten it, and to expoſe it to the air, in order to ruſt it; then vinegar has no action upon it. In order to ſhow the means of ſeparating theſe different matters, let us ſuppoſe an inſoluble reſiduum, compoſed of the five ſubſtances which we have ſaid it might contain. We muſt begin by moiſtening and expoſing

posing it to the rays of the sun: when the iron is very rusty, the residuum is digested in distilled vinegar. This acid dissolves the lime and magnesia; the solution is evaporated, and we get an acetous calcareous salt; which is distinguished from acetous magnesia by its not attracting the air's humidity. These two salts may be separated by deliquescence, or rather by pouring into their solution some vitriolic acid. This forms some selenite which precipitates: if there has been an acetous magnesia, the Epsom salt, formed by the vitriolic acid, would remain dissolved, and might be obtained by a well-conducted evaporation. In order to discover the quantity of magnesia and calcareous earth contained in the residuum, the selenite and Epsom salt, formed by pouring the vitriolic acid into the acetous solution, are separately precipitated by the mild vegetable alkali, and the precipitate is weighed. When the chalk and magnesia has been separated from the residuum, there remains only the clay, iron, and quartz. The iron and clay are both taken up by marine acid: the iron is precipitated by the Prussian alkali, and the clay by the fixed alkali; and these two substances are weighed to ascertain their quantity. The matter which remains after the separation of the clay and iron is generally quartzy; we ascertain its weight and its nature by fusing it with the fixed alkali by means of a blow-pipe. Such are the most accurate processes recommended by M. Bergman to ascertain the unsoluble residuum of waters.

2. We then take the spirit of wine, which has served for the washing of the dry residuum of the waters: it is evaporated to dryness. M. Bergman directs treating it with spirit of vitriol, like the acetous solution mentioned above; but it is to be observed, that this process serves to ascertain only the base of these salts. In order to determine the acid united with the magnesia or lime, and sometimes both, in this residuum, some drops of oil of vitriol must be poured in, which excites an effervescence,

fervescence, and disengages the marine gas, known by its smell and white vapour, when the salt which we examine is formed of marine acid. We may likewise ascertain it by dissolving all the residuum in water, and mixing with it a few drops of the solution of silver. With regard to the nature of the base, which is, as has been said, either lime or magnesia, or both, their quantity and nature are likewise known by the vitriolic acid, as we have already observed on the acetous solution.

3. The ley of the first residuum of the mineral water, with eight times its weight of cold distilled water, contains the alkaline neutral salts; such as Glauber's, marine, febrifugal, mild fixed alkali, mild soda, and Epsom salt. Sometimes also a small quantity of martial vitriol is found. These salts never are altogether in the same waters. Glauber's salts and mild tartar are found but very rarely in the waters; but marine salt is frequently found with aërated soda: Epsom salt also occurs very often, and even in very considerable quantity. When the first washing of the residuum of a mineral water contains only one kind of neutral salt, it is very easy to obtain it by crystallisation, and to ascertain its nature by its form and taste, the action of the fire, and likewise by the tests. But this case is very rare; and it is much more common that several salts are united in this ley. Their nature is then to be investigated by a slow evaporation. Even this means does not always succeed completely, however carefully this first ley is evaporated: each of the salts must be examined afresh which are obtained at different times of the evaporation. It is for the most part the mild mineral alkali, which is confusedly deposited with the marine or febrifugal salts; these may be separated by a process of M. Giaonetti. It consists in washing this mixed salt with distilled vinegar. This acid dissolves the mild soda: the mixture is dried, and washed afresh with spirit of wine, which is charged with the terra foliata mineralis, without touching the marine salt. The spirituous solution

ſolution is evaporated to dryneſs, and the reſiduum calcined; the vinegar is decompoſed and burnt: we have then no more but the mineral alkali, whoſe quantity is exactly found.

4. The ley of the firſt reſiduum of the mineral water, with 400 or 500 times its weight of boiling water, contains only ſome ſelenite: of this we are certain, by making uſe of the very pure cauſtic volatile alkali; whilſt the cauſtic fixed alkali precipitates it in abundance. By evaporation to dryneſs, we aſcertain the quantity of the earthy ſalt which was contained in the water.

§ 7. *Of artificial Mineral Waters.*

THE numerous proceſſes which we have juſt now deſcribed for the examination of the reſidua of the mineral waters after evaporation, are ſufficient to point out with the greateſt preciſion all the different ſubſtances which theſe fluids contain. A ſtep ſtill remains to be made to aſſure the ſucceſs of the analyſis; it is to imitate nature by ſyntheſis, and by diſſolving in pure water the different ſubſtances extracted by analyſis from a mineral water. If the artificial mineral water has the ſame taſte and weight, and exhibits with the teſts the ſame phenomena with the analyſed mineral water, it is the moſt complete and certain proof of the perfection of the analyſis. This combination has alſo the advantage of forming, at all times, in all places, and with ſmall expence, medicines as uſeful for the cure of diſeaſes as the natural waters, whoſe conveyance, and many other circumſtances, may tend conſiderably to alter their properties.

The moſt celebrated chemiſts think, that it is poſſible to imitate natural mineral waters. M. Macquer has obſerved, that, ſince the diſcovery of fixed air or aërial acid, and of its property of rendering many ſubſtances ſoluble in water, it is much eaſier to prepare artificial mineral waters. M. Bergman has taught us the way of compoſing waters perfectly to imitate thoſe of Spa, Seltz,

Pyrmont,

Pyrmont, &c. He has told us, that they are very ſucceſsfully employed in Swed·n. M. Duchanoy has publiſhed a work, in which he has given a ſet of proceſſes to imitate all the mineral waters employed in medicine. There is, then, all room to hope, that chemiſtry may be of important ſervice to the healing art, by furniſhing it with precious medicines, whoſe activity it may at pleaſure diminiſh or augment *.

* As the Spa, Pyrmont, and Seltzer waters are much uſed in medicine, I have added the accurate analyſis of them given by M. Bergman, in his excellent treatiſe upon the ſubject of medicinal waters. According to this analyſis, the Swediſh cantharus, a meaſure which contains 100 geometrical cubic inches, yielded the following reſult:

Seltzer water.	Grains.
Aërated calcareous earth,	17
——— magneſia,	29½
Cryſtalliſed mineral alkali,	24
Common marine ſalt,	109½
	180

Spa water.	Grains.
Aërated iron, - -	3½
——— calcareous earth,	8½
——— magneſia, -	20
Cryſtalliſed mineral alkali,	8½
Common marine ſalt, -	1
	41½

Pyrmont water.	Grains.
Aërated iron, - - -	3¼
——— calcareous earth, -	20
Vitriolated calcareous earth, -	38½
Aërated magneſia, - -	45
Vitriolated magneſia, - -	25
Common marine ſalt, - -	7
	138¾

Artificial waters may be made to imitate the natural waters in every uſeful quality. The firſt part of the proceſs is, to impregnate the water with a ſufficient quantity of aërial acid, and then to add the other ingredients in their reſpective proportions. When thoſe artificial waters are prepared with proper attention to the proportion of the materials, they prove excellent ſubſtitutes for the natural ones. However, we muſt be cautious to employ ſalts purified by cryſtalliſation from any foreign matter; and to uſe iron in the moſt metallic ſtate, as the ſmalleſt tendency to ruſt renders it no longer ſoluble by means of the aërial acid.

LEC-

LECTURE XLIV.

VEGETABLE KINGDOM.

Structure and Functions of Vegetables.

VEGETABLES are organised substances, which are confined to the surface of the earth, without either motion or sensibility. They are distinguished by their appearance and conformation. They differ from minerals; because they are nourished by intussusception, and because they prepare the juices which are destined for the increase of their bulk. The phenomena they exhibit, depending upon their organisation, are called *functions*; the principal is, their reproduction by means of seeds or eggs like animals.

Vegetables differ from one another, 1. In size: They are distinguished into trees, shrubs, herbs, mosses, &c. 2. In the place where they grow: Some grow in dry grounds, others in a humid soil; some in sands, clays, waters, upon the surface of stones, or upon other vegetables. 3. In the smell, taste, colour. 4. In their duration: plants live, are annual, biennial, &c. 4. In their use: they are employed as aliments or as medicines. A great number are employed in the arts, dyeing, &c.; others serve for the ornaments of gardens, &c.

§ 1. *Structure of Vegetables; Botanic Philosophy.*

VEGETABLES, considered externally, are formed of six parts or organs, destined for particular functions: these parts are, the root, the stem, the leaf, the flower, the fruit, and the seed; each of these differ in their form, texture, size, number, colour, hardness, and taste.

1. The root is concealed in earth, water, or in the bark of other vegetables. It is either tuberous, fibrous, or bulbous. It either hangs, as if on a pivot, or its direction is winding.

2. The stem, part of the root, and supporting the other parts: it is either solid or hollow, woody or herbaceous;

baceous; round, ſquare, triangular, or with two very acute angles. The ſtem comprehends the wood and the bark. The wood is diſtinguiſhed into wood properly ſo called, and into pulp: the bark is formed of the epidermis, of a veſicular texture and cortical layers. The ſtem is divided into branches.

3. The leaves are very variable in vegetables; *a*, in form; they are oval, round, lineal, ſagittated, lanceolated, &c.: *b*, in poſition upon the ſtem; being ſeſſile, petiolated, oppoſite, alternate, vertical, perfoliated, vaginal, &c.: *c*, in their margin; being entire, dentated, crenated, ſerrated, plaited, undulated, laciniated: *d*, in their ſimplicity or compoſition: the compound leaves are made up of ſmall ones; then they are either palmated or pinnated, *cum vel ſine impari*: *e*, in place; they are radical, caulinary, or floral: *f*, in colour, ſmell, taſte, conſiſtence, &c. Their uſe appears to be, from M. Ingenhouz's experiments, to abſorb the phlogiſticated air by their inferior ſurface, and to emit dephlogiſticated air by the ſuperior, when they are expoſed to the ſun. By this means the atmoſphere is renewed.

4. The flowers are parts deſtined to contain the organs of generation, and to defend them till parturition be accompliſhed; then they fall. In the flower two parts are diſtinguiſhed. The external parts are deſtined to envelope and to protect the internal, whoſe uſe is to reproduce the plant. The former comprehend the calyx and corolla; the calyx is external and green. Linnæus diſtinguiſhes ſeven kinds of it; to wit, the perianthium, the ſpatha, the ball, the covering, ſtrobilus, pappus, drupa. The corolla is what every body calls the flower or the coloured part: it is of one piece; and monopetal, or of ſeveral pieces, and polypetal. Upon the corolla Tournefort founded his ſyſtem. The pieces of the corolla are called *petals*. The organs, contained, and frequently hid, in the corolla, are the ſtamina and piſtils. The ſtamina are the male or generating parts; they are al-

moſt always more numerous than the piſtils. They are formed of a filament and anthera. This laſt, placed at the extremity, is a ſmall purſe, full of a generating powder: the piſtil is in the middle of the ſtamina; ſometimes it is in one flower, and ſometimes in another. On this account, ſome plants have been diſtinguiſhed into male and female. The piſtil is formed of three parts: the inferior part or ovarium, which contains the embryo, called in Latin *germen;* the filament, which is on the top of the ovarium or ſtile; and its extremity more or leſs dilated, called *ſtigma*. Linnæus, upon the number and reſpective poſition of the ſtamina and piſtils, has founded his ſexual ſyſtem. M. de Juſſieu has eſtabliſhed a ſyſtem from the inſertion of the ſtamina being either above or below the germen.

5. The fruit ſucceeds the flowers. Botaniſts diſtinguiſh ſeven kinds of fruits; the capſule, the ſiliqua, legumen, the ſtrobilus, which are dry; the drupa, the pomum, and bacca, which are ſucculent.

6. The ſeed differs greatly in form, appendices. It comprehends the corculum or little plant, the radicle, and the lobes.

Vegetables, conſidered internally, preſent five kinds of veſſels or organs, which are found in all their parts. 1. The common veſſels, deſtined for carrying the ſap. They are placed in the heart of plants and trees: they riſe perpendicularly; but they twiſt ſo as to form little holes or diviſions. 2. The proper veſſels, which convey the proper juices; the oily, the gummy, the reſinous juices. They lie under the bark: frequently they dilate into cavities or reſervoirs: they ſeem to be the excretory canals. 3. The veſſels for circulating the air which they receive from the atmoſphere. Upon cutting a young green branch, we perceive that they are ſpiral, and reſemble worms. They are frequently found filled with ſap. 4. The utricules, formed of ſacs, which contain the pith, and frequently a coloured part: they are placed in the middle of the ſtems.

5. The

5. The veſicular texture, exhibiting a ſet of ſmall cells, which are detached in a horizontal direction from the pith, and which paſſing through the ſap-veſſels, the cavities of which they fill, ſpread out below the epidermis, and form a net-work, ſtuffed up, ſimilar to the ſkin of animals. The veſicular texture of vegetables ſeems to correſpond with the cellular texture of animals.

§ 2. *Functions of Vegetables; Vegetable Phyſiology.*

ALL the organs of vegetables, of which we have juſt now given a ſhort deſcription, are deſtined to execute different motions, called *functions*. Theſe functions are, 1. The motion of the fluids, or a kind of circulation: 2. The alterations or changes of theſe fluids, which indicate a ſecretion: 3. The increaſe and evolution of the vegetable, which ſhows a nutrition: 4. The exhalation of different fluids elaborated in the different organs; and the inhalation of ſeveral principles contained in the atmoſphere, by means of the ſame organs: 5. The action of the air, and the uſe of this fluid in the veſſels of vegetables: 6. The motion made by ſome of their parts: 7. The kind of ſenſibility by which they know the contact of bodies that are uſeful to them, the light, &c.: 8. The different phenomena which ſerve for the reproduction of ſpecies, and which conſtitute the generation of plants.

Let us run over each of theſe functions in particular.

The principal fluid of vegetables, known by the name of *ſap*, is contained in particular canals, which are called *common veſſels*. Theſe veſſels, placed in the middle of the ſtems, and below the bark, are continued and prolonged from the root to the leaves and to the flowers. The ſap which they convey is a colourleſs fluid, more or leſs flat to the taſte, ſerves the ſame purpoſe with the blood of animals, of being converted into different juices for the nouriſhment and ſupport of the different organs. In the ſpring it is very copious; and its motion then appears from the evolution of the leaves and flowers. By ligature, as well as by all the phenomena of vegetation, it appears demonſtrated, that it mounts from the root to the ſtems and the branches. We do not know if it deſcends again towards the root, as ſome philoſophers have imagined. The valves, admitted in the common veſſels by ſeveral botaniſts, have not been demonſtrated, unleſs they chooſe to give this name to ſeveral filaments or hairs with which their internal ſide ſeemed ſtuck full to Tournefort and M. Duhamel. This irregular motion widely differs from the circulation of animals.

The ſap, conveyed in the utricules or proper veſſels, is elaborated in a particular manner. It gives riſe to different ſaccharine, oily, mucilaginous fluids, which iſſue out by an organic excretion, and whoſe evacuation ſeems to be of advantage to the vegetable, ſince it does not ſuffer from the great loſs which is thus frequently occaſioned. This alteration of the fluids, which is obſerved even in a remarkable degree in ſeveral organs, as in the nectarines at the extremity of the piſtil, in the pulp of the fruits, at the baſe of the calices, and of ſeveral leaves, correſponds exactly with the function which in animals goes by the name of *ſecretion*. M. Guetard has puſhed the analogy ſo far, as to deſcribe glands of ſeveral different forms at the baſe of the leaves of fruit-trees, and towards the extremity of the petals of certain flowers. It is this ſecretion which evolves the odorous principle, the colouring matter, the combuſtible ſubſtance: but it differs from the animal ſecretion in this, that this latter function entirely depends upon the organiſation of the glands, which elaborate the animal juices; whereas, in vegetables, the juices conveyed by the common veſſels are more expoſed to the contact of the air, of the light, and to the action of heat; and that their nature renders them ſuſceptible of undergoing by the action of theſe agents the motions of fermentation, which alone are capable of producing alteration upon them.

The ſap, by its remaining in the cavities of the utricles and of the cellular texture, thickens, and puts on a greater or leſs conſiſtence. This alteration renders it ſuſceptible of adhering to the ſides of the fibres, and of gradually increaſing their dimenſions. Such is the mechaniſm of the nutrition of vegetables, of their growth, and of the evolution of all their parts. It has a great reſemblance to the nutrition of animals. The veſicular texture and the utricles have the ſame ſtructure and the ſame uſes in theſe two claſſes of organic ſubſtances. They likewiſe penetrate all their organs; they eſtabliſh betwixt them an immediate communication; and both are the real foundation of nutrition.

It is a long time ſince botaniſts were convinced of exhalations from the ſurface of plants which ſpread in the air. The odorous ſpirit of the leaves and of the flowers forms round vegetables an atmoſphere, which is ſenſible to our organs, and which the contact of a burning body is ſometimes capable of inflaming; as is the caſe with the fraxinella. This exhalation ſeems to be an inflammable gas of a particular nature. Unhappy experience has even taught us, that ſeveral vegetables emit vapours mortal to animals which are expoſed to their action. Of this kind are the walnut and yew trees, and ſeveral others of the hot countries. M. Ingenhouz, in his experiments, found, that the leaves of all plants, expoſed to the ſun and to the light, mix with the atmoſphere an inviſible fluid, an air ſimilar to that procured from the calces of mercury, and improperly called *dephlogiſticated air*. Darkneſs entirely changes this property of the leaves, which then give only fixed air when they are deprived of the contact of the light. This beautiful diſcovery, firſt announced by Dr Prieſtley, demonſtrates a new property in vegetables, of purifying the air by communicating to it this portion of vivifying fluid, continually deſtroyed and abſorbed by combuſtion and reſpiration. But if vegetables continually emit vapoury fluids, which are only the laſt labour of vegetation, they likewiſe have the property of abſorbing ſeveral principles contained in the atmoſphere. The inferior ſide of the leaves abſorbs the humidity of the dew, according to M. Bonnet's experiments. Prieſtley's experiments have demonſtrated, that vegetables abſorb the gaſeous reſidua of combuſtion and reſpiration, ſince vegetation becomes more rapid and ſtrong in air altered by theſe two phenomena. Exhalation and inhalation, then, are much more extenſive in the vegetable kingdom than was imagined before the modern diſcoveries.

The gaſes abſorbed by vegetables, are conveyed into all their organs by means of veſſels called *air-veſſels*; which in their uſe and ſtructure approach to thoſe of inſects and worms. They enter the compoſition of the fluids as the air appears to do in the lungs of moſt animals; perhaps they contribute in a great meaſure to form the proper and eſſential ſalts of vegetables, theſe ſubſtances containing, as is well known, a great quantity of air. However, the air-veſſels are not ſolely deſtined to contain this fluid; they are found filled with ſap in the ſeaſons when this humour is very abundant; which makes them widely differ from the organs of reſpiration, ſo eſſential and ſo conſtant in a great number of animals.

It cannot be doubted, but that ſeveral parts of vegetables poſſeſs motion. In ſome it is ſo great, that it is ſenſible to the eye. Such are the motions of the ſenſitive plant, of the ſtamina of the opuntia, of the parietaria, &c. This motion ſeems to reſemble the function called in animals *irritability*, being occaſioned by the action of a ſtimulus, and having particular organs, that ſome botaniſts have compared to muſcular fibres. Upon this force alſo, does not the contraction of the woody fibres, occaſioned by the action of fire, depend? If that were the caſe, as M. Bucquet thought, irritability would be much longer inherent and durable in vegetables than it is in animals; ſince wood,

wood, however ancient it be, ſtill exhibits this phenomenon in a remarkable degree.

Can we refuſe likewiſe a ſort of ſenſibility to plants, when we ſee them turn their leaves and flowers to the ſun? when we obſerve that, when ſhut up in woody caſes, glazed, perforated, or ſimply, thinner in one part than in another, they conſtantly tend towards the tranſparent body, or to the opening which admits the light to paſs, or even to the ſide which is the moſt penetrable by this fluid, by its ſmall degree of thickneſs? or rather, ought not this degree of ſenſibility to be conſidered as the effect of the force of affinity of the tendency to combination ſubſiſting between vegetables and the light? It is fully demonſtrated, that this fluid evolves in plants, either by percuſſion or by combination, their colour, taſte, and combuſtible property; ſince plants raiſed in darkneſs are white, inſipid, aqueous, and contain nothing inflammable; whilſt vegetables, expoſed in hot climates, in the middle of the day, to the ſun's rays, acquire a conſiderable colour, and are charged with bitter and reſinous parts, and are ſingularly combuſtible. However ſtrong we may ſuppoſe this affinity, we cannot conceive how it ſhould be capable of exciting ſo great a motion in the branches and in the leaves of vegetables. It is then neceſſary to admit of a particular ſenſation; a feeling, it is true, very different from that of animals, by which vegetables are diſpoſed to chooſe the moſt bright places, or thoſe through which the light has the freeſt acceſs.

The means which nature employs to reproduce the ſpecies in vegetables, have great affinity to thoſe which ſhe makes uſe of for the production of animals. The ſexes, and their reunion, are requiſite in the moſt of plants. From the works of the celebrated Linnæus, it is found, that a remarkable analogy ſubſiſts between the organs deſtined to this function, in theſe two claſſes of organic ſubſtances. The ſtamina anſwer to thoſe of the male, and the piſtil is compoſed of three parts, analogous to the genital parts of female animals. M. Decemet, physician of the faculty of Paris, has also imagined, that he discovered a striking resemblance in the external form of the parts of generation of the apocynum and other vegetables, and that of animals. The embryo is evolved by the action of the generating power; without which it is impossible to produce a new individual, as is every day observed with regard to birds. But without this analogy, which it would be useless to prosecute farther, vegetables being of a much more simple structure than animals, and all their parts being composed of the same organs, each of them is capable of producing a new individual similar to itself. This is the reason of the reproduction of plants by means of suckers, buds, slips, layers, and likewise of the alteration of the fluids by the operation of grafting, either natural or artificial. There is still a new analogy between vegetables and this class of animals, which are produced by slips; as the polyps, crustaceous insects, some worms, &c.

All the functions of vegetables, the whole of which give rise to great relations between them and animals, are susceptible of alterations which produce diseases. These diseases, which depend most commonly either upon the abundance or defect of the sap, as well as upon its bad qualities, are very analogous to those of animals: their causes, their symptoms, their cure, depend absolutely upon the great principles of medicine, and form a part of agriculture, little advanced, it is true, but susceptible of great progress, when it shall be prosecuted upon the plan laid down by some famous writers on husbandry; at the head of whom may be placed Messrs Duhamel and the Abbé Tessier. The latter has thrown great light upon the diseases of grains, in his Observations described in the volumes of the Royal Society of Medicine, and deposited in part in the Royal Academy of Sciences.

LECTURE XLV.

OF VEGETABLE JUICES.

THE humours of vegetables are of two claffes; the common and proper juices. The firft conftitute the fap, which is found in all plants. This fluid feems to fupply the function of blood in vegetables. It is contained in the common veffels: it flows naturally from their furface; but a greater quantity is got by incifion. The fap is not an aqueous fluid; it contains falts, extracts, and mucilages. When a certain quantity is wanted for the examination of its properties, or for medicinal ufe, the plant is bruifed in a mortar, and ftrained through a linen rag: if, in this way, the plant does not give out its juice eafily, we fubmit it to the prefs.

The fucculent vegetables furnifh their juice by fimple expreffion: thofe whofe juice is vifcous or lefs abundant, require dilution with water; fuch are borage and the dry aromatic plants. This juice being extracted by a ftrong expreffion, contains a portion of the folids of the vegetables which have been bruifed by the peftle: it muft therefore be purified. The purification of the juices is effected, 1. By fimple repofe, when they are very fluid, as thofe of purflane and Barba Jovis. 2. By the white of an egg, which collects the feculent parts by means of coagulation, as is done to thofe of borage, nettles, &c. 3. By fimple heat, which coagulates and precipitates the filamentous parts, as M. Baumé directs in the treatment of juices containing volatile principles; fuch as thofe of the cochlearia, creffes, &c. The phial containing the juice is plunged into boiling water, being ftopped up with a perforated piece of paper: the juice, when it becomes clear, is drawn off; then it is immerfed into cold water, and filtrated. 4. By fimple filtration for thofe which are very fluid. 5. By the addition of fpirit of wine, which coagulates the feculent parts. 6. By the vegetable acids, as prefcribed by the London

don pharmacopœia, for the juices of the cruciform plants.

Of Extracts.

The juices of plants hold, in solution, substances which, when separated from the aqueous vehicle, form kinds of extracts. These substances are distinguished into three species; the mucous extracts, the soapy, and the extracto-resinous. The name of mucous extracts is given to those which easily dissolve in water, very sparingly in spirit of wine, and which pass to the spirituous fermentation; such is the rob of gooseberries, which is prepared by evaporating the juice of this fruit. The soapy extracts are characterised by their dissolubility in water, and partly in spirit of wine, and by their disposition to grow mouldy, rather than to pass to the spiritous fermentation. The thickened juice of borage is an extract of this nature. These are the extracts properly so called. The extracto-resinous dissolve in water and in ardent spirits: they are inflammable, because they contain an oily principle; and they are altered in no way by the air. The inspissated juice of the wild cucumber, called *elaterium*, is of this kind. Incisions are made in the root of this plant; it is expressed; the juice is allowed to clarify itself, and then evaporated in a balneum mariæ to dryness.

In commerce, extracts of these three different kinds are prepared, by evaporating the juice of several plants: such, among others, are,

a, The juice of acacia, brought from Egypt, by shaving the fruit of this tree, expressing its juice, and evaporating it by the heat of the sun: the juice of the German acacia is prepared with the juice of prunes by a similar process.

b, That of hypocistis, which is made like the preceding, with the fruits of this parasite plant.

c, Opium, a very important medicine, whose nature ought to be completely known. It is extracted from the

2

the white poppy in Perſia. There flows through inciſions which are made in the green capſules, a white juice, which is dried into brown tears: that is the true opium. That in commerce is formed by preſſing theſe capſules after they are moiſtened with water: the juice is dried, and ſent away in flattened circular loaves, covered with leaves, and mixed with many impurities. In order to purify it, it is diſſolved in the ſmalleſt poſſible quantity of water by means of heat; the liquor is ſtrongly expreſſed, and evaporated in a balneum mariæ. This is the extract of opium. This ſubſtance contains a reſin, a ſolid eſſential oil, an odorous principle, poiſonous and narcotic, an eſſential ſalt, and a ſoapy extract. As the odorous, poiſonous, and narcotic part, is frequently nauſeous, the method of having the extract of opium free of it has been ſought for. M. Baumé, who has ſtrictly examined this medicine, volatiliſed this principle at the ſame time with the eſſential oil, and ſeparated likewiſe the reſin by a digeſtion of ſix months. M. Bucquet diſcovered, that this ſame extract may be obtained, calming and not narcotic, by diſſolving the opium in cold water, and evaporating the ſolution in a balneum mariæ. Our knowledge of the principles of opium goes no farther. M. Lorry has made ſome very excellent experiments upon the ſubject: he found that opium fermented, gave by diſtillation a calming, not poiſonous, water, which he employed with much ſucceſs. He obſerves, that the odorous principle of this medicine cannot be deſtroyed by any means.

When the plants, whoſe extracts we want, are dry and woody, we employ, in order to get their principle, maceration in water, infuſion, or decoction, according to the ſtate of the ſubſtances; maceration often ſuffices. Decoction extracts too much ſubſtance, and ſeparates the reſinous part: it forms a very impregnated thick fluid, frequently loathſome. Infuſion may ſuffice in all

caſes:

cases: this is the opinion of the most celebrated chemists and physicians.

Extracts, differing from one another, are extracted by means of water, as those which the thickened juices yield. Thus juniper-berries yield with water a mucous extract; Peruvian bark a soapy one, which is obtained in small transparent and as if saline scales, if the solution is evaporated in very flat vessels; from rhubarb an extracto-resinous substance is obtained.

In commerce, also, extracts are prepared in great quantity by means of water: such as,

a, The juice of liquorice root, yellow by the first infusion, and black by the strong decoction.

b, The terra Japonica, which is extracted from the infusion of the seeds of a kind of palm: the infusion is evaporaed, and the residuum is formed into flattened cakes. The terra Japonica in laboratories is purified by solution and evaporation.

From these considerations, it is easy to conceive, that the name of extract is given in general to all substances soluble in water, and separated from this fluid by evaporation. However, as some eminent chemists, and particularly M. Rouelle senior, gave this name to a particular substance, which he considered as one of the ultimate principles of vegetables, it becomes a matter of importance to fix our ideas on this subject. There are only the soapy and extracto-resinous extracts, which properly deserve the name of extract. M. Rouelle distinguished this last into extracto-resinous and resino-extractive. The extracto-resinous burns only after it has been dried: it seems to contain more extract, properly so called, than resin. The resino-extractive burns much better than the former: it seems to contain more resin than extractive substance. This clear distinction proves, that these two kinds are only mixtures of extracts in a different dose with a resinous principle. There are therefore no more extracts, properly so called: and this name

name ſhould peculiarly belong to the ſoapy matter. On this account, it is the properties of this ſubſtance which we muſt examine.

Pure extract, taken in the ſenſe of M. Rouelle, is a dry ſolid ſubſtance, coloured brown, or of a dirty green, which does not burn by itſelf, which emits much ſmoke, and in which is formed a greater or leſs quantity of eſſential ſalt. Its taſte is almoſt always bitter. It yields in diſtillation an inſipid phlegm: with a gentle fire this phlegm gradually becomes coloured and acid, according to M. Rouelle: for the moſt part, however, this phlegm is alkaline, as is obſerved in the Peruvian bark, the extract of borage. This volatile alkali is formed by the heat; then ſome empyreumatic oil paſſes over: the charcoal is light, contains ſome alkali, and almoſt always ſome neutral ſalts. The extract, when expoſed to the air, becomes mouldy, dries, or attracts the air's humidity, according to the nature of the ſalts which it contains. Theſe ſalts cryſtalliſe and ſeparate from the extractive part; frequently they are altered and entirely decompoſed. It diſſolves in water, and then forms a ſtrong infuſion. The acids decompoſe this ſolution as they do ſoaps, and occaſion a more or leſs oily precipitate. Metallic ſolutions alſo precipitate it, and are themſelves decompoſed. The proſecution of its chemical properties has been carried no farther; and hence it has been conſidered as a kind of ſoap.

In medicine, extracts are employed as aperient, deobſtruent, diuretic, ſtomachic remedies; and they are every day obſerved to produce the moſt excellent effects.

Of Eſſential Salts in general.

Essential ſalts of plants is given as a name to the ſaline ſubſtances held in ſolution by their juices, or by the water of their infuſion. They are extracted by cooling theſe fluids, when evaporated, to the conſiſtence of ſyrup. As theſe ſalts are impregnated with mucilages and

and fatty matters, they must be purified by means of lime and the whites of eggs. If these salts are acid, we should not make use of lime, which neutralizes them, but of pure white clay in powder. After this first process they are still impure; they are then dissolved in distilled water, and made to crystallise several times till they be white.

The essential salts of plants are of different natures; and should be distinguished into two classes.

Class 1. *Essential Salts.*

THE first class comprehends those which resemble the mineral salts. The principal kinds are, 1. The mild fixed alkalis, which are got from almost all plants by macerating them in acids, as demonstrated by Messrs Margraaf and Rouelle junior: the vegetable alkali is the most common; the mineral is found in the marine plants. 2. Vitriolated tartar from the millefolium, old borragines, astringents, and aromatics, thymelæa, and pulp of olives. 3. Glauber's salt, from the tamarisk. 4. Nitre from the borragines, turnsol, tobacco, &c. 5. Sylvius's febrifugal salt from marine plants. 6. Selenite, from rhubarb, discovered by M. Model.

In vegetables, no doubt, several other salts might be found resembling the mineral, if an accurate analysis were made of a great number of plants. The volatile alkali, or rather mild volatile sal ammoniac, has also been imagined to exist ready formed in the class of cruciform plants; because these plants submitted to distillation, yield, upon the first impression of the heat, a phlegm containing a little of this salt in solution. On this account, the ancient chemists gave these the name of *animal plants*. But M. Rouelle junior has demonstrated that it is not to be got ready formed in them; and that it is the reaction of these principles, occasioned by the fire, which produces it. M. Baumé pretended, that the

the volatile principle of cruciform plants was merely some sulphur.

Naturalists have entertained different opinions about the mineral salts which are found in plants. Some have thought that they have been conveyed from the heart of the earth by the water, and thus passed without alteration into the vegetables. Other have imagined, that vegetation formed these saline substances. It is certain, that two very different plants, as the borage and millefolium, growing in the same soil, furnish each the same salt, which is proper to them; *viz.* the former nitre and the millefolium vitriolated tartar. One experiment alone, of which much has been said, and which has not yet been made with proper accuracy, might decide this question. It is this, To raise in a well-washed earth, plants which yield one kind of salt, as nitre, and to moisten the earth with a solution of marine or of any other salt; if they still furnished nitre and no marine salt, the conclusion might be drawn, that the salt in that state does not enter the plants, and that the salt which is peculiar to them is formed by the operation of vegetation.

Class 2. *Of Essential Salts.*

The second class comprehends the salts particular to vegetables. These truly essential salts are always formed of an acid united with an alkali and an oil. Frequently the acid is uncovered: sometimes its taste is covered by other substances; and therefore these salts should be distinguished into acid and sweet.

§ 1. *Essential acid Salts.*

The essential acid salts of vegetables are found in a great number of plants; and, in general, all those of a sourish taste contain them: such are sorrel, acid fruits, lemons, oranges, &c. The salt the most known in this class, is that called in commerce *salt of sorrel*, which is got from the oxalis. This plant is much cultivated in Switzerland and Germany. The false salt of sorrel forrel, or the falt of oxalis, is in white irregular cryftals; it has a four tafte, and reddens the blue colours of vegetables. M. Baumé, who has examined it, difcovered in it the following properties. It diffolves very well in water, and it may be cryftallifed without lofing its acid; when heated in a crucible, it emits a fharp acid fmell; it bubbles up on burning coals; it becomes charry and kindles; it burns blue like fpirit of wine; it leaves after its combuftion a white falt, which, with fpirit of falt, forms marine falts. Upon diftillation, an ounce of this falt afforded three gros and a half of an acid liquor, which was colourlefs, fmelling a little of marine acid. No oil came over. The refiduum was fuliginous. This falt precipitates white the nitrous folution of mercury, as well as the acid does, which it gives to diftillation. This acid, mixed with the nitrous, did not diffolve gold-leaf. M. Bergman has placed the acid of forrel as a particular acid in the thirteenth cocolumn of his Table of Affinities. He differs from M. Baumé in fome points, although he correfponds in the greateft number, as we fhall fee; but he did not fay whether it was the falt of forrel in commerce, or the true effential falt which he employed. His doctrine upon this fubject is as follows: The falt of forrel is vegetable alkali, fuperfaturated with a particular acid. M. Scheele has found a very good method of obtaining this falt: he mixed the acid of forrel, faturated with volatile alkali, with a folution of terra ponderofa in nitrous acid: by dint of double affinity, the principles of thefe two compounds reciprocally changed their combination; and that of the terra ponderofa with the acid of forrel is precipitated, becaufe it is very infoluble. This precipitated falt is decompofed by acid of vitriol, which has the ftrongeft affinity to it of any fubftance hitherto known: the acid of forrel fwims above the fpatum ponderofum formed by this decompofition, which may be poured off. This falt feems to refemble the acid of fugar more than it does that of tartar. It differs from both: for,

for, combined to superfaturation with vegetable alkali, it forms the falt of forrel analogous to tartar; but it decrepitates and fufes in the fire, turns a little black, and is fufceptible of being decompofed entirely by aërated lime, properties not to be found in tartar: and farther, the vegetable alkali, combined with the acid of fugar, refembles neither tartar nor falt of forrel. The acid of forrel prefers lime to the alkalis; but it is ftill doubtful, whether it does fo to terra ponderofa and magnefia: it decompofes felenite, having more affinity with lime than the vitriolic acid has. If a ftrong heat is applied to the acid of forrel it is deftroyed; but it fwells and blackens in a lefs degree than the acid of tartar. It gives to diftillation a much more acid phlegm than tartar does treated in the fame way. From thefe details, we fee, that M. Bergman differs from M. Baumé only in admitting the vegetable alkali in this falt, whilft the latter calls it the mineral. Perhaps it is the true falt of forrel which M. Bergman talks of.

All the acid falts in plants have not yet been examined, although we know a great number. That of citrons fhould be feparated from its mucilage by reft, and concentrated by means of freezing. It has been fuppofed analogous to the acid of tartar: however, in its more ftrong tafte, it feems to refemble rather that of forrel and of oxalis. Stahl fays, that this acid, faturated with crabs-eyes, and digefted upon fpirit of wine, gradually affumes the nature of vinegar. It has been remarked by M. Bergman, that the fparry, phofphoric, arfenical acids, and thofe of borax, fugar, tartar, forrel, and citrons, refemble each other in this, that when combined with earths they are almoft infoluble, and that they become foluble only by an excefs of acid; whereas this property is not found in other acids. However, felenite and heavy fpar, two earthy falts formed by the vitriolic acid, have almoft no folubility.

The fruits, which are at firft four, and which turn fweet by ripening, yield a falt, whofe acid is more covered than that in the preceding. This ſalt is in a medium between the ſtrong acid eſſential ſalts and thoſe which are quite ſweet; it reſembles the tartar of wine; it is got from apples, pears, quinces, tamarinds, &c. They have been carefully examined by M. Rouelle junior. We ſhall attempt a pretty full account of them when we come to the ſpirituous fermentation.

LECTURE XLVI.

§ 2. Essential Saccharine Salts.

THE eſſential ſaccharine ſalts are found in a great number of vegetables. The mapple, the birch trees, red beet, parſnips, raiſins, wheat, Turkiſh corn, &c. all contain ſome of them. M. Margraaf extracted them from the greateſt number of theſe vegetables.

The ſugar-cane (*arundo ſaccharifera*) contains the greateſt quantity of it; from which it is got with the greateſt profit. Theſe canes, when ripe, are ſqueezed between two iron cylinders placed perpendicular. The expreſſed juice falls upon a plate placed below. It runs into a veſſel, where it is boiled along with aſhes and lime; the ſcum is taken off; and thus boiled and ſcummed in three other veſſels; it then gets the name of ſyrup. Then it is again ſtrongly boiled with lime and alum: when ſufficiently boiled, it is poured into a cooling baſon; when it grows ſo cool as to be handled, it is put into barrels laid upon ciſterns, and having ſeveral holes ſhut up with canes. The ſyrup turns ſolid in the barrels, a portion runs into the ciſterns. The ſugar thus rendered concrete is yellow and thick. They refine it in the Weſt Indies by boiling, and then pour it into inverted earthen cones. The ſugar which cannot concrete, runs through holes into a pot placed below: it is called *thick ſyrup*. The baſe of the ſugar-loaves is taken off, and ſome white powder of ſugar ſubſtituted in its place, which is made hard by ſtriking. The whole is covered over

over with wetted pure clay. The water of the clay fil-
:res through the ſugar, and carries along with it a por-
:ion of the mother-water of the ſugar, which flows thro'
·he holes, and is received into new pots. It is called *fine*
ſyrup, being more pure than the firſt. A ſecond layer
of clay is applied after the firſt is dry, and the water is
illowed to filtre a ſecond time; and when the water has
ill run through, the loaves are carried to a ſtove to dry.
n eight or ten days the loaves are broken, and the different ſugars thus formed are ſhipped for Europe, where
hey are refined to form the various kinds of it. The
efining conſiſts in boiling the ſugar in lime-water along
vith ox blood, taking off the ſcum two or three times,
iltering the liquor, and running into the form of loaves:
hen the loaves are covered with wet clay, and allowed
o filtre. This is repeated till the ſugar be quite white;
he loaves are carried to a ſtove, and after eight days covered with paper, and tied up for exportation. The ſy-
·ups, which cannot be farther cryſtalliſed, are diſpoſed
of under the name of *melaſſes*. All chemiſts have
hought that theſe different operations ſeparated a fat
natter from the ſugar, and rendered this ſalt ſuſceptible
of cryſtalliſation. M. Bergman ſuppoſes the uſe of the
ime to be, to carry off the exceſs of acid which prevents its aſſuming ſolidity. As it undergoes a ſtrong
:vaporation in theſe proceſſes, it turns into a granulated
and irregular maſs, juſt as we have remarked happens
at Goſlar in the preparation of white vitriol.

Sugar is formed of a particular acid united with a
mall quantity of alkali, and changed by a great quan-
ity of oily matters. It cryſtalliſes into truncated hexaë-
lral priſms, ſo as to form a kind of two-ſided pyramid.
n this ſtate it is called *ſugar-candy*. It gives to diſtilla-
ion an acid phlegm, and a few drops of an empyreu-
natic oil. A ſpongy light coal remains, containing
ixed alkali. This ſalt is inflammable; laid upon burn-
ng coals, it fuſes and bubbles up; it emits a pungent
vapour; it turns of a brown yellow, and becomes a ſu-

gar useful in coughs; it is very soluble in water; it gives it a great degree of consistence; and forms a strong saccharine mucilage, called *syrup*. This, diluted with water, is susceptible of fermentation, and of giving ardent spirit.

M. Bergman has extracted an acid of a particular nature from all saccharine matters, and particularly from sugar. We put into a retort one part of powdered sugar with six parts of aquafortis: a gentle heat is applied. After the passing over of the red vapours, the solution is left to cool, and white needled crystals are precipitated, or they are prisms with four sides terminated by two-sided tops. The liquor poured off, and again heated with three or four parts of the same nitrous acid, yields by a fresh crystallization prisms of the same form. The same process may be repeated upon the second mother-water of these crystals. One ounce of sugar yields about three gros of a prismatic salt, which dissolves in warm water, and may be crystallized by cooling to have it very pure.

The acid of sugar has a very pungent acid taste. Diluted with water, it forms a sourish agreeable liquor. It reddens all the blue colours of vegetables. Exposed to a gentle heat, it becomes opaque, as if efflorescent; it turns into powder, and it loses three-tenths of its weight by the evaporation of the water of its crystals. The water may be collected by distillation. Heated more strongly, it fuses; it turns brown; an acid phlegm passes over into the receiver, resembling the salt itself in all its characteristic properties. A part sublimes in form of a white crust; the retort contains almost no residuum. What remains is grey or brown, and comes, according to M. Bergman, only to about the fiftieth of the salt employed. A very considerable quantity of a gas is also produced: half an ounce of acid of sugar gave M. Bergman one hundred cubic inches of a gas, consisting half of aërial acid, and half of inflammable gss, which burnt blue. M. l'Abbé Fontana, who repeated this experiment,

ment, obtained from an ounce of this falt four hundred and thirty-two inches of a gas; a third of which was aërial acid, and the reft inflammable gas mixed with common ai·. Upon a repetition of the experiment, I had a refult very nearly fimilar to that of this laft chemift. What is moft fingular, is, that the fublimed portion, again diftilled twice over, gives no coaly matter, and leaves only a whitifh grey refiduum. This falt, heated in an open fire, emits a very pungent vapour, and its refiduum is quite white.

The acid of fugar expofed to the air effforefces in length of time. Cold water diffolves half of its weight of this falt, and boiling water its own weight. This falt cryftallizes as the folution cools.

Acid of fugar diffolves clay, the bafe of alum. By evaporation the folution yields a yellowifh, tranfparent, fweet, aftringent mafs, which attracts the air's moifture, and reddens turnfol. This falt bubbles up when put into the fire; it parts with its acid, and leaves the clay brown. It is decompofable by the mineral acids.

This acid combines with the faline-earthy and alkaline fubftances. It forms, 1. With terra ponderofa, an infoluble falt, yielding cryftals of an acid tafte: hot water, which wafhes off the fuperfluous acid, renders them opaque, powdery, and infoluble. 2. With magnefia, a white powdery falt, decompofable by the fparry acid and terra ponderofa. 3. With lime, a falt infoluble in water, powdery, decompofable by fire alone, on account of the immenfe attraction of thefe two fubftances, the acid of fugar carrying off this bafe from all the other acids. M. Bergman on this account propofes the acid of fugar to detect the prefence and quantity of lime when contained in mineral waters, or combined with any acid. This falt turns the fyrup of violets green.

It unites with the vegetable fixed alkali, and is fufceptible of cryftallifation when there is an excefs of either of its principles. This falt, which is very foluble in in water, is decomposed by the action of fire and by the mineral acids.

Combined with two parts of fixed mineral alkali, the acid of sugar forms not a very insoluble salt, which dissolves better in hot water, and turns the syrup of violets green.

With the volatile alkali an ammoniacal salt is formed, which crystallises by slow evaporation into quadrilateral prisms; which is decomposed by fire, and furnishes some mild volatile alkali, formed by the destruction of the acid of sugar.

The acid of sugar is soluble in the mineral acids. It turns the oil of vitriol brown, and is decomposed by the spirit of nitre.

The acid of sugar combines in general more easily with metallic calces than with the metals. It forms, 1. With white arsenic, crystals; prismatic, very fusible, very volatile, decomposable by heat. 2. With cobalt, a powdery salt, of a clear rosy colour, not very soluble. 3. With the calx of bismuth, a powdery white salt, very insoluble in water. 4. With the calx of antimony, a salt in crystalline grains. 5. With nickel, a salt of a white or of a greenish yellow, insoluble. 6. With manganese, a powdery white salt, which blackens by the fire. 7. With zinc, whose solution is accompanied with an effervescence, a white powdery salt. 8. It dissolves the calx of mercury, and reduces it to a white powder, which blackens by the contact of light: this acid decomposes mercurial vitriol and nitre. 9. At first it blackens tin, which is afterwards covered with a white powder. The salt which it forms with this metal is of an austere taste: it crystallises into prisms by a well-conducted evaporation. If in the evaporation a strong heat be used, it yields a transparent mass resembling horn. 10. It tarnishes lead, but dissolves its calx better. The liquor, when saturated, deposites small crystals; which are also obtained by pouring the acid of sugar into a solution of nitrous or marine lead, and

likewise

likewiſe into a ſolution of ſaccharum Saturni. 11. It attacks iron filings, and produces ſome inflammable gas. This ſolution is ſtyptic. It yields priſmatic cryſtals of a greeniſh yellow, decompoſable by heat. Crocus martis, united with this acid, affords a yellow powder, reſembling that which is obtained by pouring the acid of ſugar into a ſolution of martial vitriol. 12. It acts upon copper, and entirely diſſolves its calx; the ſalt which it forms is of a clear blue, inſoluble: this ſalt may alſo be had from the precipitation of the vitriolic, nitrous, marine, and acetous ſolutions of copper, by means of the acid of ſugar. 13. The calx of ſilver, precipitated by the fixed alkali, diſſolves in ſmall quantity in this acid. The beſt method of preparing this ſalt, called by M. Bergman *argentum ſaccharatum*, is to precipitate the nitrous ſolution of ſilver by means of the acid of ſugar: a white depoſition falls, difficultly ſoluble in water, which turns brown by the contact of the light. 14. This acid has but a very inconſiderable action upon the calx of gold. 15. It diſſolves the precipitate of platina, occaſioned by the mineral alkali. This ſolution is a little yellow, and yields cryſtals of the ſame colour. Theſe are the phenomena mentioned by M. Bergman reſulting from the combination of the acid of ſugar with the metallic ſubſtances.

It might be imagined from the proceſs directed by M. Bergman for procuring the acid of ſugar, that this ſalt is owing to the nitrous acid employed in the operation. This learned chemiſt does not think this conjecture admiſſible; becauſe the acid of ſugar has none of the properties of the acid of nitre, and becauſe it differs from it in all its combinations. In fact, the nitrous acid does not ſeem to make a conſtituent part in the acid of ſugar; but, however, the great quantity of nitrous gas which exhales during the proceſs, announces the decompoſition of the ſpirit of nitre. Now, as, from M. Lavoiſier's experiments, the nitrous acid yields no gas, unleſs it part with the pure air which, combined with this gas, conſtitutes this acid, might we not imagine, that a part of this pure air combines with the combuſtible matter of the ſugar to form the acid which is extracted from it? We have ſeen, that in this way we might account for the production of the arſenical and dephlogiſticated marine acids.

M. Bergman aſcribes the cryſtalline form of the acid of ſugar to a portion of phlogiſton which remains combined with it. M. de Morveau, in an excellent note, taken notice of in the laſt paragraph of M. Bergman's Diſſertation upon the Acid of Sugar, has obſerved, that many acids become more fluid by the addition of phlogiſton, and that this property in all the acids depends upon the degree of their fuſibility.

Sugar is of immenſe uſe. It is an aliment, a great quantity of which is in danger of heating the conſtitution. Great uſe is made of it in pharmacy; forming ſyrups, lozenges, boluſes, &c. It ſerves well to ſuſpend in water reſins, oils, &c. It is uſed to preſerve the juices of fruits made into jelly: it may be conſidered even as a medicine, being expectorant, aperient, ſlightly tonic, and ſtimulant: there are likewiſe ſome facts recorded of ſome diſeaſes depending upon obſtructions being cured by the continual uſe of ſugar.

Of Manna.

THERE are ſome juices which flow from plants, and have a ſweet taſte. Manna and nectar are of this kind. Manna is produced by the leaves of the pine, of the holm, of the juniper, of the willow, of the fig, and of the maple trees. The aſh-tree, which is very plentiful in Calabria and Sicily, produces that of commerce. It flows naturally from theſe trees; but it is obtained in greater quantity by inciſions in the bark. What is collected upon the ſtraws or ſmall ſticks which are introduced into the artificial openings, forms kinds of ſtalactites, pierced in their middle; it is called *manna in tears*. Manna *en ſortes* flows upon the bark, and contains ſome impurities. The fat manna contains a great many ſtrange matters: it is formed from the impurities of the two former; it is always moiſt, and often changed. The taſte of manna is ſweet and inſipid. That which the larch-tree, very copious in Dauphiné, affords, and that of the alhagi, which grows in Perſia about Tauris, is of no uſe: this laſt is called *tereniabin*. Manna is ſoluble in water; it gives the ſame products to diſtillation as ſugar does. By means of lime and whites of eggs, a matter reſembling ſugar is extracted; and when treated with nitrous acid, it gives an acid ſalt, of the ſame nature with acid of ſugar.

In the doſe of an ounce, or of two or three ounces, it is uſed as a purgative; or in the doſe of a few gros, when diluted with water, if it is given as a deobſtruent.

Of Gums and Mucilages.

ANOTHER kind of proper juice is that called *gum* or *mucilage*. This is inſipid, ſoluble in water, to which it gives a thick and viſcous conſiſtence. This mucilage, when evaporated, becomes dry, tranſparent, and friable. Mucilage burns without flame; it gives to diſtillation a great quantity of acid phlegm, a little empyreumatic oil, and ſome volatile alkali: its coal, which is very bulky, contains vegetable fixed alkali.

There are three kinds of gum. 1. The gum of Pays, which flows from the apricot and pear trees, &c.: it is white, yellow, or reddiſh. 2. The Arabic gum, which flows from the acacia in Egypt and Arabia. The gum of Senegal is of the ſame nature. It is employed in medicine as a ſoftening relaxing remedy. It is uſed in ſeveral arts. 3. The gum tragacanth, which flows from the tragacanth in Crete; *tragacantha Cretica*. It is uſed like the preceding.

Mucilages of the ſame nature are extracted from a great number of plants. The roots of the wimote, linſeed, quinces, give out, by maceration in water, viſcous fluids, which, when evaporated, yield true gums.

All theſe ſubſtances, conſidered chemically, ſeem at

firſt

firſt ſight to be bodies not very compound; ſince chemical experiments frequently preſent ſubſtances whoſe gelatinous form approaches to that of gums and mucilages. However, from theſe products of vegetation, which ſeem to ſerve the purpoſe of an excrementitious humour, water is extracted, ſome acid liquor, ſome aërial acid, an oily principle, and ſome fixed alkali, in the charry reſiduum. This reſiduum contains a fixed earth, whoſe nature is not yet known. Their inflammable part is not conſiderable, ſince they are not combuſtible. M. Bergman ſays, that he obtained from gums, treated with the nitrous acid, an acid ſimilar to that of ſugar.

LECTURE XLVII.

Oily Juices.

OILS are proper juices, fat and unctuous, fluid or ſolid, inſoluble in water, combuſtible with flame, volatile in different degrees of heat; they are contained in proper veſſels or particular veſicles. They are diſtinguiſhed into fat and eſſential oils, into fluid and concrete oils. Theſe bodies are found in two ſtates in vegetables: either they are combined with other principles, as they are found in the extracts, mucilages, &c.; or they are free and detached. It is theſe we ſhall examine.

Chemiſts have thought a ſimple oily principle, like a primitive ſalt, exiſted. This oily principle, combined with different ſubſtances, and modified by theſe combinations, conſtitutes, according to them, the different kinds of oils which are obtained in the analyſes of vegetables. This ſimple and primitive oil is very fluid, volatile, colourleſs, inodorous; it burns with flame and fume. This fume condenſed has all the characters of coal: it does not unite with water: it is thought to be formed of an acid united with earth and phlogiſton. It is certain that the oils yield in their decompoſition inflammable gas united to some aërial acid. They also contain a good quantity of water. The earth is in the smallest quantity, since they leave only a very small quantity of fixed charry residuum. This theory of the oily principle ought to be considered merely as an hypothesis. The oils are always formed by organic substances; and all the bodies which show their characters in the mineral kingdom, owe their origin to vegetable or animal life. The oily juices of vegetables are distinguished into fat and essential oils.

§ 1. *Of fat Oils.*

The fat oils are very unctuous; they have for the most part a mild insipid taste, without any smell: for their volatilisation, they require a higher degree of heat than that of boiling water; and kindle only when exposed to the heat requisite for their volatilisation. This is the use made of the cotton of lamps for kindling the fat oils. The oil is heated to the point of volatilisation.

The most part of the oils are fluid, and require a very considerable cold for their conversion to a solid state: some concrete with the slightest cold; others, again, are almost always solid; these are named *butters*.

The fat oils do not flow from the surface of vegetables: they are contained in the almonds, quinces, and emollient seeds. The cells which contain them being bruised, they come out by expression.

The fat oils being exposed to air, alter and turn rancid: their acid is evolved; they lose their properties, acquire new ones, in which their nature approximates to the essential oils. Water and spirit of wine, by carrying off the evolved acid, deprive them of their strong taste, but never reduce them to their former state.

They give to distillation an acid phlegm of a pungent odour, a light oil, a thick oil, and a great quantity of inflammable gas mixed with some aërial acid. They afford little charcoal. By the redistillation of these

pro-

products, ſome phlegm and oil, becoming more and more light, is obtained. This oil is named *oleum philoſophorum*. The alchemiſts prepared it, by diſtilling ſeveral times a fat oil, with which they had impregnated a brick. It is not exactly known whence this decompoſition ariſes, although it may have been ſaid, that a fat oil can be reduced into a detached inflammable oil, an acid phlegm, air, and earth.

Water does not alter the fat oils; it purifies them by taking away a part of their mucilage, which likewiſe precipitates during combuſtion; and to which they owe their fermenting property, or that of becoming rancid.

The fat oils do not combine with the quartzy and vitreſcent earths: they afford with clay a ſoft paſte, which is employed in chemical manipulations under the name of *fat lute*.

By particular proceſſes they combine with magneſia, which reduces them to a ſoapy ſtate.

Lime unites with them, but not in a remarkable manner, when they are immediately combined.

The alkalis eaſily combine with the fat oils, and form a compound called *ſoap*.

In its preparation, the oil of olives or of ſweet almonds is triturated with mineral alkali, rendered cauſtic by lime; and hence called *ley of the ſoap-boilers*. The mixture does not thicken till it has ſtood a few days, and then forms the medicinal ſoap. That in commerce is made by boiling the ley with ſome altered oil: it is then white: orpiment is made uſe of to marble it. The green ſoap is made with the reſiduum of the olives after expreſſion and potaſh.

Soap is ſoluble in pure water. Heat decompoſes it, drives off phlegm, ſome oil, and ſome volatile alkali, ariſing from the decompoſition of the fixed alkali and oil. The coal contains much fixed alkali. Lime-water, M. Thouvenel ſays, decompoſes ſoap; and then a calcareous inſoluble ſoap is formed, which turns gru-

 mous,

mous. The acids poured upon ſoap diſengage its oil a little altered.

The volatile alkali does not unite but with difficulty with the fat oils; however, by a long trituration, the mixture acquires a little conſiſtence, and becomes opaque.

The fat oils unite with the acids, and form particular kinds of ſoap. Meſſrs Achard, Cornette, and Macquer, have ſtudied theſe compounds. M. Achard made them, by pouring gradually ſome fat oil upon vitriolic acid. By a continued trituration of this mixture, a brown maſs reſults, ſoluble in water and ſpirit of wine. The oil, which is extracted from it by means of the alkalis, is always more or leſs concrete, juſt like that which is obtained by diſtillation. M. Macquer here directs pouring the acid upon the oil; but he remarks, that an acid ſoap made in this way has little ſolubility in water. That which is prepared by the trituration of ordinary alkaline ſoap with oil of vitriol is more ſoluble.

Nitrous acid inſtantly conſumes the fat oils, and kindles thoſe that are dry. Thoſe which are not dry, cannot be kindled, unleſs by a mixture of ſpirit of nitre and oil of vitriol, as M. Rouelle ſenior has remarked in his Memoirs upon the Inflammation of oils, Académie, anno 1747.

The marine and aërial acids have but a very weak action upon the fat oils. The former, however, in its ſtate of concentration, combines, according to M. Cornette, with them to a certain degree.

The action of the other acids upon the fat oils is not known. The oils do not ſeem to combine with the neutral ſalts. Several of theſe, and particularly all the calcareous ſalts, decompoſe the alkaline ſoaps. In this decompoſition, particularly that effected by ſelenite and Epſom ſalt, which frequently are found united with mineral waters, the vitriolic acid unites with the fixed alkali of the ſoap, and forms Glauber's ſalt: the lime

lime or magnesia combines with the oil, and forms a kind of very insoluble soap, which swims in whitish lumps upon the water. This is the cause of the phenomenon exhibited by the waters, which curdle the soap without dissolving it.

The action of inflammable gas upon the fat oils has not yet been examined.

The fat oils dissolve sulphur by means of a boiling heat; and this solution is of a deep red colour inclining to brown: it has a very fetid smell; it gradually deposites some crystallised sulphur. If this combination is distilled, the sulphur is decomposed, and not an atom of it is to be afterwards found. This experiment would merit a particular examination. Some sulphureous gas is also obtained in this decomposition.

The fat oils do not seem susceptible of uniting with the pure metallic substances, except with copper and iron, upon which they exert a remarkable action: but they combine with their calces, and form thick concretes, which have a soapy appearance, as is observed in the preparation of ointments and plasters. These combinations have not yet been examined chemically; it is known only, that each of the metallic calces are reduced in the formation of plasters, as the calx of copper in the emplastrum divinum, and the litharge in the unguentum martis. In docimasia, or in the art of essaying, the fat oils are made use of in the reduction of the metallic calces. M. Berthollet has given an ingenious and simple process for the formation of a combination of a fat oil with any metal, or for the formation of a metallic soap. It consists of pouring a metallic solution into a solution of soap; the acid of the solution seizes upon the alkali of the soap, and the metallic calx precipitates with the oil, to which it gives its colour.

The fat oils dissolve the bitumens, and particularly amber; but in this solution they require heat. They form kinds of fat varnishes, which dry but with difficulty.

The fat oils should be distinguished into three kinds.

The first comprehends the pure fat oils, which fix by cold, thicken slowly, form soaps with the acids, and kindle only by the addition of those of nitre and vitriol. Such are, 1. Oil of olives; which is extracted from the pulp of this fruit squeezed between two stones, and then submitted to the press in linen rags. That which comes out first is called *virgin oil;* that which is obtained from the residuum, wetted with water, is less pure, and deposites a ley; that which is extracted from unripe olives is the omphacine of the ancients. Oil of olives congeals at 10 degrees above 0 of Reaumur's thermometer, and turns rancid only in about twelve years. 2. Oil of almonds, extracted without fire, turns rancid very quickly; it congeals at about 10 degrees below 0. 3. That of cabbage, which is extracted from the grain of a kind of collyflower called *colsat.* 4. That extracted from almonds, which come from Egypt and Arabia; it is very bitter and unodorous; it congeals easily.

The second kind comprehends the dry oils, which thicken readily, do not fix by cold, which inflame by the nitrous acid alone, and which form kinds of resins with vitriolic acid. Such are, 1. The oil of lin, which is extracted from the grain of roasted lin by expression. It is employed in fat varnishing and in painting. 2. That of nuts, which serves the same purpose. 3. That of artichokes or of poppy seed, containing nothing narcotic, as is very well demonstrated by M. l' Abbé Rozier. 4. That of hemp-seed, which is very dry.

In the third genus, we comprehend the concrete fat oils, or the butters. 1. That of cocao, extracted from the almonds of the cocao tree. Four kinds of cocao are distinguished; the great and small, berbeche, and that of the islands. The butter is extracted by torrefaction and boiling in water: it is purified by liquequefying it with a very gentle heat: it owes its concrete ſtate to an acid. 2. The cocoa furniſhes a ſimilar butter. 3. The wax of vegetables is of the ſame nature, only it has more ſolidity. It is extracted from the gall-nut in China: yellow, white, and green candles, are made of it, according to the method made uſe of to extract the wax. The berries of the birch and poplar trees may likewiſe yield a ſmall quantity of a ſimilar wax.

The uſe of the fat oils is great in the arts and in medicine.

As medicines, they ſerve as emollient, relaxing, calming, and laxative: ſome even are purgative, as the oleum ricini; which has alſo the virtue of killing and expelling the vermis ſolitaris. They enter a great number of compound medicines; as the balſams, the ointments, and plaſters; and they are uſed as aliments on account of the mucilage which they contain *.

§ 2.

* Although M. Fourcroy ſeems to be of opinion that the fat oils have no power to act upon any of the metallic ſubſtances except copper and iron, there is much reaſon to believe that they are alſo capable of exerting their action upon mercury in a metallic ſtate. We have an inſtance of the combination of mercury with a fat of a ſimilar nature in the common preparation of mercurial ointment, where theſe two ſubſtances are intimately blended together by long friction in a marble mortar. In this ointment, the particles of the mercury do not merely ſeem to be interſperſed and diſtributed among the particles of the fat without any adherence or chemical union: on the contrary, the oily matter of mercurial ointment very quickly becomes rancid; and we know that rancidity is always the conſequence of the combination of oil with ſome other ſubſtance. When the mercurial ointment is old, if we rub a portion of it between two bits of paper, the whole of the oil is abſorbed, without leaving any globules of mercury viſible behind it: but when we treat mercurial ointment recently prepared in the ſame manner, we can very readily perceive a great number of metallic particles quite diſtinct. M. Baumé took equal quantities of mercurial ointment; one of which was new made, and the other had become ſlightly rancid by keeping. He kept both of them in a ſtate of liquefaction during eight days, in a degree of heat much below what could poſſibly decompoſe the fat. The new made ointment allowed three

drams

§ 2. *Eſſential Oils.*

The eſſential oils differ from the fat oils in the following characters: Their odour is ſtrong and aromatic; they are ſo volatile as to diſtil with the heat of boiling water; their taſte is very acrid; they are much more combuſtible than the former.

Theſe oils are found in almoſt all odorous plants. They are contained either through the whole of the

drams of mercury to ſeparate; the other, which was rancid, only one dram and a half. All thoſe obſervations do not allow us to doubt of the reality of the combination; they pointedly prove, that what we call the extinction of mercury in fat, is not purely the effect of a mechanical diviſion, ſince thoſe two ſubſtances exert a ſlow ſpontaneous action upon one another, from which a more intimate union at length reſults. This is much confirmed, by obſerving the difference in colour and conſiſtence between old and new ointment. New-made ointment is of a very light colour, and extremely ſoft; while what has been kept for ſome time is much darker in colour, and much firmer in conſiſtence: A ſufficient proof of ſome change in the intimacy of their union.

We are in the next place to inquire in what ſtate the mercury unites with the fat, whether in form of a metal or in form of a calx. When old mercurial ointment is converted into a ſaponaceous compound by the addition of cauſtic alkali, there is always a quantity of fluid mercury ſeparated from the mixture, the fat forſaking the mercury to unite with the alkali. Mercurial ointment is alſo decompoſed by the action of ether upon it. When a ſmall quantity of good mercurial ointment is put into a flaſk, which is two-thirds full of ether and diſtilled water, and the mixture frequently ſhaken, the mercury ſoon begins to precipitate, carrying a ſmall portion of fat along with it, which gives the mercury the appearance of a calx; but this fat ſoon diſappears, and the mercury unites in the form of metallic globules, by ſimply drying it upon bibulous paper. By this analyſis we collect almoſt the whole of the mercury in a fluid ſtate.

In reviewing all thoſe facts carefully, it ſeems probable, that the mode in which mercury combines with fat, more reſembles the amalgamation of the metals with mercury, than their diſſolution in acids, as the mercury is taken up in a metallic ſtate, and not calcined; the fatty matter ſerving the purpoſe of a ſolvent to the mercury in the preparation of mercurial ointment, in the ſame way that mercury itſelf ſerves the purpoſe of a ſolvent to the other metals in the combination of the different amalgams.

plant, as in the Bohemian angelica; or in the root alone, as in the alder-tree, the oris, the white dictamnus, &c.; or in the ſtem, as in ſaunders, ſaſſafras, pines, &c.; or in the bark, as in the canella. Sometimes the leaves conceal it, as in the meliſſus, peppermint, wormwood. In other plants it is found in the calices of the flowers, as the roſe. The petals of camomile and of the orange are filled with it. At other times it is fixed in the fruit; as in cubebs, pepper, juniper berries: And, laſtly, many vegetables contain it in their ſeed; as nutmeg, aniſe, fennel, and moſt of the umbelliferous plants.

They differ from one another, 1. In the quantity, which greatly varies according to the ſtate or age of the plant. 2. In the conſiſtence. There are ſome very fluid; as thoſe of lavender and rue. Some congeal by cold; as that of aniſe, fennel: Others are always concrete; as thoſe of roſes, parſley, and alder. 3. In the colour: Some have none; ſome are yellow, as that of lavender; of a deep yellow, that of canella; blue, that of camomile; aqua marine, that of mille-pertuis; green, that of parſley. 4. In weight: Some ſwim upon water, as moſt of our own country oils: others go to the bottom; as that of ſaſſafras and cloves, and moſt of thoſe from foreign plants. This property, however, is not conſtant, relatively to the climate; as the eſſential oils of nutmeg, mace, pepper, are lighter than water. 5. In ſmell and taſte: This laſt property is often very different in the eſſential oil from what it is in the plant; for inſtance, pepper gives a ſweet oil, and wormwood a bitter.

Eſſential oils are got, 1. By expreſſion, from the citron, bergamot, orange, cedar trees, &c. 2. By diſtillation. The plant is put into an alembic of copper along with ſome water; the water is made to boil; the oil paſſes over with the water, from which it is ſeparated by means of a proper veſſel.

The eſſential oils are adulterated, either by the fat oils,

oils, and the fraud is then detected by trying if they ſtain paper; or by the oil of turpentine, which is known by its ſtrong ſmell, and which remains after the evaporation of the eſſential oil; or by means of ſpirit of wine; then water, by rendering them turbid, diſcovers the fraud.

The eſſential oils loſe their colour by a gentle heat; and as they are very volatile, the fire cannot decompoſe them. By heating them in cloſe veſſels, a great quantity of inflammable gas eſcapes. When they are heated with the contact of air, they kindle readily, and emit a very thick fume, which condenſes into a charry matter, very light and fine: they leave, after inflammation, very little coal, being ſo volatile, that the charry part riſes in the volatiliſed portion.

Expoſed to the air, they thicken as they grow old, and appear like a reſin. Needled cryſtals are depoſited, reſembling thoſe of ſublimed camphor, which Geoffrey junior obſerved in that of motherwort, marjoram, and in that of turpentine. Their ſmell alſo approaches to that of camphor, according to the ſame obſerver, Acad. 1721, p. 163.

They unite difficultly with lime and the alkalis: the acids alter them; the concentrated vitriolic acid changes them into bitumens; and if it is weak, it converts them into kinds of ſoaps. They kindle into flame with the nitrous acid, and the marine acid makes them into ſoaps.

They have no action upon the neutral ſalts. They eaſily combine with ſulphur, and form compounds called *balſams of ſulphur*; in which the ſulphur is ſo changed, that it never can be wholly ſeparated, nor its preſence ſcarce made to appear.

Mucilages and ſugar render them ſoluble in water *.

 In

* The eſſential aromatic oils have not the ſmoothneſs, ſlipperineſs, and unctuoſity of the expreſſed oils. They are all volatile in the heat of boiling water, or nearly in that degree of heat; and the ordi-

In medicine they are employed as cordial, ſtimulant, antiſpaſmodic, emmenagogue, &c. Exhibited externally, they prove frequently antiſeptic, and retard the progreſs of carious bones.

ordinary method of obtaining them is, to macerate the plant in water, and afterwards to ſubject the whole to diſtillation, when the volatile oil ariſes along with the vapour of the water. The only purpoſe of adding the water is to give an eaſy method of adjuſting the heat; for if the vegetable was to be put into the ſtill without any addition of water, ſome parts would be ſcorched and deſtroyed before others were ſufficiently warmed to exhale the eſſential oil, as it is not poſſible to diſtribute the heat equally without the aſſiſtance of a fluid medium. The water which comes over is impregnated with the taſte and flavour of the oil, for water is capable of diſſolving a ſmall portion of eſſential oils; and in this ſtate is called a *ſimple diſtilled water*. Many ſuch are prepared for common uſe from roſes, cinnamon, peppermint, and other aromatic plants. They all owe their virtues to a quantity of the eſſential oil of the vegetables; as we know they may be prepared in a different manner, merely by mixing ſome eſſential oil of the vegetable with pure water, by agitation. The ſolution of theſe oils in water does not ſeem to be very perfect, as there is commonly ſome degree of milkineſs and want of tranſparency; which is proof ſufficient of an imperfect ſolution. This diſſolving power of water diminiſhes the yield of oil to a certain degree, ſince we do not yet know of any way by which it can again be ſeparated. But this inconvenience is not very great in practice, as the water is ſoon ſaturated, and will diſſolve no more; ſo that all the redundant oil ſeparates from the water, and either ſinks to the bottom or riſes to the top, according to the relative ſpecific gravity. If therefore the pure oil be the object of our purſuit, we have only to pour water, fully ſaturated by a former diſtillation, upon a freſh parcel of materials, when the whole of the oil may be eaſily collected without loſs, as no more can be diſſolved in the water. Some vegetables are ſo tender, that the ſimple maceration in water impairs the delicacy and fragrance of their odours. In ſuch caſes, the vegetable is only expoſed to the ſteams of boiling water, which penetrate every where, and carry the eſſential oil along with them. To accompliſh this end, we extend a piece of hair-cloth over the water of a common ſtill, and lay the vegetable upon it, and raiſe the heat to the degree which we wiſh. For as ſome vegetables do not require a boiling heat, but are greatly injured by expoſure to one ſo great, we can by this contrivance temperate the heat at pleaſure, and employ the exact degree which is requiſite for the particular plant, and never allow the fire to excite a greater. Roſemary and

lavender,

lavender, diſtilled with this precaution, give over an oil much ſuperior in fragrance to what is obtained by a boiling heat.

Spirit of wine diſſolves the eſſential oils more thoroughly than a watery menſtruum; and as the ſpirit alſo riſes in a lower degree of heat, all the aromatic properties of the plant are extracted in higher perfection. There are, however, but a ſmall number of plants which are volatile in the heat which raiſes ſpirit of wine; ſo that, if we wiſh the ſpirit to be impregnated with ſuch oils, we either diſſolve directly ſome oil which we have obtained by a different proceſs; or we add the ſpirit to the ſimple water, which, in the courſe of diſtillation, carries off the aromatic odorous principle; and thus we obtain a ſpirituous impregnation of the eſſential oil.

The aromatic parts of vegetables are contained in ſeparate ſpaces, diſtinct from the general matter of the plant. A good microſcope will ſhow a number of ſhining particles compoſed of this matter; and if theſe be all carefully picked out, the plant will no longer yield a ſingle drop of aromatic oil.

Some chemiſts have ſuppoſed that the lily, violet, jeſſamine, narciſſus, hyacinth, and other very delicate flowers, derived the fragrance of their odour from the preſence of ſome principle different from the eſſential oil, becauſe thoſe chemiſts had always failed in their attempts to collect the oil in any conſiderable quantity. This opinion is, however, ſo far erroneous, that Neuman ſaw eſſential oils obtained from all thoſe plants in different parts of the warmer countries of Europe. The product was indeed extremely trifling; ſo that the very powerful effluvia of theſe flowers muſt ariſe from the extreme volatility, pungency, and diffuſive nature of their eſſential oils. And in this conſiſts the only difference between them and other aromatic vegetables.

As it is ſometimes difficult to collect a large enough quantity of rare flowers to prepare oil from them, ſome have propoſed to ſalt the aromatic parts, and keep them till a ſufficient collection be made. When this advice has been followed with roſe-leaves, the diſtilled water was found to contain a ſmall portion of marine acid; which appeared from the effect of the roſe-water in diſturbing the nitrous ſolution of ſilver.

Another practice is recommended in particular caſes where a ſlight degree of fermentation increaſes the produce, or in ſome meaſure ſeems to create the oil. To ſuch plants a little yeſt of ſome kind is added, to promote the fermentation. The veronica, a plant which yields ſcarcely any oil when in a freſh ſtate, will give over a great deal by being allowed to ferment for ſome days. Whether this incipient proceſs only ſerves to extricate what was already formed, or really alters the combination of the different principles, is a matter which has not yet been fully examined.

LECTURE XLVIII.

Of CAMPHOR.

CAMPHOR is a white, concrete, cryſtalline ſubſtance, of a ſtrong ſmell and taſte, which approaches to the eſſential oils in ſome of its properties, but widely differs from them in others.

This ſubſtance is extracted from a kind of laurel, growing in China, Japan, and in the iſlands of Borneo, Sumatra, Ceylon, &c. The tree which produces it ſometimes contains ſo great a quantity, that cleaving it is ſufficient to obtain very large and very pure tears. It is obtained, however, by diſtillation. The roots or other parts of the tree are put into an iron alembic along with ſome water; they are covered with a head, in which ſome rice-ſtraws are placed; and the whole is heated. The camphor ſublimes in ſmall greyiſh grains, which are united into larger pieces. This coarſe camphor is very impure. The Dutch purify it by ſublimation in kinds of balloons, by the addition of an ounce of lime to the pound of camphor. It may alſo be extracted from the roots of the canella tree, zedoary, thyme, roſemary, ſage, and from all the labiated ones, either by diſtillation or decoction, as Meſſrs Neuman and Cartheuſer have remarked; but the camphor is in very ſmall quantity, and it always has a ſlight ſmell of the plant from which it was extracted. This ſingular ſubſtance ſeems to be found in combination with the eſſential oils of theſe vegetables, ſince Geoffroy obſerved that theſe oils depoſited camphorated needles.

Camphor is much more volatile than the eſſential oils, as it ſublimes with the gentleſt heat; it cryſtalliſes into hexaëdral laminæ, adhering to a filament. If it is briſkly heated, it fuſes before it is volatiliſed. It ſeems to be incapable of decompoſition in this way: however, if it is ſeveral times diſtilled, it yields a reddiſh and manifeſtly acid phlegm; which indicates, that by a frequent repetition of this proceſs it might be decompoſed.

Its

Its own temperature is sufficient to volatilise it: exposed to the air, it dissipates entirely; included in a vessel, it sublimes i. hexagonal pyramids, or into polygonal crystals; which were observed and described *anno* 1756 by M. Romieu, (Acad. 1756, p. 443). It emits a strong smell, unsupportable by some persons: it kindles very rapidly, burns with a great smoke, and leaves no charry residuum.

It does not dissolve in water; it communicates to it its smell; it burns at its surface. M. Romieu observed, that some little pieces of camphor, of a third or of a fourth of a line in diameter, put in a glass of pure water, are moved, turn round, and dissolve in half an hour. He suspects that this motion is an effect of electricity; and he remarks, that it ceases when the water is touched with a conductor, as a wire of iron; and, on the contrary, that it continues when touched with an insulated body, as glass, resin, sulphur, &c.

The earths, saline-earthy, and alkaline substances, exert no action upon camphor; the effect, however, of the caustic alkalis has not been tried.

The acids dissolve camphor when they are concentrated; oil of vitriol dissolves it by means of heat: this solution is reddish. The nitrous acid dissolves it without noise: the solution is yellow. As it swims above the acid as oils do, it has been called *oil of camphor*. The marine acid in the state of gas dissolves camphor, and likewise the sulphureous and sparry gases. If some water is added to these solutions, they grow turbid; the camphor separates in flocci, which swim a-top, and have undergone no alteration. The alkalis, the saline-earthy, and metallic substances, likewise precipitate these solutions.

The neutral salts have no action upon camphor. The effect of sulphur and the bitumens is not known, though it is probable that they are capable of uniting with it.

The fat and essential oils dissolve camphor by means of heat. These solutions, when cool, gradually deposite crystals, like vegetation, resembling those which are found in the solutions of sal ammoniac; that is, made up of a moderately sized side, to which very fine filaments, placed horizontally, adhere. These kinds of plumage, viewed with a microscope, are very beautiful and regular. This nice observation is owing likewise to M. Romieu, (Acad. 1756, p. 448). We shall afterwards take notice of the solution of camphor in spirit of wine, which is much better known and much more used than the preceding, that it exhibited a somewhat different crystallisation, which he obtained by a particular process.

We must not forget to observe, that a physician, as illustrious for his extensive knowledge as for his great conception of the action of medicines, M. Lorry, considers camphor as a principle very diffused in vegetables; and he places its spiritus rector at the head of a class of very strong smells, to whose effects upon the animal-œconomy all chemists and physicians ought to attend.

Camphor is one of the most powerful remedies which physic possesses. Applied to inflamed tumours, it soon dissipates them. It is employed as an antispasmodic and antiseptic in contagious diseases, in malignant fevers, and in all diseases accompanied in general with nervous and putrid affections. In France it is administered scarcely in the dose of a few grains; in Germany they augment the quantity to several gros a-day. It is also important to know, that camphor calms ardors and pains in the urinary passages, often as if by enchantment. Triturated with yolks of eggs, sugar, gums, or in the state of oil of camphor, it is put into some convenient drink. The surgeons employ the *spiritus vini camphoratus*, to be described afterwards, in external gangrenes, whose progress it frequently retards and stops.

Spiritus Rector.

BOERHAAVE has given the name of *spiritus rector of plants* to their odoriferous principle; only very little is known

known about the properties of this singular substance, altho' so interesting in its effects upon the animal œconomy. The spiritus rector seems to be very volatile, very dissipable, and attenuated; it is perpetually flying off from plants, and forms round them an odorous atmosphere, which is diffused to a greater or less distance. All plants differ from one another in the quantity, power, and nature of this principle. Some are abundantly provided with it, and even lose it only partially upon exsiccation; so that it hence appears to possess a certain degree of fixity; such as, in general, the odorous woods, and all the odorous parts of vegetables, dry and woody. In others it is so dissipable and so volatile, that though they have a strong smell, the principle is with difficulty fixed. And there are some plants whose smell is insipid, and not very sensible: they have been named *inodorous;* these having only the smell of an herb, their principium rector has been called *herbaceous*.

The slightest heat is sufficient to dissipate the spiritus rector from plants. To obtain it, we must distil the plant in a balneum mariæ, and receive its vapours in a cold head, which condenses them, and makes them run in a liquid state into the receiver. The product is a limpid water charged with smell, which has been called *essential* or *distilled water*. This liquor should be considered as a solution of the odorous principle. As this principle is more volatile than the liquid in which it is dissolved, if the spiritus rector is heated, it gradually loses its smell, and becomes insipid: exposure to air produces the same alteration: it deposites very light, apparently mucilaginous, tufts, and even acquires a mouldy smell.

The principle of odour unites with the oily juices, and also seems to make one of the elements of the essential oils; since, 1. These always contain it. 2. The plants which have a tenacious odour constantly yield more essential oil than those do, whose odour is volatile, and which yield frequently none at all. To retain the spiritus

ritus rector of these last, as of the jessamine, tuberose, &c. it must be combined with the fat oils. The flowers are put into a tin cucurbit with cotton, impregnated with some oil of benzoin: the flowers and cotton are laid alternately; the cucurbit is shut up, and exposed to a gentle heat. The spiritus rector is disengaged, and combines with the oil, and fixes itself in a durable manner. 3. The plants which have no smell never yield a drop of essential oil. 4. The vegetables from which the spiritus rector has been extracted in a balneum mariæ yield no more of this kind of oil, unless they still retain a little of the smell; in this case they yield, too, only a very small quantity. 5. An essential oil which has lost its smell easily regains all its properties, when it is distilled upon the fresh plant from which it was at first extracted.

The action of the saline substances upon the spiritus rector has not been examined.

The nature of this principle is not identical, and it seems to differ according to the genera of plants to which it belongs. M. Macquer thinks with Boerhaave, that it is in general composed of an inflammable substance and a saline matter; but he observes, that it sometimes partakes more of the saline nature, whilst in other plants it approaches more to the oily matters. The spiritus rector of the cruciform plants appears to him to be saline, and he gives it the characters of being pungent and penetrating without affecting the nerves. That, on the contrary, which is insipid or strong, but without being pungent, and which affects the nerves so as to produce or allay the fits which depend upon their impression, as those of the aromatic and narcotic plants do, participates greatly of the oily nature, according to this celebrated chemist. Some facts support this assertion. The fraxinella emits an odour, which forms an inflammable atmosphere round the plants, which an ignited combustible body kindles; then this vapour burns from top to bottom, along the stem which supports the flow-

ers

ers. The ſpiritus rector of the fraxinella ſeems then to be of an oil" nature. M. Venel, a chemiſt of Montpelier, and ſcholar of Rouelle, extracted from the cheſnut, with a gentle heat, an acid ſpiritus rector: and the late M. Roux profeſſor of chemiſtry, who examined this product, found that it does not redden the blue colours of vegetables, but that it ſaturates the alkalis. The nature of the ſpiritus rector of cruciform plants is not determined. Some ſuppoſe it acid, others alkaline. It ſeems from the works of Meſſrs Deyeux and Baumé, that ſulphur is found combined with the odorous principle of the antiſcorbutic plants. As to what remains. we have already taken notice of theſe beautiful diſcoveries elſewhere.

There are ſtill two important reflections to be made upon ſpiritus rector. The firſt is what M. Macquer very properly ſuſpects, that this principle is perhaps a gas of a particular nature; its inviſibility, its volatility, the manner in which it is diffuſed in the atmoſphere, its expanſibility, and ſome experiments of Dr Ingenhouz upon the noxious gas of flowers, all go to ſupport this opinion. There only remains to make upon the ſubject experiments, which indeed require great care and accuracy, but which alſo promiſe brilliant and uſeful diſcoveries. Boyle has already ſaid much about the ſmells, their alterability, their reciprocal combination; and the ſubject has been continued with the greateſt ſucceſs by M. Lorry. This learned man has proſecuted the alterations reſulting from their mixture, from fermentation, the action of the fire, the air, and of different menſtrua. His intereſting memoir will appear in the next volnme of the Royal Society of Medicine. We cannot, without deviating from our plan, enter upon theſe details; but we ſuppoſe his firſt diviſion of the odours may be given. M. Lorry divides theſe bodies into five claſſes; the camphorated ſmells, the etherial, the poiſonous or narcotic, the acid, and alkaline. All the ſmells may be referred to theſe five primative claſſes. In

gi-

giving a reason for this division, this philosopher confines himself to the effect the smells have upon the organs of smell and upon the nerves in general, without entering upon their chemical nature. But it is more than probable, as he thinks himself, that those of every class approach to one another in their chemical nature, as we know they do in their action upon the animal œconomy.

The second reflection with which we shall terminate the chemical history of the odorous principle is, that though the plants which have been called *inodorous*, are considered destitute of this principle, it is, nevertheless, now fully demonstrated, that with the most gentle heat of a water-bath, a slightly odoriferous water may be extracted, but which plainly discovers the plant from which it was produced. I am assured, having often tried it along with M. Bucquet, that the plants deemed the most inodorous, as the mercury, plantain, &c. yield to the water-bath a water, which so sensibly emits their smell, that the one cannot be distinguished from the other. It is true, that these insipid spiritus rectors are very quickly decomposed, and part with the small odour which characterises them. They are altered, ferment, and also pass to the acid or alkaline state according to their quality.

There is an art founded upon the method of extracting the odorous part of vegetables, preserving them, and fixing them in different substances, called *the art of perfuming*. The most part of its properties is entirely chemical.

Medicine makes great use of distilled waters. They have virtues different according to their nature: it is only those distilled with a naked fire which are used, as is done in the process for obtaining the essential oils. We may observe, that this manipulation is proper in the case of the aromatic essential waters, but that it is defective in regard to those commonly called *inodorous*. It is our opinion, that distillation in a balneum mariæ is in-

indefpenfably requifite: this precaution not particularly attended to, they have a burnt or empyreumatic fmell, without being charged with that of the plant. If the virtue of thefe waters refides only in the fpiritus rector, however weak it may be, it is certain that the manner in which they are prepared may deprive them of all the properties of which they might be poffeffed.

Of Refinous Inflammable Juices.

THE name of refins has been beftowed upon dry inflammable matters, immifcible with water, foluble in oils and in fpirit of wine, running fluid from the trees which produce them; thefe matters are only oils which have become concrete by drying. The difference between balfams and refins is not determined. Some give the name of *balfams* to fluid inflammable fubftances; fome of them, however, are dry. Others alfo call the inflammable fubftances the moft odorous. M. Bucquet has thrown great light upon this fubject, by giving the name of *balfams* only to thofe of thefe combuftible fubftances which have a fweet fmell, which they can communicate to water, and which particularly contain the odorous acid falts, which may be obtained concrete by fublimation, or by decoction in water.

§ 1. *Balfams.*

THE principal kinds of balfams may be reduced to the three following:

1. Benzoin. Two kinds of this are diftinguifhed; the amygdaloide, formed of white tears fimilar to almonds, adhering by means of a brown juice. The common benzoin is brown and without tears; it emits a a very fweet fmell when it is heated or rubbed. The tree is unknown which furnifhes it. This balfam is imported to us from the kingdom of Siam and from the ifland of Sumatra. It yields but very little effential oil, on account of its folidity. It gives out to water an acid falt in needles, whofe fmell is ftrong, and which cryftallifes

lizes by cooling. It is extracted alſo by a ſublimation. It is then called *flowers of benzoin*. This operation is done into two glazed earthen veſſels, placed one above the other. The fire muſt be very gentle, otherwiſe the ſalt is brown. The paſteboard cone, which was formerly employed, allows the diſſipation of much of the flowers. The ſmell of this ſalt is ſtrong, exciting coughing: its taſte is acid: it reddens the ſyrup of violets, and effervefces with the mild alkalis. Benzoin in a retort yields a very acid phlegm, a concrete brown ſalt of the ſame nature, a brown and thick oil; the charcoal remaining contains fixed alkali.

Benzoin diſſolves in ſpirit of wine; and its tincture, precipitated by water, gives the lac virginale. The ſalt of benzoin is employed as a very good expectorant and aperient in pituitous diſeaſes of the lungs, kidneys, &c. Its oil is reſolvent; it is uſed externally for paralytic members.

2. Balſam of Tolu, Peru, Carthagena. It is brought or contained in the cocoa nut, or in yellow tears, or in a fluid ſtate: it flows from the toluifera, placed by Linnæus in the decandria monogynia. It may be extracted from the ſhells by heating them in boiling water, which renders it fluid. It comes from South America, in a country ſituated between Carthagena and the Name of God, which the inhabitants call *Tolu*, and the Spaniards *Honduras*. When analiſed, it gives the ſame products as benzoin, and particularly a concrete acid ſalt: it is employed in diſeaſes of the lungs: a ſyrup is made with it.

3. Storax is in red and ſmooth, or brown and greaſy tears. It has a very ſtrong ſmell; it flows from the oriental liquidambar, a plant little known. M. Duhamel ſaw it flow from the aliboufier, a juice of a ſimilar ſmell. Newman analyſed ſtorax; he got very little eſſential oil, a concrete acid ſalt, and a fat oil. Its uſe is the ſame with that of benzoin; it is employed chiefly as a perfume.

LECTURE XLIX.

§ 2. Resins.

RESINS differ from the balſams in their leſs agreeable odour, and particularly in giving no concrete acid. The principal kinds are the following.

1. The balſam of Mecca, Judea, Egypt, Grand Cairo. It is liquid, white, bitter, of a very ſtrong citron-ſmell. It flows from a tree called *amyris opobalſamum*, and claſſed by Linnæus in his Octandria Monogynia, and diſcovered in Arabia Felix by M. Forſkahl. This balſam yields eſſential oil to diſtillation: it is uſed as a vulnerary when incorporated with ſugar, yolks of eggs, &c.

2. The balſam of copahu, brown or yellow, which flows from a tree called *copaiba*, and called by Linnæus *copaifera*, and placed by this botaniſt in his Decandria Monogynia. The common kind, and likewiſe that of the balſam of Tolu, is a mixture of true balſam of copaiba and of turpentine, according to Cartheuſer. It is employed in ulcers of the lungs and of the bladder, like the former.

3. Turpentine of Chio flows from the turpentine tree, which produces the piſtachò nuts; it is of a white colour, or of a yellow inclining to blue. It yields a very fluid eſſential oil in a balneum mariæ; that which it gives with the naked fire is leſs fluid. The turpentine is afterwards more yellow: if it is diſtilled with water, it is white and ſilky. In this ſtate it is called *terebinthina cocta*. This is rare, and of almoſt no uſe.

4. Venice turpentine, or the reſin of the larch-tree, is what is commonly employed in medicine. It is uſed in its natural ſtate, or combined with the fixed alkali. This combination is called *Starkey's ſoap*. The Paris diſpenſatory preſcribes, for its preparation, to pour upon half a pound of nitre, fixed by tartar, and ſtill hot, four ounces of eſſential oil of turpentine; to agitate the mixture with an ivory ſpatula, and to cover the veſſel with paper. The oil is gradually added till the

the whole forms into a white mafs. As this procefs takes up feveral months, chemifts have endeavoured to find out a more expeditious procefs. M. Rouelle, by triturating the alkali drop by drop with the oil, and adding a little water near the end, prepared in three hours a very confiderable quantity of foap. M. Baumé directs bruifing upon a porphyry one part of alkali of tartar, dried with a heat nearly fit to fufe it, and gradually to add two or three times its weight of effential oil of turpentine. When the mixture has acquired a foft confiftence, it is put into a glafs cucurbit covered with paper, and fet in a moift place. In five days the deliquefcent alkali forms a layer of a liquor at the bottom of the veffel, the foap is in the middle, and a portion of oil, which has acquired a red colour, fwims a-top. M. Baumé thinks, that the alkali unites only to the portion of oil which is in the ftate of refin. M. le Gendre extends this idea, by propofing to faturate in the cold the fixed alkali diffolved with thickened oil of turpentine, or turpentine itfelf. This foap has a certain degree of folidity, which becomes gradually more confiderable: it is formed of cryftals, which have been confidered as the combination of the acid of the oil with the alkali; but which, according to the Academicians of Dijon, is only the alkali faturated with aërial acid and cryftallifed. As this foap is very difficult to make, and very alterable, M. Macquer thinks, that when we wifh to unite the properties of the effential oils with thofe of foap, it is better to incorporate with the white medicinal foap a few drops of the effential oil, whofe effects are known. The pure volatile alkali, triturated with turpentine, forms a folid foapy compound, which diffolves very well in water, and renders it milky and frothy.

5. The refin of the fir-tree is called *turpentine of Strafbourg*. It is collected, by piercing the veficles of the bark of the fir-tree, which is very copious in the mountains of Switzerland.

6. Pitch is the juice of a kind of fir called *picea*. It is got by incisions made in the bark. It melts with a gentle fire: it is squeezed through linen cloths, and received in casks. This is the Bourgogne pitch, or white pitch: mixed with the *noir de fumée*, it gives the black pitch. When it is long kept melted with vinegar, it dries, becomes brown, and forms hard resin. Its grossest parts are burnt in a furnace, whose chimney ends in a small apartment terminated by a cone of cloth. In this cone the smoke is condensed, and forms a fine soot called *noir de fumée*.

7. The resin of the pine which yields the pine-apple. This tree is cut towards the bottom; the resin flows out into troughs. These incisions are continued upwards when the last yields no more resin. When it runs fluid, it is called *galipot*; that which dries upon the tree in yellow masses is called *barras*. These juices are liquefied in pots, and then they are thickened by the heat: they are filtered through mats of straw, they are run into holes made in sand, they are formed into loaves called *black pitch*. If some water is interposed, the matter becomes white, and forms resin or pitch-resin. The people of Provence distil in great the galipot; from it they extract an oil, which they call *oil of pitch*. It is from the roots and trunks of the pine that the tar is prepared; which is only the empyreumatic oil of this substance. The wood of the tree is put in a heap and covered with turf, and the fire applied. The oil which the heat disengages not being able to pass through the turf, is conveyed into a bucket by means of a gutter, and is collected and sent to market by the name of *tar*.

8. Tacamahaca, resina elemi, resina animata, are little in use: the tree which yields the first is not known. Elemi comes from a kind of amyris: the oriental resina animata or copal, whose origin is not known; the west animata or courbaril, which flows from the hy-

menæa, a tree of South America, are employed in making of varnishes.

9. Mastich is in white tears like corn, of a weak smell: it flows from the turpentine and mastich trees. It is used as an astringent and aromatic. It enters into the composition of dry varnishes.

10. Sandarac is in white tears, more transparent than those of mastich. It is extracted from the juniper-tree between the wood and the bark: it is also called *varnish*, because it is greatly used in these preparations. It is used for putting upon scraped paper, to prevent its sucking in the ink.

11. The resin of guaiac, which is greenish, is employed in the gout. It flows from the guaiac-tree by incisions.

12. The labdanum, or resin from Candy, is blackish. The peasants collect it with a rake, to which are fixed several skins of animals, which they rub upon the trees: they form it into kinds of cylinders, called *labdanum in tortis*. It is impure from a great deal of blackish sand. It is used as an astringent.

13. Sanguis draconis, which is extracted from the dracæna draco, and from several other similar trees. It is in flat or round leaves, or in small spheres inclosed in red leaves, and knit together like a chapelet. It is used in medicine as an astringent.

§ 3. *Gum-resins.*

THE gum-resins are juices mixed with resin and extractive matter, which has been taken for a gummy substance. They flow by incision, and never naturally, from trees or plants, in the form of fluids like an emulsion, white, yellow, or red, which are dried more or less easily. Water, spirit of wine, wine, vinegar, all dissolve only a part of the gum-resins; they differ in the proportion of resin and extract, and their analysis yields very various results. The most important kinds to be known are the following.

1. Oli-

1. Olibanum is in yellow tranſparent tears, of a very diſagreeable ſmell. What tree produces it is not known. Diſtillation furniſhes a little eſſential oil and an acid ſpirit, and leaves a very conſiderable remainder, owing to the extractive part which it contains. It is uſed as a reſolvent fumigation.

2. Galbanum is a thick juice, of a brown yellow and of a nauſeous ſmell. In Syria, Arabia, at the Cape of Good Hope, it flows from inciſions made in a plant of the ferulaceous kind, called *bubon galbanum*. Diſtilled in an open fire, it yields a blue eſſential oil, which after that becomes red, an acid ſpirit, a weighty empyreumatic oil. It is a good deobſtruent and antiſpaſmodic.

3. Scammony is of a blackiſh grey, of a ſtrong and nauſeous ſmell, of a bitter and very acrid taſte. We diſtinguiſh that of Aleppo, which is the pureſt; and that of Smyrna is weighty, black, and mixed with foreign ſubſtances. It is extracted from the convolvulus ſcammonia of Linnæus. The root of this plant, cut and expreſſed, furniſhes a white juice, which is dried and becomes black. Scammony contains a various quantity of extract and of reſin, according to the different ſpecimens; the cauſe of very different effects on different conſtitutions. It is employed as a purgative in the doſe of four grains to twelve: mixed with a ſoft extract, as that of the liquorice, it forms the ordinary diagrede. The juice of quinces is alſo uſed for this purpoſe. It is generally adminiſtered triturated with ſugar and ſweet almonds.

4. Gammboge is yellow, reddiſh, without ſmell, of a very acrid and corroſive taſte. It comes from Siam, China, Iſland of Ceylon: it is extracted from a large tree little known, called in the country *coddam pulli*. It contains a great deal of reſin, which renders it ſtrongly purgative in the doſe of four or ſix grains. It ought to be uſed internally with the greateſt attention.

5. Euphorbium is in yellow tears, rotten or carious,

without ſmell. It flows from inciſion of the euphorbium, which grows in Ethiopia, Libya, and Mauritania; it contains a very acrid reſin, and is ſtrongly purgative. It is little employed, unleſs externally in carious bones.

6. Aſafœtida is ſometimes in yellowiſh tears, and moſt commonly in loaves, formed of different agglutinated pieces. It is remarkable for its very fetid ſmell of garlic, and its bitter and nauſeous taſte. It is extracted from the root of a kind of ferula, which grows in Perſia, in the province of Chorafin, and which Linnæus has ſurnamed *aſafœtida*. The root of this plant is fleſhy and ſucculent; it furniſhes by expreſſion a white juice of a hideous ſmell, which the Indians eat as a ſeaſoner, and which they call *meat of the gods*. It is uſed internally as a powerful antiſpaſmodic, and externally as a diſcutient.

7. Aloes is a juice of a deep red and even brown colour, and conſiderably bitter.

Three kinds of it are diſtinguiſhed; ſuccotorine aloes, hepatic aloes, and caballine aloes; they differ only in purity. The firſt kind is the pureſt. M. de Juſſieu has ſeen the different aloes prepared at Morviedro in Spain with the leaves of the common aloes. Inciſions are made, the juice is allowed to flow, decanted from above the feculæ, and inſpiſſated in the ſun's heat. It is exported in leather ſacks, under the name of *ſuccotrine aloes*. The leaves are expreſſed, the juice purified by repoſe and dried: and this is the hepatic. Laſtly, the ſame leaves are more ſtrongly expreſſed, and the juice mixed with the leys of the two preceding; and this is called *caballine aloes*. The firſt aloes contains much leſs reſin than the two laſt, which are much more purgative. The firſt is uſed in medicine as a draſtic purge, and recommended as a provoker of the menſtrual and hemorrhoidal fluxes. It is recommended likewiſe as a very excellent hydragogue.

8. Myrrh is in reddiſh brilliant tears, of a ſtrong very

very agreeable ſmell, a bitter taſte, preſenting in their fracture white lines of the form of claws. Some of theſe tears are entirely gummy and inſipid. Myrrh comes from Egypt, and particularly from Arabia, from the country of Troglodytes. The plant is not known from which it is obtained: it contains much more extract than reſin. It is uſed as a very good ſtomachic, antiſpaſmodic, and cordial. M. Cartheuſer recommends it to the learned, who have a delicate ſtomach, to chew it, and ſwallow it diluted in the ſaliva. It is uſed in ſurgery as a detergent, and to ſtop the progreſs of gangrene.

9. Gum ammoniac is ſometimes in white tears internally, and yellow tears externally, and frequently in maſſes reſembling thoſe of benzoin. Their white colour and fetid odour eaſily diſtinguiſh them. It is ſuſpected, that this gum-reſin, which is imported to us from Africa, is extracted from an umbelliferous plant, on account of the ſeeds that are in it. The phenomena of the ſolution of this ſubſtance by water and ſpirit of wine, and particularly its inflammability, aſſimilate it to the reſino-extractives of M. Rouelle.

Gum ammoniac is uſed in medicine as a very good deobſtruent in obſtinate obſtructions. It is exhibited in the doſe of a few grains in pills or emulſions: it enters alſo into the compoſition of ſeveral diſcutient and reſolvent plaſters.

Of the elaſtic Gum or Caoutc-houc.

Elastic gum is one of thoſe ſubſtances whoſe nature it is difficult to determine: Although its combuſtible property, part of which is extracted in America to give light, ſeems to aſſimilate it to the reſins, its elaſticity, ſoftneſs, and inſolubility in the menſtrua which generally diſſolve theſe ſubſtances, are ſo many characters which diſtinguiſh it from them.

The tree which yields it grows in ſeveral places of America. Long inciſions are made in its bark, ſo as to penetrate the wood: the white, more or less fluid, juice, is received into a vessel, and then different utensils are formed: it is applied in different layers upon moulds; it is dried in the sun or at a fire. Different forms are marked upon it with a steel-rod: they are exposed to smoke; and when they are very dry, the moulds are broken. Such is the formation of the elastic gum bottles brought into Europe.

The vessels which are made with this substance are capable of containing water and other fluids without its being affected. When it is cut asunder, and the sides of the pieces applied together, they join and unite very firmly.

We are unacquainted with the action of the fire upon elastic gum; we know only that it softens and kindles it.

It is insoluble in water: we do not know what action the saline substances have upon it. M. Macquer, who tried to dissolve it in different menstrua, found that spirit of wine had no action upon it, as Messrs Condamine and Fresneau had already announced, Academy 1751; but that the oils dissolved it by means of heat. However, as his intention was to convert it into a liquid state, so as to be capable of being made use of, but to restore its properties by the evaporation of the solvent, M. Macquer was obliged to have recourse to another menstruum than the oils, because these matters, however volatile, always altered it, and remained fixed, so as to deprive it of its elasticity and force. Very highly-rectified ether, in which he easily effected its solution, entirely removed these objections by its evaporability, Acad. 1768; and though this liquor was very expensive, he thought it right to show this method of forming very useful utensils; such as flexible catheters, by applying successive layers of this solution upon a wax mould, till they are of the thickness desired. When the catheter is dry, it is plunged into boiling water, which liquefies the wax; and in this manner the wax is sepa-

ſeparated fror. the mould. The ſoftneſs and elaſticity of this inſtrument render it very uſeful to ſuch perſons as are forced to carry about it continually.

Such was the knowledge acquired of elaſtic gum, when, in the month of April 1781, M. Berniard, who is celebrated for his accurate performances, inſerted in the Phyſical Journal a very good paper upon this ſingular ſubſtance. From his reſearches, this chemiſt concludes, that elaſtic gum is a kind of particular fat oil, coloured by a matter ſoluble in ſpirit of wine, and defiled by the ſoot of the ſmoke to which each layer is expoſed in the drying. Water does not at all alter it. Spirit of wine takes its colour from it by means of ebullition. The cauſtic fixed alkali has no action upon it. The oil of vitriol reduces it to a coaly ſtate, and is itſelf blackened, aſſuming the odour and volatility of ſulphureous acid. The nitrous acid acts upon this gum as upon leather, and turns it yellow. Spirit of nitre very readily deſtroys it. Marine acid alters it in no ſhape. Rectified vitriolic ether did not diſſolve it. This fact would appear ſingular, as the author ſays, to all thoſe who are acquainted with the accuracy and veracity of M. Macquer. Nitrous ether diſſolves it. This ſolution is yellow, and gives, by evaporation, a tranſparent friable ſubſtance, ſoluble in ſpirit of wine: in a word, a true reſin, formed, according to the author, by the action of the nitrous acid upon the elaſtic gum. Eſſential oil of lavender, and that of ſpikenard and turpentine, diſſolved it by means of a gentle heat: but they formed gluey fluids, which ſtuck to the hands more or leſs; and therefore could be of no uſe. A ſolution of elaſtic gum in oil of ſpikenard, mixed with ſpirit of wine, depoſited ſome white flocci, which were inſoluble in hot water, ſwam upon its ſurface, and by cooling became white and ſolid like wax; in a word, a true fat concrete oil.

Oil of camphor diſſolves elaſtic gum by mere maceration. In evaporating the ſolution, the camphor is volatilised, and there remained an amber-like matter, of a solid consistence, and almost not gluey, which easily dissolved in spirit of wine. The fat oils boiled upon elastic gum dissolve it; wax likewise dissolves it. This substance does not fuse with the heat of boiling water; but when exposed to the fire in a silver spoon, it is reduced into a black fat oil: it emits white vapours; it then remains fat and gluey although exposed to the air for several months, and does not resume the dryness and elasticity, which are so useful for the purposes to which it is destined. Lastly, M. Berniard finished his experiments upon elastic gum with exposing it to a naked fire. From an ounce of this matter he obtained very little phlegm; an oil, at first clear and light, then thick and coloured; and some volatile alkali; but does not describe the quantity. A coal of twelve grains weight remained, similar to that of the resins. This chemist attributes the presence of volatile alkali to the soot which colours the elastic gum.

We shall observe, upon this analysis, that it does not very exactly demonstrate the nature of elastic gum, as the acids in their action upon it do not show the same upon the fat oils, which is much more rapid; as the caustic alkalis do not convert it into soap; as it fuses only with a heat, much stronger than that which is necessary to melt the most solid fat oils; as no fat oil becomes elastic, and never dries like it.

Besides, the author advances, in the fifth experiment, that this gum is composed of two distinct substances, which he has not demonstrated; and he concludes with considering it as a product of human industry. From all these reflections, and many others which it might be possible to make upon this subject, otherwise very excellently handled by M. Berniard, it is our opinion much still remains to be known, as he has said himself, before its properties are fully understood, and before its nature is fully determined.

LECTURE L.

OF THE FECULÆ AND FARINÆ.

THE juices of vegetables, elaborated in their veſſels, thicken, and are gradually depoſited upon the ſurface of their fibres, for their nutrition and growth; or they accumulate in a more or leſs ſolid form in the different organs which compoſe them. After having ſpoken of the fluid parts of theſe organic ſubſtances, it is neceſſary to examine the ſubſtance which forms the texture of their different ſolids. It is alſo of great importance to know the nature of all the ſolid matters which compoſe the texture of vegetables: however, the knowledge acquired upon this ſubject ſeems to announce, that theſe organs, treated by the proceſſes which we are going to deſcribe, are reduced into a dry, powdery, inſipid, white or grey ſubſtance, or of different colours, inſoluble in cold water, and as if earthy, called *fecula*.

This ſubſtance is obtained, by reducing in a mortar a root, ſtem, leaf, or any ſeed, into a pulp. When theſe parts are ſucculent, they may be treated in this way without addition of water; but, in general, this fluid is uſed to facilitate the ſeparation of the fibres, and to carry off the divided powdery portion of their texture. Theſe parts thus reduced into a pulp are expreſſed: the juice or the water, which the force of the preſs ſqueezes out, is turbid, white, or coloured, and it gradually depoſites by repoſe a flocculent matter, in part fibrous, ſometimes powdery; which is the true fecula of the vegetable. Some parts of vegetables appear entirely formed of this matter; ſuch are the ſeeds of the graminous and leguminous plants, the tuberous roots, &c. Theſe parts furniſh in general the fineſt and moſt copious fecula. As to the tender ſtems and leaves, their more fibrous texture, when treated by the above-mentioned proceſs, never yields any but a

coarſe

coarse deposition, which is coloured, filamentous, and is called *coarse fecula.* If, after having dried them well, they are powdered, and this powder is washed, the water carries off a much more fine fecula, perfectly resembling that of the tuberous roots and graminous seeds. To the eyes of a chemist, then, these two genera of feculæ differ only in this, that the former comes from a less fibrous, less organised part, as if formed of cells, in which nature has deposited the dry or farinaceous juice; whilst the second, of a fibrous texture, needs to be disorganised and attenuated by art.

All the solids of vegetables may with rigour furnish a kind of fecula; but as some of it is prepared in the arts, pharmacy, and for our food, it is that chiefly which we shall consider. The fecula of briony, pomme de terre, cassave, sagou, salep, and starch, are chiefly made use of.

1. In order to prepare the fecula of briony, the roots of this plant are taken fresh, the bark is taken off, they are filed down, and submitted to the press. The juice which flows is white, and deposites a very fine fecula. The juice is decanted off in twenty-four hours; the fecula is dried, as it contains a certain quantity of extract which the juice has left in it: it is very acrid, and purges violently: if it is washed before drying, it becomes finer and whiter, but it loses at the same time its purgative virtue. This manner of preparing the fecula of briony furnishes only a very small quantity of it; but more of it may be got by applying more water to what remains in the press, by passing this water thro' a hair-sieve, to separate the coarse fibrous parts, and allowing the fluid to settle. When this second fecula is deposited, the water is decanted, and the deposition dried. It must be observed, that this fecula, obtained by washing the residuum of the press, is not purgative like the first; because the water has carried away the extractive part possessing this virtue. M. Baumé has observed, that the fecula of briony, well washed, is ab-

abſolutely ſimilar to ſtarch, and that a powder for powdering the hair might be made from it, which would make great ſaving of wheat. In the ſame way, for the uſe of medicine, is prepared the fecula of the roots of colt's-foot and ſword-graſs.

2. The pommes de terre are one of the moſt uſeful alimentary ſubſtances, on account of their abundance and fertility. A very great quantity of very fine white fecula is eaſily procured from them, which furniſhes a light food by baking and boiling in water. This fecula is obtained by filing ſome of them above a ſieve, and pouring upon them a great quantity of water. This fluid carries along the moſt fine and divided portion of the fecula, and is allowed to depoſite by repoſe: the water is decanted, the fecula dried with a gentle heat; then it is in an extremely fine very white and light powder.

3. The Americans extract from the root of a very acrid plant, called *maneſe*, a very ſweet nouriſhing fecula, which they call *caſſave*. They take off the ſkin from the root; they file it, and put it into a conical ruſh baſket, of a very looſe texture, which they ſuſpend by means of a ſtick placed upon two wooden reſts. To the extremity of this baſket they fix a very weighty veſſel; which by its weight expreſſes the root, and receives the juice which flows out. This is a very acrid and dangerous poiſon. When the fecula is well preſſed, and deprived of all the juice which it contained, it is expoſed to ſmoke to dry it, and paſſed through a ſieve: it then forms the caſſave. This farina is ſpread upon a hot iron-plate in order to be baked; and it is turned, in order to give its two ſurfaces the reddiſh yellow colour which announces the baking: in this ſtate it is called loaves of *caſſave*. By heating it in a baſon, and agitating it from time to time, it aſſumes, upon drying, the form of grains, which are called *couac*. From the expreſſed juice a very fine ſweet fecula is preci-

cipitated, called *mouſſache*, which is employed in paſtry-work.

4. The ſagou is a dry fecula, reduced into grains by the action of the fire, which is imported from the Molucca Iſlands, Java, and the Philippines. It is extracted from a kind of palm, called *ladan*, in the Moluccas. The trunk of this tree contains a ſweet marrow, which the inhabitants extract after the trunk is felled. They bruiſe this marrow, and put it into a kind of cone or funnel, made of the bark of a tree, and laid upon a hair-ſieve; they waſh it with plenty of water, which carries through the holes of the ſieve the fineſt and whiteſt portion of the marrow; the fibrous portion remains upon the ſieve. The water which is charged with the moſt attenuated part is received into pots, and gradually depoſites the fecula which diſturbed its tranſparency. The clear water is decanted off, and the depoſition paſſed through perforated copper plates, which give it the form of ſmall grains called *ſagou;* the red colour at their ſurface is owing to the action of the fire, which ſerved to dry them. This kind of fecula diſſolves in hot water, and forms with milk or broth a ſort of light agreeable jelly, which is much recommended in phthiſical complaints.

5. Salep, ſalop, ſalab, &c. is the root of a kind of orchis, prepared in the Eaſt Indies. They pick out the fineſt bulbs of this plant, take off their bark, macerate them in cold water, and boil them; then they are tied together when they are well drained, and dried in the air. M. Jean Moult gave another proceſs for the preparation of ſalep, which may be made with all kinds of orchis. The roots are rubbed to dryneſs, or in water with a bruſh, to carry off the external pellicle, and then ſet to dry in an oven: they become very hard and very tranſparent. However, they may very eaſily be reduced into powder; and this powder, waſhed in hot water, forms a nouriſhing jelly, of whoſe virtue Geoffroy greatly boaſted in all the diſeaſes depending upon an acrimony

acrimony of the lymph, and particularly in the phthisis and bilious dysentery.

Starch, properly so called, is a fecula absolutely like the preceding; but as the farina of wheat, of which it makes one of the constituent parts, is one of the most important substances, and worthy of chemical notice, we shall dwell a little longer upon this fecula than upon the rest.

What is called *farina* is in general a dry, friable, insipid substance; susceptible of taste and solubility by the action of fire, and formed of matters easily separated from each other. This substance resides in the graminous seeds; and particularly in wheat, rye, oats, rice, Indian corn, &c. The leguminous seeds seem also to contain a compound analogous to the farina: however, it is only the farina of wheat which enjoys the truly desirable properties of this substance, because it alone contains a just proportion of the different substances whose mixture gives rise to these properties. Although the œconomical use of the farina of wheat be established as the best nourishment from time immemorial, it is but a short time since it was attempted to be chemically analised. M. Beccari, an Italian physician, and M. Keyssell Meyer in Germany, are the first who endeavoured to separate its different component parts. Messrs Rouelle, Spielman, Malouin, Parmentier, Poulletier de la Salle, and Macquer, have resumed the task, and pushed the matter much farther than they had done. From their different researches, and from our own experiments, we shall deliver what is to be said upon the subject.

Water is the most proper agent, which is the least capable of altering the different matters which it carries along with it, or which it separates according to the laws of their solubility. This fluid is also the most successfully employed in obtaining the different matters of which the farina of wheat is composed. For this purpose we form a paste with some farina and water. We squeeze this paste above a vessel, and placed under another vessel,

vessel, which lets fall upon it a continued stream of water: this fluid, in its descent, carries along with it a very fine white powder, which renders it milky: the process is thus continued, till the water which falls upon the paste goes clear into the vessel. Then the farina is found separated into three substances; a greyish elastic matter, which remains in the hand, called the *glutinous part*, or *vegeto-animal part*, on account of its properties; a white powder, deposited by the water, that is the fecula or starch; and a fine matter in solution in the water, which appears to be a kind of mucous extract. Let us proceed to the properties of each of these three substances.

§ 1. *Of the glutinous part of the Wheat.*

The glutinous part is a tenacious, ductile, and elastic matter, of a whitish grey. When it is drawn out, it extends about twenty times the length it was before, and seems to be composed of fibres or filaments, adhering sidewise, according to the direction in which it has been drawn. Upon letting it loose, when extended, it resumes, by its elasticity, its former volume. By increasing its dimensions, it may be made to imitate sufficiently well, by means of its polished surface, the texture of animal-membranes. In this state it firmly adheres to dry bodies, and forms a very tenacious glue, employed by some in the junction of broken porcelain, a long time before the chemical means of obtaining it was known.

Its smell is sweet, as if mucous; its taste is flat: exposed to a fire which is capable of drying it quickly, it bubbles up prodigiously. It dries very well with a dry air or with a gentle heat; then it becomes semitransparent, hard, and brittle, like common glue; it breaks like it, clean and with a crack.

If in this state it is put upon a burning coal, or above the flame of a candle, it exhibits all the marks of an animal-substance; it leaps about, bubbles up, liquefies,

is

is agitated, and burns like a feather or horn, emitting a ſtrong fetid ſmell. Diſtilled in a retort, it yields, as all animal-ſubſtances do, ſome alkaline ſpirit, concrete volatile alkali, and an empyreumatic oil: its coal is very difficult of incineration, and gives no fixed alkali.

Freſh gluten, expoſed to a hot moiſt air, changes and putrefies in the ſame way as the animal-parts: When it ſtill retains a little ſtarch, this laſt paſſes to the acid fermentation, retards and modifies the putrid fermentation, and converts it into a ſtate very like cheeſe. In like manner, M. Rouelle prepared with gluten a cheeſe, remarkably ſimilar in ſmell and taſte to thoſe of Gruyre and Holland, of a moderate fineneſs.

Water has no action upon the glutinous part: when it is boiled in this fluid, it becomes ſolid; it loſes its elaſticity and gluey quality; but it acquires neither taſte nor ſolubility in the ſaliva. We obſerve, however, that it is to the water which forms the paſte that the gluten owes its ſolidity and elaſticity; and in the farina this vegeto-animal portion, ſuſceptible of aſſuming a ſolid elaſtic form, was powdery, and not coheſive: but as ſome water was poured upon the farina, and it is mixed, theſe molecules, which ſhould poſſeſs the glutinous portion, abſorb this fluid, turn gluey by this means, and form at laſt the kind of elaſtic ſolid which is called *gluten*. Water, then, contributes greatly to the formation of this ſubſtance; and this ſubſtance ſhould be perhaps conſidered as a particular compound, which is ſaturated with water, and which can abſorb no more. It is true, that by depriving it of water by exſiccation, it totally loſes its elaſtic and gluey property.

The moſt part of the ſaline ſubſtances have a more or leſs remarkable action upon gluten. The fixed alkali by means of heat diſſolves it. This ſolution is troubled, and depoſits ſome gluten, not elaſtic, by the addition of acids.

The mineral acids diſſolve the gluten. The nitrous acid diſſolves it with great activity, emitting a great quantity of nitrous gas. This ſolution is yellow; that

by

by the marine and vitriolic acids is of a violet brown. A kind of oily matter ſeparates from theſe ſolutions: the gluten is in a true ſtate of decompoſition.

M. Poulletier, to whom we owe all theſe nice experiments, diſcovered that ammoniacal ſalts might be extracted from theſe combinations diſſolved in water, ſpirit of wine, digeſted and evaporated in the open air. This diſcovery ſeems to prove, that the volatile alkali is ready formed in the gluten.

From what has been ſaid it appears, that it is entirely different from any ſubſtance we as yet know in vegetables; and it approaches in many characters to the fibrous part of the blood. It is this gluten which gives the farina of wheat the property of forming a very pliant paſte with water, and the facility with which it riſes. It ſeems not to exiſt, or at leaſt to exiſt in very ſmall quantity, in other farinas; ſuch as thoſe of rye, oats, French corn, rice, &c. all of which form ſolid paſtes, brittle, and not ductile, and which riſe only very little, or not at all, when expoſed only to the temperature which makes the paſte of the farina of corn to riſe. In this laſt alone, then, are to be really found all the neceſſary qualities for making good bread.

M. Rouelle junior ſays, that he found a glutinous ſubſtance, analagous to that of the farina of wheat, in the green feculæ of plants, which, in their analyſis, yield volatile alkali, and empyreumatic oil, as the glutinous part, of which we have juſt now ſpoke.

§ 2. *Of the Starch of Wheat.*

Starch, or the amylaceous matter, is the moſt copious part of the farina; it is it which precipitates from the water which carries it along with it, when we waſh the paſte with a view of obtaining the gluten pure. This ſubſtance is very fine, ſoft to the touch: it has no ſenſible taſte. It is of a grey white colour, and dirty when extracted in the manner deſcribed; but the ſtarch-makers render it extremely white, by leaving it among

which they name *aqua certa*. From the
I. Poulletier it appears, that the fermen-
excited in this fluid whitens and purifies
attenuating, and even deſtroying, the
tive part which precipitates with it in
. Starch, chemically conſidered, is a
particular nature. This mucilage, which
lered falſely as an earth by ſome che-
lifiers from the glutinous part. It burns
g an empyreumatic odour like this laſt.
open fire, it yields an acid phlegm of a
and a very thick empyreumatic oil to-
f the proceſs. Its coal is very eaſily in-
which ſtate it is found to contain ſome

ſoluble in cold water; but when it is
water, it forms a glue, or rather what is
This laſt, expoſed to the moiſt air, gra-
conſiſtence, ferments, paſſes to the acid
nd is covered with mouldineſs.
f the ſaline ſubſtances upon ſtarch has
amined; but the account now given of
ay ſuffice to diſtinguiſh it from the glu-
If the white feculæ above deſcribed are
ure with the amylaceous matter.
ns the greateſt part of the farina, there
t but that it is the principal alimentary
ad. M. Beccari found, that the quan-
nous part was from a fifth to a third,
and that this quantity varies according
the farina, which depend upon the kind
iſhed it, and according to the qualities
eceives from the earth and from the at-

mucous extractive part of the Farina.

ating the clear water which ſerved to
and depoſite the ſtarch, M. Poulletier

obtained a matter of a brown yellow, viscous, gluey, pitchy, whose taste was very weakly saccharine. This substance, which this learned man called *mucoso-saccharine*, exhibited in its combustion and distillation all the phenomena of sugar. It is it which excites the acid fermentation in the water which swims above the starch; since, as M. Macquer has very well observed, this last is by no means soluble in cold water. The mucoso-saccharine matter is only in very small quantity in the farina of wheat: perhaps there are other farinæ in which it is more copious. It cannot be doubted, however small the quantity of this substance be in the farina of wheat, but that it has a principal share in the particular fermentation which takes place in the paste, and which makes it rise. This motion, necessary for the formation of good bread, is still little known in its nature. It seems that this can be only a beginning of the putrid fermentation in the gluten, the acid in the starch, and perhaps the spirituous in the mucous saccharine matter; perhaps from these three commencing fermentations, which mutually oppose each other, the mixture little known arises, which is much lighter than the paste, and which by baking forms bread. This is so far certain, that in bread, the three substances just now examined are found combined together, and so altered that they cannot be separated. The action of heat is even sufficient without the motion of fermentation to combine and alter these substances, in such a way that the baked bread, without being raised, furnishes no more glutinous part, according to M. Malouin.

From these details, we see how far the other farinæ, except those of wheat, and still more, how far the fruits or the leguminous and farinous seeds, such as the beans, pease, chesnuts, &c. are from possessing all the qualities necessary for the formation of good bread.

Of

Of the Colouring Parts of Vegetables.

Vegetables contain colouring parts in all their organs. Theſe parts greatly differ from one another; frequently a vegetable matter, which has no apparent colour, aſſumes a very remarkable one by certain menſtrua. The art of dyeing, all whoſe proceſſes are entirely chemical, is founded upon the ſolubility of the colouring parts in the different menſtrua, upon the manner of applying them to the ſubſtances to be dyed, and of rendering them fixed and tenacious in them. In examining the properties of each colouring matter, we ſhall have occaſion to ſpeak of the principles of this important art, of which Meſſrs Hellot, Macquer, le Pileur d'Apligny, Hequet d'Orval, Quatremere Dijonval, and l'Abbé Mazeas, have given ſo excellent deſcriptions.

It appears, that the colouring-matter of vegetables, properly ſo called, is not yet known. M. Rouelle ſuppoſed that the green part, ſo copious in the vegetable kingdom, was analogous to the gluten of the farina: but it is certain, that this matter preſents different chemical characters, according to the baſe to which it is united. It is then this baſe, rather than the colouring part itſelf, which we would wiſh to talk of, in ſaying, that ſuch or ſuch a colour is extractive, ſuch another reſinous. The real ſubſtance which colours each of the vegetable parts employed in the arts, is, no queſtion, a very ſubtile body, and perhaps as divided as the principle of ſmell. One might be led even to believe it is only particularly modified in the ſolid and liquid parts of vegetables.

It is requiſite to relate here, that the colour of vegetables depends entirely upon the contact of light. Might we ſuppoſe then, that it is a portion of this fluid which combines with the humours of vegetables? Phyſics and chemiſtry are ſtill very inadequate for the reſolution of this problem. However it be, as it is impoſſible to ſeparate entirely the colouring part of the vegetable baſe, to which it adheres, it is agreed to take both these substances together for the colouring part.

M. Macquer has, more distinctly than all the chemists, described the colouring part of vegetables, considered relatively to dyeing; and his theory upon the application and fixation of the colours in the substances to be dyed, is without contradiction the most satisfactory. As our intention is to unite this theory with the history of the colouring parts of vegetables, we shall consider them relatively to these last properties.

1. A great number of vegetable parts, which are extractive or soapy, dissolve very easily in water. Madder, dyer's-weed, logwood, Indian-wood, Brasil-wood, furnish yellow or red colours of this kind. It is known that stuffs dyed with these colours should lose their dye in water: for that reason, to render these colours durable, a matter is used capable of fixing them by decomposing them; as an acid salt, such as red tartar, alum, and several others. These salts are said to to be eating. An uncombined acid would have the same effect, but it would alter the colouring part. The superabundant portion of acid in the alum unites with the alkali of the colouring soapy extract, and precipitates, upon the body which is dyeing, the resinous part which is insoluble in water. However, this colouring portion, rendered insoluble by alum, is of two kinds; the first is very solid, and resists in the free air the soaps and all the proofs used in dyeing.

This first colour gets the name of *good* or *great dye*. The other is altered by the air, and particularly by the action of the proofs: it is called *false* or *small dye*. It must be observed that it is wool which assumes the best colour; and then silk, cotton, hemp, and flax, are the substances which are dyed with more and more difficulty, and which retain less strongly the colouring substances.

The authors who have treated of dyeing, have entertained different opinions upon the manner in which the colour-

colouring parts are applied to the ſubſtances expoſed to their contact. Several have imagined, that this took place only on account of the pores of the ſubſtances to be dyed, which are more or leſs great, and more or leſs numerous; and that wool aſſumed more colour than ſilk, only becauſe its pores were more open and more numerous. But M. Macquer thinks, that this application which is more or leſs eaſy, depends on the nature of the colouring part relative to the matter to be dyed; and that the colouring is a true painting, whoſe ſucceſs and adherence is owing to an affinity, and to an intimate union between the colour and the dyed ſubſtance. This celebrated chemiſt adopted this opinion from the great number of experiments which he made upon this art, which is much indebted to his diſcoveries.

2. There is another claſs of colouring ſubſtances, which ſeem to be compounds of ſoapy extracts and reſins. M. Macquer calls them *reſino-earthy*. When theſe ſubſtances are boiled in water, the reſinous matter which they contain diffuſes through the fluid by means of the heat and diſſolved ſoapy portion; but it is precipitated in proportion as the decoction or bath cools. When, then, any wool, or any other matter, is plunged into the decoction of a colouring mixture of this kind, the reſin precipitates by cooling, and is applied without any other preparation to theſe ſubſtances. As it is not ſoluble in water, it gives a colour of a good dye. The colouring parts of this nature are extracted from almoſt all aſtringent vegetables; ſuch as the ſhell of walnuts, the root of the walnut-tree, ſumac, the bark of the alder-tree, ſaunders, &c. Theſe colours are all yellow: the dyers call them *colours from roots*. They ſerve for the moſt part to form a very good coat upon which the other more brilliant colours are applied. It muſt alſo be remarked, that the colouring ingredients, which require no preparation neither for themſelves nor for the matters

ters to be dyed, furnish the kind of dyeing which is the most simple and most easily practised.

3. The colouring principle of several other substances resides in a purely resinous matter, insoluble in water. Some of these substances are not even soluble in spirit of wine, but all are in the alkalis, which convert them into a kind of soapy state, and render them soluble in water.

The principal colours of this nature which are employed in dyeing are the following:

a. Annotto, a kind of fecula, which is extracted by maceration from the seeds of the urucu purified in water. This fecula is deposited during putrefaction: it is first red, but in time becomes of a brick colour. This paste is washed in water with the alkali of ashes of tartar, which we shall know very soon, and the substances to be dyed are plunged into this bath. A fine gold yellow or an orange colour is produced.

b. The flower of safflower or bastard saffron, gives a very fine red colour by the same process. This flower contains two distinct colouring parts; the one purely extractive and soluble in water; the other resinous. To obtain this last, first the soluble part must be extracted by accurate washings in water; then it is mixed with tartar-ashes or soda; the mixture is washed, and thus used for the tincture. But as the alkali alters and tarnishes its colour, the dyed matter is steeped in water rendered acid by the juice of citrons: this acid seizes upon the alkali, and leaves the colouring matter, which it enlivens and converts to a red.

c, Archil is a paste which is prepared from mosses, which are macerated in urine and lime. This last disengages the volatile alkali, which evolves the red colour. The archil diluted in water gives a tincture without any other preparation: the alkalis extract a violet colour; but it is a false teint: it is altered by the air, and the acids turn it yellow.

d, Indigo, whose blue is deep, violet, and as if coppery,

pery, is a fecula, which is prepared at St Domingo and the Antilles, by macerating in ſtone-troughs filled with water the ſtems of the indigo-tree oranillo. The water becomes blue; the ſtems are ſtruck ſtrongly, and the fecula precipitates. The indigo, precipitated from the water, is put into ſtraining-bags to drop through; it is then dried in ſmall wooden boxes, and broken into pieces when dry. It is conſidered as good when it ſwims upon water; and it burns entirely away upon a red iron. The colouring part is extracted by the alkalis, and it is applied to the ſtuffs, requiring no kind of preparation: the colour cannot be enlivened by the acids, which would alter the colour of the ſtuffs.

4. There are ſeveral colouring parts that are ſoluble in oils. Alkanet, or the red root of a kind of bugloſs, communicates its colour to oil. Spirit of wine alſo diſſolves ſome: the green feculæ are diſſolved in it as well as in oil.

It is eaſy to conceive, that no uſe of theſe colours can be made in dyeing, becauſe it is impoſſible to employ the ſubſtances which are neceſſary for extracting them.

Such is our principal knowledge of vegetable colours. Hence it reſults, that all the immediate principles of vegetables may be the baſe of theſe colouring parts; ſince ſome are found ſoapy, reſinous, and extractive. Some ſeem to be even of the nature of fat oils, ſince they are inſoluble both in water or in ſpirit of wine, whilſt they diſſolve very well in the alkalis. Laſtly, there are ſome analogous to the glutinous part, according to M. Rouelle. There is great room to believe, that experiments made upon this ſubject will bring to light ſeveral other properties in theſe matters, which are very abundant in vegetables; and they will contribute to the progreſs of dyeing, one of the arts to which chemiſtry may be of the greateſt ſervice.

LECTURE LI.

THE ANALYSIS OF PLANTS WITH THE NAKED FIRE.

AFTER having examined all the matters which may be extracted from vegetables by simple means, incapable of producing in them any change, and after having considered these matters as their immediate principle, it is necessary to take a view of the alterations which they may undergo from the fire.

Ancient chemists knew little more than this sort of vegetable analysis; and all their investigations of the nature of these substances consisted in determining how much spirit, oil, and volatile salt, they gave to distillation. At present we have lost all confidence in this means; we know only that almost all plants yield nearly the same products: and the distillation of a great number of these bodies performed by chemists, in other respects very valuable and interesting, has served only to deceive us in this analysis. For how is it to be conceived, that the action of the fire, which is exerted upon alll the different principles contained in vegetables, such as the extract, mucilage, oil, resin, saline matter, and which decomposes each of these principles in a particular manner, is able to throw any light upon the nature and quantity of these principles; particularly when we observe that the products of these different decompositions mutually combine and produce a new body, which did not exist in the vegetable before? The analysis of vegetables in a retort, is then a complicated, false, and deceiving analysis.

However, as in the chemical examination of any body, no means should be neglected which art furnishes in the discovery of its nature, recourse may be had to this analysis, with a view of observing its effects, well aware not to trust too much to this kind of investigation. It sometimes even happens, that when, in processes performed in order to know its chemical properties, we compare

pare the effects of the aqueous, oily, and ſpirituous menſtrua upon it, with the alterations which they undergo on the part of the fire, theſe agree with the action of the ſolvents; and by the product of diſtillation, indicate the matter contained in the vegetable in more or leſs quantity, the nature of its ſalt, &c. But in order to perform this part of the analyſis with the naked fire, we muſt, 1. Fully know the action of the fire upon each immediate principle of vegetables, ſuch as the extract, the mucilage, the ſaline matter, the oily juices, fluid or dry. 2. We muſt compare the products of diſtillation of the vegetable entire, with thoſe which the immediate principles commonly yield when treated in the ſame way. 3. We muſt analyſe at the ſame time the vegetable by menſtrua, in order to diſcover the immediate principles, and to draw uſeful inductions upon the changes which the fire produces.

The proceſs neceſſary for diſtilling vegetables with the naked fire is very eaſily performed and ſimple. Into a retort of glaſs or earth, we put a given quantity of the dry vegetable: we take care to fill it only a half or two-thirds; we place it in a reverberatory furnace, and adjuſt a proper receiver. A receiver perforated with a ſmall hole was formerly recommended, with a view of giving vent to any air which might be diſengaged from the vegetable, and which endangers the rupture of the veſſels. Now we know, that the aëriform fluid which eſcapes from theſe bodies in diſtillation, is almoſt never any other air but aërial acid or inflammable gas. Therefore as theſe fluids are as much the product of the vegetable, decompoſed by the action of the fire, as the phlegm, the oils, and volatile ſalts, it is as important to collect them as theſe laſt: for this end we ſhould uſe a perforated receiver with a crooked ſyphon adapted to it, whoſe extremity is received in a veſſel full of water or rather of mercury. By this means the liquid products are collected in the recipient, and the aëriform fluids in the pneumato-chemical apparatus. When the ſubſtance which

which we are diſtilling is ſuſceptible of furniſhing a concrete ſalt, we adjuſt between the retort and the recipient another veſſel, upon whoſe ſides the ſalt ſublimes. In this kind of diſtillation the fire ſhould be gradually and cautiouſly raiſed to obtain the products in the order of their volatility, and to prevent our confounding them. We begin with a few coals placed under the retort, and gradually augment the heat till the retort is red and nothing more paſſes over. We leave all to cool, and then unlute the apparatus to examine the products obtained. Though the diſtillation of vegetables with the naked fire always yields us products upon which we cannot fully rely; theſe products, however, differ from one another materially enough to demand a careful diſtinction. What paſſes firſt is an aqueous liquor charged with ſome odorous and ſaline principles. This phlegm aſſumes from leſs to more colour and ſaline properties. To it a coloured oil ſucceeds, whoſe colour deepens as the diſtillation advances, and which at the ſame time acquires conſiſtence and weight. This oil is ſometimes light and fluid, at other times weighty and ſuſceptible of ſolidity. It conſtantly emits a ſtrong empyreumatic ſmell; at the ſame time a greater or leſs quantity of elaſtic fluids is diſengaged, which are either aërial acid or inflammable gas, and moſt commonly both mixed together. At this time alſo, the volatile ſalt ſublimes when the vegetable is of a nature to furniſh it. When all theſe matters are paſſed, the vegetable is reduced to a charry ſtate. Let us reſume each of theſe products, and ſee what their nature is, and to what ſubſtances they owe their formation.

The phlegm is owing to the water in the compoſition of the vegetable, and in part to the water of vegetation, particularly when the analyſed body is not entirely dry; it is more or leſs copious, in proportion as the vegetable has ſuffered a greater or leſs degree of exſiccation. This phlegm is coloured more or leſs red, from the ſmall quantity of oily matter which it carries over

with

with it, and which is in a ſoapy ſtate, by means of the ſalt, which it generally holds in ſolution. The ſaline matter which is united with it is moſt commonly acid; on that account it generally reddens the ſyrup of violets, and effervefces with mild alkalis. This acid belongs to the mucilages and to the oils. Sometimes the phlegm is alkaline; as in the diſtillation of the nitrous cruciform plants, emulſive and farinous ſeeds. Frequently it is ammoniacal, becauſe the volatile alkali which ſucceeds the acid combines with it. This fact is confirmed by throwing a little fixed alkali or quicklime into the phlegm. When it is ammoniacal, a briſk ſmell of volatile alkali flies off. Although the acids of vegetables appear not to be all of the ſame nature, thoſe which are obtained in ſolution exhibit the ſame external marks; but they have not yet been ſufficiently examined to determine the nature of their properties exactly. The oils of vegetables obtained by diſtillation in a retort are all very odorous, high-coloured, and preſent quite the ſame properties. The parts of vegetables which contain a great quantity of theſe inflammable fluids, ſuch as the emulſive ſeeds, yield in their analyſis a great quantity of oil. The odorous plants furniſh one which retains a ſmall portion of their ſmell in the beginning of diſtillation, but which very ſoon aſſumes the characters of all theſe oils, that is, the colour, weight, and empyreumatic ſmell, which diſtinguiſh them. All theſe fluids are very inflammable; nitrous acid kindles them; they diſſolve in ſpirit of wine; and all reſemble the vegetable from which they are procured. By rectification they may all be rendered very fluid and light, colourleſs; in a word, in the ſtate of etherial or eſſential oils.

As to the volatile ſalt, which is only ſome mild volatile alkali, it is obtained from only a few vegetables; but we muſt not believe, as ſome chemiſts have advanced, that it is extracted only from the cruciform plants. In general, all the plants which contain a great quantity of nitre,

nitre, fixed alkali, and oil, furnish it more or less. It is very rare, however, that we obtain a certain quantity of it in a concrete state; frequently it is dissolved in the last portions of phlegm. This salt appears to be owing to the reciprocal action of the oil and of the fixed alkali; and on this account it for the most part does not pass over till about the end of the distillation. It seems also, that that which is carried over by the phlegm in the analysis of some plants, as the cruciform ones, the poppy, rue, is always the product of a new combination; since M. Rouelle junior has demonstrated, that the first contain none of it in their natural state.

The elastic fluids which are disengaged during the distillation of vegetables, should be reckoned also among the number of products. It appears that their nature depends upon that of the vegetable. A plant which contains a great deal of oily combustible fluids yields inflammable gas. Mucilages, again, yield aërial acid. We have, in the article Sugar, observed, that Messrs Bergman and Fontana extracted from its acid a great quantity of fixed air; and this last chemist believed that the vegetable acids were formed in great part of this elastic fluid. It is, then, not astonishing that the mucilages, in which M. Bergman found the same acid as in sugar, yield in their analysis aërial acid. These aërial fluids pass over only towards the end of the distillation, because they are disengaged only the instant that the vegetable is entirely decomposed. Hales, who was unacquainted with their nature, observed, that the quantity of air disengaged during the distillation of vegetables, was so much the greater the more solid the vegetables were; and consequently he considered this element as the cement and the cause of the solidity in bodies. From these considerations we have just now delivered, it is evident how much we should value this hypothesis.

Of

Of Vegetable Charcoal.

Charcoal is the black refiduum which vegetables leave, after they have undergone a complete decompofition of their volatile principles in clofe veffels. The property of yielding charcoal belongs only to the organic matters, which contain the combuftible fubftance called *oil*. To the decompofition of this laft fubftance we afcribe exclufively this formation of this fubftance.

The coal is in general black, brittle, fonorous, and of little folidity. It retains the form of the vegetable, when this has been very confiftent, and when it contained but little fluid matter. If, on the contrary, we decompofe a tender plant, and one which contains a great deal of juice, this, in its difengagement, deftroys the organic texture, and yields a friable coal, which does not exhibit the form of the decompofed vegetable. The different vegetable matters furnifh coals more or lefs copious, according to the folidity or form of their texture. The woods yield much more than the herbs, the gums more than the refins, and thefe laft more than the fluid oils.

Charcoal is a fubftance enjoying very fingular properties, which are in general little known. Although it is very important in chemiftry, and prefents exceedingly particular phenomena, no chemift has yet undertaken to inveftigate its nature. Stahl confidered it as the principal focus of phlogifton; and he is the chemift who has faid moft upon the fubject. Our knowledge of the properties of charcoal belongs almoft entirely to the œconomical ufe which we are obliged to make of it; and the works of philofophers upon it offer nothing complete.

Charcoal, as to its phyfical properties, differs according to the ftate and nature of the vegetables employed for its formation. It is fometimes hard, and then preferves a part of the organifation of the vegetable; at other

other times it is friable, and powdery. The pure oils yield one which is in very fine, as if porphyrised, grains: it is the black of smoke. From the same circumstances also, it varies in weight. When it is well made, it has neither a sensible taste nor smell. Its colour also follows the varieties of its other physical properties. In a word, it is of a more or less deep black, brilliant or unpolished. But the most important examination of this product of fire concerns its chemical properties.

Charcoal, exposed to the most violent fire in close vessels, is nowise altered. If it is heated in a pneumato-chemical apparatus, it yields some particular inflammable gas, of which we shall take more notice afterwards: it is in no degree altered in this experiment. If, on the contrary, it is heated with the contact of the air, then it burns, and is consumed into cinders; but with particular phenomena, which it is requisite carefully to distinguish from those of the other combustible bodies. As soon as it kindles, it reddens and flames: it presents a white flame, which is the larger the greater the mass. It emits no kind of smoke: in the atmosphere, of which it absorbs a part of the pure air, it emits a vapour or destructive gas, which appears to be only aërial acid: lastly, it gradually consumes, and leaves a more or less white cinder, in part saline, and in part earthy. The different charcoals vary in their inflammability; and this is also the most useful distinction in the arts: some easily burn with flame, and are soon consumed; others are kindled with difficulty, consume only slowly, and are converted into cinders only after they have been red a very long time.

The combustion of charcoal has been explained very differently by different chemists. Stahl thought that it was owing to the disengagement of phlogiston. M. Macquer has adopted this opinion; adding, that the pure air is absorbed at the same time that the phlogiston is dissipated. What is more striking in this phenomenon is, the ne ſmall quantity of reſiduum which is left after combination: it ſeems that all this ſubſtance entirely evaporates; and this is probably what happens. M. Lavoiſier, no doubt, ſtruck with this important fact, thought he might advance, in a very good paper upon Combuſtion, Acad. 1777, that charcoal appropriates to itſelf the baſe of the air, and forms with it an acid *ſui generis*, which is fixed air or aërial acid; whilſt the portion of the matter of fire, which was the ſolvent of this baſe, is diſengaged, and preſented under the form of flame and of light, which is proper to it. According to this opinion, the aërial acid is a combination of charcoal with the baſe of air *. With regard to our own opinion of the matter, we adopt a theory which we think more ſimple: The inflammable gas, diſengaged in great quantity from the charcoal, combines with the dephlogiſticated air of the atmoſphere, and forms the aërial acid, which ſeems to be only a combination of theſe two bodies. Thus inflammable gas is to aërial acid what ſulphur is to the vitriolic acid, and nitrous gas to the nitrous acid. This agrees ſo much the better with the facts, as inflammable gas of charcoal precipitates lime-water when it is made to burn above this fluid, and appears really converted into aërial acid, by being combined with the pure air during its combuſtion. This fact, which I have often obſerved, and communicated to M. Bucquet, appeared to me very proper to favour the doctrine which I have preſented upon the nature of inflammable gas and of aërial acid; and I have had the pleaſure of ſeeing it confirmed by M. Lavoiſier, in his paper upon Pyrophyrus, Acad. 1777, of which I had no knowledge before its publication. However, let the cauſe be what it may, of the phenomena of this combuſtion, and the hypotheſis adopted in conſequence of this cauſe, the facts preſented are

* We muſt here refer to this chemiſt's theory, who conſiders pure or dephlogiſticated air as formed of a baſe united with the matter of fire, which gives it the aggregation of an elaſtic fluid.

are no less true; and these alone must command attention.

Charcoal exposed to the air attracts its humidity; probably because it is very porous, and perhaps also on account of the salts which it contains, although these salts are not uncombined. Water alters it in no manner whatever; nor does it suffer any change from the earthy matters: but almost all the saline substances are susceptible of acting upon it. M. Rouelle found, that the fixed alkali dissolved a considerable quantity of it by fusion.

The vitriolic acid decomposes it in close vessels, according to M. Baumé, when a mixture of them are distilled to dryness.

The nitrous acid acts upon it in a much more rapid manner. Dr Priestley observed, that a great quantity of nitrous gas was disengaged. M. Macquer observed the nitrous acid made a very sensible effervescence with this body by means of a certain degree of heat. M. Proust set on fire some charcoal by means of a nitrous acid, which weighed one ounce four gros and twenty-three grains in a bottle which contained an ounce of distilled water. The experiments of this chemist are so new and so important, that I think it proper to give his own words of them, as he has described them in his Observations upon the pyrophori of Alum, inserted in the Journal of Medicine July 1778.

" A charcoal, extracted from the wild saffron, reduced to powder, and recently calcined, occasioned a vivid detonation with nitrous acid, and the rapidity of the inflammation raised the powder with a very pretty effect. I calcined some very fine powder of ordinary charcoal, and the detonation succeeded very well.

" I introduced about a gros of powder of charcoal into a dry retort of glass. I then poured into it about a gros of nitrous acid: it had no sooner got to the bottom of the retort, than the detonation took place with the greatest rapidity: from the neck of the retort, whilst in in my hand, iſſued a flame more than four inches long, which brought with it ſome powder, and very deep coloured vapours of nitrous acid. Theſe vapours condenſed into a green liquor, not very fuming: it was the nitrous acid, weakened by the water, which entered into the compoſition of that which firſt detonated. I poured ſome freſh nitrous acid upon the charcoal which remained in the retort; I likewiſe kindled it when I had poured on all the quantity.

"I repeated this experiment with the black of calcined ſmoke: the ſame thing happened; only a very ſmall portion of cinder was found in the retort, ſometimes half vitrefied, and adhering to the bottom of the retort."

All charcoals are generally charged with a great quantity of humidity. It appeared to me, that calcined charcoal, kept from night to morning, was not more proper for theſe detonations; becauſe it was ſenſibly moiſtened in this ſpace of time. But what is moſt ſingular, is, that theſe experiments are capricious, and do not always ſucceed, although with the ſame charcoal, the ſame acid, and the ſame proportion. The beſt way of aſſuring ſucceſs ſeems to be, if we pour ſome acid upon the middle of the powder, it does not kindle; if, on the contrary, the acid is allowed to flow round the ſides of the crucible or capſule, and it reaches the bottom, the detonation goes from this point, the powder is raiſed, and kindled by the nitrous acid: when the nitrous acid comes to fail, the detonation ceaſes of itſelf, and the coal which ſurrounds it remains black.

We are not acquainted with the action of the other acids upon charcoal.

This body decompoſes by means of heat all vitriolic ſalts, and forms ſulphur with their acid.

It makes nitre detonate; which conſumes it by means of the pure air which it furniſhes by the action of the acid.

In chemiſtry and pharmacy, a preparation is made

called *nitre fixed by charcoal*. Two parts of nitre and one of powdered charcoal are mixed together: the mixture is projected into a red hot crucible; a vivid detonation ensues. When all is over, a white mass remains, which attracts humidity from the air, and is nothing else but the fixed alkali of the nitre and of the charcoal: by washing this substance, the water dissolves the fixed alkali, and only an earthy matter remains.

Liver of sulphur dissolves charcoal with great facility, both by the dry and humid way: It is the substance which combines with it the most easily. For this discovery we are indebted to M. Rouelle.

The metals do not unite with charcoal, but their calces pass to a metallic state when they are heated more or less strongly with this body. In the article of the Metals, we have seen how we should judge of this phenomenon, and what are the different opinions of chemists about it.

The action of the vegetable substances upon charcoal has been little attended to. We only know, that when this substance is mixed with the fat oils, they may by this means be rendered inflammable by the nitrous acid; which confirms the elegant theory of M. Rouelle, of the inflammation of the oils by this acid.

All we have said upon the known properties of charcoal, tends to prove, that it is composed of a combustible body, saline substances, earth, and a little water. Perhaps the combustible matter, which makes more than three-fourths of the charcoal, is inflammable gas. This is at least what facts seem to show. With regard to the rest of its properties, a greater number of experiments must be made, before a decisive opinion is given of its nature and properties.

Of the fixed Salts of Vegetables.

When a vegetable charcoal has been burnt, there remains a grey blackish or white matter, according to the

the charcoal: this is called *cinder*. It is
ded: when it is properly made, it con-
e and earthy matters, frequently mixed
1: when the charcoal was not very com-
netimes ſtill contains a little inflammable
Lavoiſier, upon examining the cinders
nployed by the ſaltpetre-makers, found
e and reſino-extractive matters in them.
fixed ſalts of plants has been given to
ſtances, which are extracted by waſhing

ration of vegetables is uſed to obtain
ſalts, of importance to be known:
from which the vegetable fixed alkali is
prepared in the north by burning the
grows there in great abundance.
very impure; it frequently contains com-
which change its whiteneſs; many neu-
as ſelenite, vitriolated tartar, Glauber's
ſalt, marine ſalt, and a little mineral
ron and earthy ſubſtances. In order to
, and to extract the vegetable alkali from
f purity, it is diſſolved in the ſmalleſt
ty of cold water. This fluid is charged
and ſome neutral ſalts; and it is ſepa-
ion from the earth, coal, iron, and ſele-
taſhes frequently contain. This ſolution
o a pellicle, and by repoſe and cooling
rm the cryſtals of the different neutral
contains. When this ley, after ſeveral
ſtalliſations, and evaporations, yields no
alts, it is evaporated to dryneſs and cal-
lkali is then vegetable alkali very pure: it
ns, however, ſome neutral ſalts, which
parated, by ſetting by a very charged ſo-
arating by filtration the depoſition which
may then be employed with certainty in
ite chemical experiments.

U 2 2. Soda,

2. Soda, from whence the mineral alkali is extracted, is the residuum of combustion of the plants which grow at the side of the sea. It is prepared at Alicant in Spain, at Languedoc, Cherbourg. It is made by burning different sorts of plants. At Alicant they employ the kali; at Cherbourg, sea-ware of all kinds, known by the name of *sea-weed;* the former contains much more mineral alkali than the other, which gives almost none. These plants are burnt very dry in a pit dug for the purpose. At Cherbourg, when the combustion is advanced, and the cinders are very hot, they are agitated and struck strongly with large sticks. By the motion, this substance, which is hot enough to undergo a kind of semivitrification, is made into hard and solid pieces, which are sent to market under the name of *salligot, la marie, alun catin.* The names which distinguish it best, and which announce its state, are those which it receives in the countries where it is manufactured, or from the plants which yield it. The soda of Alicant, also called soda of *barilla*, is the best for chemistry and all the arts, which require a great abundance of mineral alkali. The soda of Cherbourg or sea-ware contains the least alkali, and ought to be rejected in chemistry; though in glass-making it is employed with great success. Soda, considered chemically, is a compound of mineral alkali, of vegetable alkali, though in small quantity, of vitriolated tartar, of Glauber's salt, marine salt, charcoal; iron in the state of Prussian blue, according to the observation of Henckel; and of earth, in part free, in part combined with the fixed alkali, as that of Cherbourg. In order to separate these two substances, and to obtain the mineral alkali pure, it is washed with cold distilled water; the ley is filtrated to separate the earth, the iron, and charry matters; then it is evaporated, as we have directed upon Potashes. This alkali is not so easily purified as that of potash, because, as it crystallises more easily,

in

in its cryſtalliſation it mixes with the neutral ſalts which it contains.

Fixed ſalts are prepared in pharmacy under the name of *Takenius*. This chemiſt's proceſs conſiſts in putting into an iron veſſel the plant from which we want to extract the ſalt: the veſſel is heated till the bottom be red; the plant, which is continually ſtirred, exhales a great deal of fume: it kindles; then the veſſel is covered with a top, to allow the fume to diſſipate, but to ſuffocate the flame. By this means the plant is gradually conſumed; when it is reduced to a kind of black coal, it is waſhed with boiling water, the ley is evaporated to dryneſs, and a yellowiſh or brown ſalt is obtained. This ſalt is frequently alkaline: it contains a great deal of extractive matter which colours it, and which is found mixed with all the neutral ſalts which the plant contained. It is in a ſort of ſoapy ſtate; on which account it is employed in medicine with much ſucceſs: but we muſt not ſuppoſe that it has the ſame virtues with the plant from which it was extracted, as combuſtion neceſſarily alters its principles. Laſtly, it would be of importance to examine, by a chemical analyſis, the different fixed ſalts of plants prepared after the manner of Takenius, with a view of diſcovering the different ſaline and extractive ſubſtances which they contain, and with a view to determine their virtues, and the doſe in which each of them ſhould be given.

Of Vegetable Earths.

When, by the ablution of the cinders of vegetables, all the ſaline matter has been carried away which they contain, no more remains but a powdery ſubſtance, more or leſs white or coloured, inſipid, inſoluble in water, which hitherto has been regarded as an earth. Some iron may be got by the application of a magnet. This metal was ready formed in the vegetable; and ſome naturaliſts have thought, that upon it the colours of vegetables depended. M. Baumé, who in his Memoirs upon the Clays mentioned the earthy residuum of vegetables, asserts, that it forms alum with the vitriolic acid and a kind of selenite, a little different from that which is produced from pure calcareous earth: the other acids yield with this residuum sparry salts and a few martial salts. M. Bergman thinks from this, that the earth of vegetables is formed of clay, and of an earth approaching to the calcareous earths; although, according to him, it sensibly differs from it, since it does not form quicklime by the action of fire. He thinks that the clay is formed in these substances by the collisions which the vitrefiable earth suffers, and by the action of the acids with which it combines; that the clay, once formed, passes to the state of calcareous earth by the new elaborations which it undergoes in the vegetable vessels.

We may be allowed to observe, that the discoveries made in Sweden upon the saline nature of animal bones, which are to these substances what the fibrous texture of vegetables seems to be, seem to announce, that the residuum of these last is nothing else than an earth. Perhaps an accurate analysis, which as yet has not been made upon this subject, might inform us, that what has been taken for an earthy matter, is only a calcareous phosphoric salt; we are allowed at least to suspect so far from the works of M. Margraaf, who extracted phosphorus from mustard and several other seeds.

LECTURE LII.

Of FERMENTATION *in general.*

AFTER having considered vegetables just as nature presents them to us, we must be acquainted with the changes and alterations which they may undergo in different circumstances. These alterations, which entirely depend upon their nature, are owing always to a phenomenon which is called *fermentation.*

Fer-

Fermentation is a ſpontaneous motion which is excited in a vegetable, and which totally changes its properties. This motion is proper to the fluids of the organic ſubſtances; and it is only the ſubſtances elaborated by the principle of vegetable or animal life which are ſuſceptible of it. Chemiſts have not ſufficiently inſiſted upon this important truth; the application of which to the phenomena of organiſed ſubſtances is ſingularly uſeful to medicine.

There are ſeveral circumſtances neceſſary to every kind of fermentation. Such as,

1. A certain degree of fluidity. In fact, dry ſubſtances ſuffer no ſuch alteration.

2. A more or leſs ſtrong heat. The degrees of heat vary in every ſpecies of fermentation; but cold impedes all.

3. The contact of air. On that account bodies are very well preſerved, and without alteration, in a vacuum.

Chemiſts have diſtinguiſhed, after Boerhaave, three kinds of fermentation: the ſpirituous, which furniſhes ardent ſpirits; the acetous fermentation, which yields the acid of vinegar; the putrid fermentation or putrefaction, which produces volatile alkali. It muſt be obſerved, that there are ſeveral fermentative motions which ſeem not to belong to theſe three kinds; ſuch are, perhaps, the bread-fermentation, that of inſipid mucilages, that which evolves the colouring parts. It has been ſuppoſed, that fermentation followed the orders juſt now mentioned; but there are ſome bodies which become acid without having previouſly undergone the ſpirituous fermentation, and there are others which putrefy without undergoing the two firſt.

Of the Spirituous Fermentation.

The ſpirituous fermentation is that which furniſhe ardent ſpirit. For the thorough underſtanding of this fermentation, we ſhall conſider, 1. The conditions neceſſary

ceſſary to produce it: 2. The phenomena which accompany it: 3. The different matters which are ſuſceptible of it: 4. The product which it yields.

§ 1. *Of the Conditions neceſſary to the Spirituous Fermentation.*

EXPERIENCE has taught chemiſts, that all vegetable matters are not ſuſceptible of the ſpirituous fermentation; and that, in order that it may take place, it is neceſſary that ſeveral circumſtances be united: theſe are different objects, which we ſhall conſider as conditions neceſſary to the ſpirituous fermentation.

Theſe conditions are, 1. A ſaccharine mucilage. This matter alone is ſuſceptible of paſſing to the ſpirituous fermentation.

2. A fluidity, a little viſcous. A too fluid juice does not ferment more than a too thick one.

3. A heat from ten to fifteen degrees of Reaumur's thermometer.

4. A large maſs in which a rapid motion may be excited.

5. Acceſs of air, without which it cannot take place.

§ 2. *Of the Phenomena of the Spirituous Fermentation.*

WHEN the five conditions juſt now mentioned are united, then the ſpirituous fermentation takes place; and it ſhows conſtant phenomena which characteriſe it. It has been obſerved,

1. It excites in the liquor a motion, which increaſes till the fermentation be well eſtabliſhed.

2. The volume of the mixture is much augmented; and this augmentation follows the progreſs of the motion.

3. The tranſparency of the liquor is troubled by opaque filaments, which are agitated and carried throughout the fluid.

4. A heat is produced, which increaſes to eighteen degrees, according to M. l'Abbé Rozier.

5. The

5. The ſolid parts mixed with the liquor are raiſed and ſwim a-top, on account of the air which is diſengaged.

6. A great quantity of aërial acid is diſengaged. This gas forms above the fermenting veſſel a layer, which is eaſily diſtinguiſhed from the air. In this air Dr Prieſtley and the Duc de Chaulnes made their excellent experiments. In it candles are extinguiſhed, animals die, lime diſſolved in water is precipitated, cauſtic alkalis are perfectly cryſtalliſed. It is this acid, contained above the fermenting veſſels, which expoſes the workmen who work about it to ſo great danger.

7. The diſengagement of this gas is accompanied with the formation of a great number of bubbles, which are owing only to the viſcous liquor which the acid is obliged to penetrate.

All theſe phenomena ceaſe, as the liquor, however mild and ſaccharine it was, becomes briſk, pungent, and capable of intoxicating.

§ 3. *Of the different Matters ſuſceptible of the Spirituous Fermentation.*

Necessity has induced men to prepare fermented liquors from a great variety of different ſubſtances; but experience has convinced us, that it can be formed by ſaccharine matters only. Among theſe laſt, thoſe of which we have made the moſt uſe, and which it is conſequently requiſite to examine, are the following.

1. The juice of grapes produces wine properly ſo called; the beſt of all the fermented liquors. In order to underſtand the art of cultivating the vine, the object of which is very important to the neceſſaries of life, it muſt be examined, 1. The nature of the ſoil where the vine grows. It is known, that a dry parched ſoil is in general very good for this plant, and that it does not agree with a fat ſtrong earth. 2. The treatment and culture of this vegetable: it is cut, and the branches are lopped, to ſtop the courſe of the ſap; care muſt be taken

taken that it be exposed to the sun, and particularly to the reflection of its rays from the earth, &c.; no manure is furnished to it, &c. 3. The history of the vegetation of the vine; of its exposition; of its flourishing; of the formation of the grape; of its maturation. 4. The accidents to which it is exposed; such as cold, too much rain, moisture. 5. The time of vintage, which ought to be dry and hot. This preliminary knowledge once acquired, the art of making wine should be considered; which consists in putting the grapes taken from the stalk into a vessel; in exposing them to a heat from fifteen to sixteen degrees; in bruising, trampling, and agitating them: then the fermentation is excited, and all its phenomena begin. The juice of the grape, or the must, ought to be neither too fluid nor too thick; in the first case it is thickened by boiling, in the second it is diluted with water. Upon this subject, we should consult the works of M. l'Abbé Rozier, and of M. Maupin. When the wine is made, it is taken out from below, and put into open barrels. It undergoes a second insensible fermentation, which more intimately combines its principles: a fine lee precipitates, and a salt known by the name of *tartar*, which we shall afterwards examine. In order to preserve it, linen rags impregnated with sulphur are burnt in the vessel in which it is contained.

It is also of importance to be acquainted with the different wines. France produces a great number of excellent wines. Those of Bourgogne are the best for daily use. Their principles are perfectly combined, and there is none which prevails. The wines of Orleannois possess qualities very similar to those of Bourgogne, when time has dissipated a little of their greenness, and has fixed the ardent spirit, which they contain to excess. The red wines of Champagne are very good and very delicate. The white not musty wine of this country is much better than the musty, whose pungent and sourish taste, as well as the property of lathering, depends upon

upon the aërial acid which has been ſhut up when it has been put into the bottle before the end of the fermentation. The wines of Languedoc and of Guyenne are of a deep colour; great tonics and ſtomachics, particularly when they are old. The wines of Anjou are white, very ſpirituous, and very readily intoxicate.

With reſpect to the foreign wines, thoſe of Germany called *Rheniſh* and *Moſelle wines* are white, very ſpirituous: their taſte is cool and pungent; they very readily intoxicate. Some Italian wines, ſuch as thoſe of Orviette, Vicence, Lacryma Chriſti, are well fermented, and pretty much imitate the French wines. Thoſe of Spain are in general boiled, ſweet, little fermented, and very unhealthful. Thoſe, however, of Rota and Alicant muſt be excepted, which are juſtly accounted very uſeful ſtomachics and cordials.

2. The apples and pears yield cyder and perry: theſe kinds of wines are very good; and very good aquavitæ may be extracted from them, as M. d'Arcet has demonſtrated.

3. The cherries furniſh another good wine; from which may be extracted an aquavitæ, called by the Germans *kirchenwaſſer*.

4. Apricots, peaches, prunes, yield not ſo good wines.

5. Sugar diſſolved in water ferments eaſily: from this wine an aquavitæ may be procured called *taffia, rum, guildive*.

6. The graminous ſeeds, and particularly barley, furniſh a kind of wine called *beer*. The art of brewing conſiſts in the following proceſſes: The barley is macerated in water during thirty or forty hours, in order to ſoften it: when put up in a heap, it is left to ſhoot; it is dried at a furnace, terminated by a board, upon which the barley is laid; then it is ſearched, to ſeparate the ſhoots called *tournillons;* it is ground into a farina called *malt:* this farina is waſhed with hot water, which diſſolves the mucilage; this water is called the *firſt water:* ſome freſh water is poured upon the malt after its being

being heated, and it forms the ſecond water: it is boiled, and ſet to ferment with hops and yeſt, in a veſſel called *guilloire:* when the fermentation is over, it is agitated: the beer is taken out in caſks: the ſecondary fermentation raiſes a ſcum called *yeſt*, which ſerves to excite the fermentation of the decoction of malt. The ſhooting evolves in the barley a ſaccharine matter, to which it owes the property of forming wine: the ſame might be made from other graminous ſeeds.

§ 4. *Of the Product of the Spirituous Fermentation.*

THE product of all theſe fermented ſubſtances is a particular liquor, more or leſs coloured, of an aromatic ſmell, a hot and pungent taſte; which animates the weakened fibres when they are taken in ſmall doſes, and which intoxicates when drunk immoderately: this is what is univerſally underſtood by wine.

1. Grape-wine, which we take for our example, is a compound of a great quantity of water, ardent ſpirit, eſſential ſalt called *tartar*, and of a colouring extracto-reſinous matter, to which the red wines owe their colour.

Before we ſhow the means of ſeparating theſe principles, the properties of wine, entire and not altered, and its uſes, muſt be known. Wine is capable to diſſolve many bodies, by means of the water, ardent ſpirit, and eſſential acid ſalt, of which it is formed. It unites with extracts, reſins, certain metals, &c. Upon theſe properties the preparation of the medicinal wines is founded; ſuch as the emetic wine, made by macerating two pounds of good white wine with four ounces of crocus metallorum: the liquor is filtrated; or it is employed, in a turbid ſtate, as a very good ſtimulant in apoplectic and paralytic caſes. 2. The chalybeate wine, made by digeſting an ounce of ſteel-filings with two pounds of white wine: it is an excellent tonic and aperient. 3. The vegetable wines, which are prepared either with red wine, in which aſtringent aromatic plants are digeſted; or with white wine, which is generally employed with the antiſcorbutic plants; or with Spaniſh wine: the vinum ſcilliticum is made with this wine, and alſo the laudanum liquidum of Sydenham. This laſt is made by digeſting ſeveral days two ounces of opium cut into ſlices, an ounce of ſaffron, a gros of canella, and cloves, bruiſed in a pound of Spaniſh wine. This medicine is a very good anodyne in the doſe of ſome drops, particularly when we obſerve that the opium does not weaken the patient, or does not ſtop ſome uſeful evacuation.

In order to decompoſe wine, and to ſeparate its different principles, the action of the fire is generally made uſe of. This liquor is diſtilled in an alembic of tinned copper, to which a recipient is adapted; as ſoon as the wine boils, a white, ſlightly opaque, and milky fluid is obtained, of a pungent and hot taſte, of a ſtrong and ſweet ſmell: the operation is carried on till the vapours which are raiſed ceaſe to kindle at a candle. This product is called *aquavitæ;* it is a compound of water, ardent ſpirit, and a ſmall quantity of oil, which deprives it of its tranſparency while it is diſtilling, and which colours it yellow afterwards. We ſhould not aſcribe the colour of the old aquavitæ in commerce to this kind of oil alone, which paſſes with it in diſtillation, but rather to the extractive matter of the wood, which it has diſſolved in the barrels in which it has been kept. Aquavitæ is the liquor from which the ardent ſpirit is got, as we ſhall ſee afterwards. After having furniſhed aquavitæ, the wine is of a deep colour, of an acid and auſtere taſte; it is troubled, and a great quantity of ſaline cryſtals is obſerved, which are juſt the tartar. This fluid, then, is quite decompoſed; and its former properties cannot be reſtored to it by combining the ſpirituous product with the reſiduum. This analyſis, therefore, is complicated. If the reſiduum of the wine, from which aquavitæ has been extracted, is evaporated, it aſſumes the form and conſiſtence of an extract. The colouring

part

part may be ſeparated by ſpirit of wine, which does not touch the tartar. This kind of tincture is not precipitated by water. By evaporation to dryneſs, the reſiduum eaſily kindles, and is ſoluble in water: it is a true reſino-extractive ſubſtance, which the ardent ſpirit formed by fermentation has carried off from the covering of the grapes. From this analyſis it is evident, that wine is truly compoſed of water, ardent ſpirit, tartar, and colouring matter. We know the nature and properties of two of theſe ſubſtances, of the water and colouring extract; it now remains only to examine thoſe of ardent ſpirit and tatar.

Before we ſpeak of theſe two ſubſtances, we ſhould ſay a word of a ſubſtance which is precipitated from wine during fermentation, and called *lee*. It is a compound of ſtones and pieces of raiſins, of coarſe tartar and vitriolated tartar. Some aquavitæ is obtained by diſtilling it in an open fire. If it is treated in a retort, it gives ſome acid phlegm, oil, volatile alkali: its coal contains vitriolated tartar and fixed alkali. The incineration of the lee of wine, made in the open air, furniſhes a cauſtic vegetable alkali, mixed with ſome vitriolated tartar, which is known in the arts by the name of *lees of wine*. What we are going to ſay upon the properties of ſpirit of wine and of tartar, will complete what we have to ſay upon the lee.

LECTURE LIII. LIV.

AQUAVITÆ, which is obtained by diſtilling wine with a naked fire, is a compound of ardent ſpirit, water, and a ſmall portion of oily matter. In order to ſeparate theſe ſubſtances, and to obtain the ardent ſpirit pure, diſtillation is made uſe of. There are ſeveral proceſſes for diſtilling ſpirit of wine. M. Baumé directed to diſtil aquavitæ in a balneum mariæ a ſufficient number of times, in order to get off all the ſpirit. He recommends ſeparating the firſt fourth of the product of the firſt diſtillation, and likewiſe to ſet apart the firſt half of the product of the following diſtillations: all theſe firſt products are mixed together, and rectified with a gentle heat. The firſt half of the liquor which paſſes over in this rectification is the moſt pure and ſtrong ardent ſpirit, called *alkohol;* the remainder is a leſs ſtrong ſpirit, but ſtill very good for ordinary uſe. M. Rouelle preſcribed drawing over by diſtillation in a balneum mariæ the half of the aquavitæ employed; this firſt product is common ſpirit of wine: by rectifying it twice, and reducing it to about two-thirds, a ſtronger ſpirit is uſed, which is diſtilled a-freſh with water according to Kunckel's proceſs; water ſeparates the ſpirit from the oil which altered it. This ſpirit diſtilled with water is rectified, and then there is the greateſt certainty of having it perfectly pure. The reſiduum of the diſtilled aquavitæ is only a water charged with ſome colouring part, with a particular kind of oil floating a-top.

It is manifeſt, that this liquor, after the different proceſſes employed, may be of different degrees of ſtrength and purity. It is a long time ſince different methods of aſcertaining its purity were known. It was at firſt ſuppoſed that ſpirit of wine, which eaſily kindled, and which left no reſiduum, was very pure: but it is now known that the heat excited by its combuſtion is ſtrong enough to diſſipate all the phlegm which might be contained in it. A proof by gunpowder was propoſed: the ſpirit being kindled, when placed upon gunpowder in a veſſel, if the powder does not kindle it is conſidered bad; if, on the contrary, it takes fire, it is thought pretty good. But this proof is falſe and deceiving: for upon putting much ſpirit of wine upon powder, the water which it affords in its combuſtion moiſtens the powder which will not kindle, whilſt we might inflame it by making burn upon its ſurface a very ſmall quantity of phlegmatic ſpirit. This method, then, is not more certain than the former. Boerhaave has given a

very

very good procefs for difcovering the purity of this fluid. It confifts of throwing into the fpirit fome very dry powder of fixed alkali: this falt unites with the fuperabundant water of the fpirit of wine, and it forms a more weighty and more coloured fluid than the ardent fpirit, which does not mix with it but fwims below. Laftly, M. Baumé, confidering that fpirit of wine is fo much lighter than water the purer it is, devifed an aréometer, by whofe means the degree of purity of this fluid and of all fpirituous liquors may be exactly determined. This inftrument being plunged into fpirit of wine, finks the deeper the purer the fpirit is. It is afcertained by well-executed experiments, that the moft pure and moft highly rectified fpirit of wine ftands at thirty-eight degrees of his aréometer. In his Elements of Pharmacy, from page 463—479, may be feen the method of conftructing this inftrument, as well as the refults which fpirit of wine, mixed with different quantities of water gave; which may ferve to fhow comparatively the fpirit of wine in the liquor which is examining.

Pure ardent fpirit, obtained by the procefs juft now defcribed, is a fluid of great tranfparency, mobility, and lightnefs, weighing fix gros forty-eight grains in a bottle which holds an ounce of diftilled water. Its fmell is penetrating and agreeable; its tafte is fharp and hot; it is extremely volatile. When it is heated even flightly in clofe veffels, it rifes and paffes without alteration into the recipient; by this means it is concentrated and feparated from the little water which may adhere to it. On this account the firft portions are the moft fweet, volatile, and pure. It is proper to obferve, that when fpirit of wine is diftilled, a large quantity of air is always difengaged: it might be fuppofed that it is the moft pure fpirituous part which is feparated from the water, and which is volatilifed in the ftate of gas. But Dr Prieftley has demonftrated, that fpirit of wine cannot put on the ftate of a permanent gas. When ardent fpirit is diftilled with the contact of air, it very foon kindles

kindles and preſents a ſlight flame, white in the middle and blue at the edges; it burns alſo without leaving any reſiduum, when it is properly dephlegmated. Several chemiſts have tried to aſcertain what burning ſpirit of wine yields. They have aſcertained that its flame is accompanied with no ſoot nor ſmoke; and that by receiving what is volatiliſed, only ſome pure water is obtained, inſipid, inodorous, and abſolutely in the ſtate of diſtilled water. It is therefore not known what becomes of the ſpirit in combuſtion. However, M. Berthollet has remarked, that when a mixture of this ſpirit and water is made to burn, the remaining fluid precipitates lime-water: this experiment ſeems to ſhow, that ardent ſpirit contains inflammable gas, which by its combuſtion or combination with the air forms aërial acid. Chemiſts, from the phenomena which this ſpirit preſents in inflammation, have adopted different opinions upon the ſubject. Stahl, Boerhaave, and ſeveral others, conſidered this fluid as a compound of a very attenuated oil, acid, and ſome water. According to this opinion, then, it is a kind of acid ſoap. Others, at the head of whom we may place Cartheuſer and Macquer, think that ſpirit of wine is formed of phlogiſton and water. Each of theſe opinions is ſupported by theory and experiment. We ſhall take notice which of them is moſt ſupported by facts.

Spirit of wine, expoſed to the air, evaporates at a temperature of ten degrees above the freezing point; and it leaves no kind of reſiduum, unleſs a little water, when it is not very much dephlegmated. This evaporation in the air is ſo much the more rapid the hotter the atmoſphere is; it produces a greater or leſs cold, according to its rapidity.

Spirit of wine unites with water in all proportions, and is perfectly ſoluble in it. This ſolution is attended with heat, and it forms kinds of aquavitæ ſo much the ſtronger as the ſpirit of wine is in greater quantity. The affinity of combination between theſe two fluids is ſo

 ſtrong,

ſtrong, that water is able to ſeparate from ſpirit of wine ſeveral ſubſtances combined with it; and that, reciprocally, ſpirit of wine decompoſes the moſt part of the ſaline ſolutions, and precipitates the ſalts. From this laſt property, Boulduc propoſed uſing ſpirit of wine, in order to precipitate the ſalts contained in the mineral waters, and to obtain them without alteration.

Spirit of wine has no action upon the pure earths. It is not known if it might be altered by terra ponderoſa and magneſia. Lime ſeems ſuſceptible of occaſioning ſome change in it; ſince, when ſpirit of wine is diſtilled upon this earthy ſubſtance, this fluid acquires ſome ſmell: but this alteration has not been proſecuted.

Fixed alkali alters its nature, and really decompoſes ſpirit of wine, as is proved by the preparation known by the name of *acrid tincture of tartar*. In order to prepare this medicine, ſome ſalt of tartar is fuſed in a crucible; it is pulveriſed quite hot, and put into a matraſs; ſome very dephlegmated ſpirit of wine is poured on it to the height of three or four fingers-breadth above it; the matraſs is ſhut up with another ſmaller one; they are luted together, and the whole digeſted in a ſand-bath till the ſpirit of wine has acquired a reddiſh colour. More or leſs alkali remains at the bottom of the veſſel, and ſome cryſtals of neutral ſalts are obſerved, which M. Baron conſidered as vitriolated tartar previouſly contained in the alkali. Neverthleſs, ſeveral chemiſts thought that this ſalt was partly formed by the union of the fixed alkali with the acid of the ſpirit of wine: but this point has not been properly examined. Upon diſtilling the acrid tincture of tartar, a ſpirit of wine is obtained of a ſweet odour little altered, and there remains in the retort a true ſoapy extract; which diſtilled in the naked fire yields ſome ſpirit of wine, volatile alkali, and a light empyreumatic oil. In this operation a little coal is formed, in which is found ſome fixed alkali. This experiment ſeems to demonſtrate, that

that spirit of wine contains an oil, with which the fixed alkali unites and forms a true soap, which is found dissolved in the undecomposed portion of spirit of wine. The lily of Paracelsus differs from the acrid tincture of tartar, only because the fixed alkali employed in its preparation seems to have been made caustic by the metallic calces along with which it has been heated. The regulus of antimony, iron, tin, and copper, are fused together in the dose of four ounces each; they are detonated with eighteen ounces of nitre and as much tartar; they are brought into fusion; the mixture is pulverised, put into a matrass, and some well dephlegmated spirit of wine is poured above it, to the height of three or four fingers breadth.

This mixture digested in a sand-bath acquires a red colour deeper than the acrid tincture of tartar, and it presents all the same phenomena: this last may be made entirely like the lily of Paracelsus by digesting some spirit of wine with caustic fixed alkali, instead of using fixed salt of tartar, which the action of the fire does not entirely deprive of aërial acid, unless it be kept red for a long time.

The acrid tincture of tartar and the lily are very good tonics and powerful deobstruents. They are employed in all the cases where the strength of patients is insufficient to favour a crisis, as in the malignant fever, bad small-pox, &c.

The action of the caustic volatile alkali upon spirit of wine has not yet been examined.

All acids exhibit with this fluid very remarkable phenomena. When very concentrated oil of vitriol is poured upon an equal part of rectified spirit of wine, a remarkable heat and hissing noise ensues: these two substances grow coloured, and at the same time a sweet smell is disengaged, resembling that of citrons or rennets. If the retort in which the mixture generally is made, is placed in a sand-bath heated, and if two large receivers are adapted, the first of which is plunged into a vessel full of cold water, there comes over, 1. A ſpirit of wine of an agreeable ſmell. 2. A liquor called *ether*, of a very ſweet ſmell and extreme volatility, and whoſe preſence is announced by the ebullition of the liquor contained in the retort, and by the large ſtriæ which line the inſide of the veſſel. The receiver muſt be careſully cooled with wet cloths. 3. After the ether a ſulphureous ſpirit paſſes over, whoſe white colour and ſmell indicate that the apparatus ſhould be unluted, in order to ſeparate the ether. 4. A light yellowiſh oil is at the ſame time alſo volatiliſed, called *ſweet oil of wine*. The fire ought to be greatly moderated after the ether has paſſed, becauſe the matter contained in the retort is black, thick, and it bubbles up conſiderably. 5. When the ſweet oil is all diſtilled, ſome ſulphureous acid alſo paſſes over, and which becomes more and more thick; and at the end is no more than the oil of vitriol black and impure. 6. By continuing this operation with a gentle fire, the reſiduum is entirely dried, and has acquired the form and conſiſtence of a bitumen. An acid liquor is extracted from it, and a dry yellowiſh ſubſtance like ſulphur, by expoſing this bitumen to a very ſtrong fire. M. Baumé, who made many experiments upon vitriolic ether, examined the reſiduum with great care; he found ſome martial vitriol in it, ſome Pruſſian blue, a ſaline ſubſtance, and a particular earth, whoſe nature he did not determine. He ſays alſo, that the yellowiſh ſublimate which it furniſhes is not ſulphur, and that it remains white and powdery without inflaming upon coals. To theſe details we ſhall add, that the reſiduum of the ether may furniſh a new ſupply of ether, by adding to it a third of ſpirit dephlegmated by ſalt of tartar, and diſtilling this mixture. Theſe diſtillations might be ſeveral times repeated; and thus, from ſix pounds of oil of vitriol and ſpirit of wine, to which mixture five pounds of this laſt have been ſucceſſively added, more than two pounds of good ether may be extracted.

The operation juſt now deſcribed is one of the moſt

singular in chemistry, from the phenomena which it presents; and at the same time one of the most important, from the light which it is capable of throwing upon the composition of spirit of wine. There are two opinions upon the composition of ether, which it is necessary to make known. M. Macquer, who, as we have already said, considers spirit of wine as a compound of water and phlogiston, thinks that oil of vitriol carries off the water of this fluid, and brings it nearer and nearer to the characters of oils. Thus, according to this opinion, some spirit of wine passes over, at first little altered; then a fluid, which is in a middle state between spirit of wine and oil, which is the ether; and lastly, a true oil; because the oil of vitriol acts with so much the more energy upon the principles of spirit of wine, as the heat employed to obtain the ether is more strong. M Bucquet, struck with a strong objection which he had made to this theory, its being difficult to conceive how the oil of vitriol, charged at the beginning of its action upon spirit of wine with a certain quantity of water which it had carried off from this fluid, could, though phlegmatic, react sufficiently upon another portion of the same spirit, so as to put it into an oily state, proposed another opinion upon the production of ether: He considered spirit of wine as a fluid composed of an oil, an acid, and water; he thought, that when the vitriolic acid was mixed with this spirit, a sort of bituminous fluid resulted, which by heat furnished the same principles as the bitumens; that is, a light, very odorous, and combustible oil; a kind of naphtha, which was the ether; and then an oil more coloured and less volatile than the former, which was the sweet oil of wine. Indeed, we shall observe, from the properties of ether which we are going to examine, that this fluid has all the characters of a very subtile oil, such as naphtha. The ether obtained by the process just now described is not very pure; it is united with some spirit of wine and sulphureous acid. It is rectified by distillation in a retort in a sand-bath along with some caustic fixed alkali. This salt combines with the sulphureous acid, and the ether passes over very pure with the most gentle heat. If the first half of this product is separated, the purest and most highly rectified ether is obtained.

Ether is a much lighter fluid than spirit of wine, of a strong smell, of a hot and pungent taste. It is so volatile, that upon being poured out or agitated it is dissipated in an instant. In its evaporation, it produces so great a cold, that it may congeal water, as M. Baumé has demonstrated in his excellent experiments. It kindles very easily, as soon as it is heated in the open air or touched with a kindled body: the electric spark likewise kindles it. It emits a very luminous white flame; and it leaves a black mark upon the surface of a body exposed to its flame. Ether dissolves in ten parts of water, according to M. le Count de Lauraguais.

The phenomena which ether might present with all the saline substances, have not been examined; the action of some of the acids only is well known. Lime and the fixed alkalis do not appear susceptible of changing it. The caustic volatile alkali combines with it in all proportions, and it forms a fluid, whose mixed smell might be very useful in asphyxies and spasmodic diseases. Oil of vitriol becomes very hot with ether; and it may convert a great part of it into sweet oil of wine by distillation. Fuming nitrous acid raises a considerable effervescence with it; and the ether in this experiment seems to become more consistent, coloured, and oily. Ether has not been combined with the other saline matters, not even with the mineral inflammable substances. It is only ascertained that it dissolves the essential oils and resins like spirit of wine; and physicians frequently employ ætherial tinctures. Ether is considered in medicine as a powerful tonic, and as a very good antispasmodic. It is employed in hysterical cases, in spasmodic colics. It readily opposes bad digestion which is caused by a weakness of the stomach. It

should

should be administred with prudence, because it is known that its excessive use is dangerous: it is also used with great success externally in pains of the head, burns, &c. Hoffman, who greatly studied the combinations of spirit of wine with vitriolic acid, made use of a medicine composed of sweet oil of wine dissolved in spirit of wine, which he called *anodyne mineral liquor*. The Faculty of Medicine of Paris added ether to this liquor; and ordered in their dispensatory to prepare it by mixing two ounces of spirit of wine, which passes over before the ether, two ounces of ether, and twelve drops of sweet oil of wine. This medicine is employed as the ether, but it has not nearly the same virtue.

Nitrous acid acts in a very rapid manner upon spirit of wine. This brisk action has hindered chemists from examining this combination so fully as that with oil of vitriol. M. Navier is the first who has given an easy process for the making of nitrous ether. According to him, a very strong bottle of Sêves is taken; twelve ounces of very pure and highly-rectified spirit of wine is poured into it and plunged into cold water, or, still better, ice; eight ounces of spirit of nitre are added at several times, and the mixture agitated each time; the bottle shut up with a cork-stopper, which is strengthened with leather, and well tied. The mixture is set in a by-place to prevent accidents from the fracture of the bottle, which sometimes happens. In a few hours bubbles rise from the bottom of the vessel, and drops collect at the surface of the liquor, which gradually form a layer of true ether. This disengagement goes on for four or six days. As soon as the motion is thought to have ceased, the cork is pierced with a sharp probe to give vent to a certain quantity of air, which, without this precaution would burst out upon unstopping the bottle, and carry off the ether, which would be lost. When the air is dissipated, the bottle is uncorked, the liquor is poured into a funnel, whose neck is stopped with the finger, the residuum is separated from the ether, which swims a-top, and is received into a separate vessel.

Mr Woulfe, a famous English chemist, has offered another process for preparing nitrous ether. It consists in employing numerous and very large vessels, to allow plenty of room to the disengaged air. A white glass vessel, from eight to ten pints, is taken, terminated by a neck seven or eight feet long: it is placed upon a tripod sufficiently high, in order that a refrigeratory may be placed below it: a tubulated head is adjusted to this vessel; to whose neck a glass tube of seven or eight feet long is adapted: this last is received by its lower extremity into a globe with two points, pierced below with a hole, to which a vessel is joined: to the third tube of this globe are added the bottles which constitute Mr Woulfe's apparatus, which we have already several times described

When all these vessels are well luted, a pound of well-rectified spirit of wine, and as much fuming spirit of nitre, is poured into the globe by the tube of the head; then the head is shut with a crystal stopper, which is tied down with leather. The mixture, as soon as made, becomes warm; vapours are disengaged, which run rapidly through the neck of the globe. This being heated till the liquor boil, the nitrous acid passes over into the globe, which serves for the recipient. This process, though very ingenious, has several inconveniences: the apparatus is long of being fixed, it is dear and very embarrassing; besides, it exposes to danger, because, in spite of the space allowed to the vapours, they are so rapidly disengaged, that the vessels often burst with considerable noise.

M. Bougues, anno 1773, published another method of making nitrous ether. He directs to mix, in a glass retort of eight pints, a pound of spirit of wine with a pound of weak nitrous acid, so as to stand no higher than 24 degrees of Baumé's aréometer; to adapt to the retort a globe of twelve pints; to give passage to the

the air, by adjuſting two quills at the junction of the luting, and to diſtil with a very gentle heat, ſinking the retort but a very ſhort way into the ſand. In this way he got ſix ounces of a very pure nitrous ether. From what M. l'Abbé Rozier has ſaid, it appears that M. Mitouard employed, ſo late as the year 1770, a proceſs very ſimilar to M. Bougues.

This chemiſt put four ounces of fuming ſpirit of nitre, with twelve ounces of ſpirit of wine, into a retort for diſtillation: the retort was ſunk but very ſlightly into the ſand. By this means, which appears the moſt ſimple of all, he obtained a nitrous ether, reſembling that of M. Navier. Laſtly, M. de la Planche, a Paris apothecary, has thought of two methods of preparing nitrous ether in a very commodious manner. The firſt conſiſts in putting ſome nitre into a tubulated free-ſtone retort, to which a large globe, or two joined together, are adapted, to pour through the tube, firſt oil of vitriol, and then ſpirit of wine. The vitriolic acid diſengages the nitrous, which reacts upon the ſpirit of wine, and almoſt inſtantly forms nitrous ether. As it might be ſuſpected that the ether thus prepared was in part vitriolic, he ſubſtituted to this firſt method another very ingenious proceſs. He adapts to a tubulated glaſs-retort, in which he has put ſix pounds of very dry nitre, a long neck and a globe, which communicates by means of a crooked tube with an empty bottle. This correſponds by means of a ſyphon with another bottle, which contains three pounds of the moſt perfect ſpirit of wine. The whole being well luted, and the retort placed upon warm cinders, three pounds of very pure oil of vitriol are put in upon the nitre through the tube of this laſt veſſel: the retort is ſhut with a cryſtal ſtopper, a boiling heat is applied, and it is kept up till the vapours ceaſe to paſs. In this experiment, the vitriolic acid diſengages the nitrous, which paſſes partly into the globe, and partly into the ſecond bottle. When the operation is finiſhed, the

globe

globe contains fuming ſpirit of nitre, the retort vitriolated tartar, and the ſecond bottle an etherial liquor. This is diſtilled in a retort with a ſingle globe, and only two thirds of the product are drawn over. This product is diſtilled with a fifth of fuming ſpirit of nitre, which is gradually poured upon it with a glaſs funnel with a long ſtem; only two thirds of it are taken. Laſtly, this ſecond product is rectified with ſome ſalt of tartar; only four ounces are drawn over with three-fourths of remainder. The four ounces are very pure nitrous ether; the three-fourths of remainder are a nitrous mineral anodyne liquor. The reſidua of the two rectifications are dulcified ſpirit of nitre.

The nitrous ether obtained by all theſe proceſſes is a yellowiſh fluid, as volatile as vitriolic ether: its ſmell bears ſome analogy to that of this laſt, although it is ſtronger and leſs agreeable; its taſte is hot and more diſagreeable. It contains a little ſuperabundant acid: it raiſes up the cork of the veſſel in which it is kept; becauſe a great quantity of air is continually diſengaged: burnt, it emits a more brilliant flame and a thicker ſmoke than vitriolic ether; it alſo leaves a more copious coal: laſtly, like the vitriolic ether, it takes gold from its ſolution, and charges itſelf with a certain quantity.

The reſiduum of nitrous ether is of a citron yellow colour; its odour is acid and aromatic; its taſte is pungent, and imitates that of diſtilled vinegar. If it is diſtilled, it yields, M. Baumé ſays, a clear liquor, of a more agreeable ſmell than that of nitrous ether, of an agreeable acid taſte: it reddens the ſyrup of violets, unites with water in all proportions, and efferveſces with mild alkali; then a yellow amber-like matter, friable, ſimilar to ſuccinum, which attracts humidity from the air, becomes pitchy, and diſſolves in water without rendering it mucilaginous, remains in the retort. M. Baumé calls this a gummy ſoapy ſubſtance. It yields upon diſtillation a few drops of an

aci-

acidulous very clear liquor, of an oily consistence, and of a slight empyreumatic odour. After distillation, there remains a spongy brilliant coal without taste, and very fixed in the fire. M. Bucquet says, that if the liquor is evaporated which remains after the formation of nitrous ether, it acquires the consistence of a mucilage, and in time saline crystals form very like hairy caterpillars, to which has been given the name of *crystals of Hierne*, from the chemist who first described them.

Nitrous ether may be employed in the same cases as the vitriolic; but it must be given in a smaller dose on account of its greater action *.

The

* M. Baumé, who tried many experiments to determine the best mode of preparing nitrous ether, recommends the proportion of two parts of fuming nitrous acid to one part of spirit of wine. The use of a larger proportion of acid never fails to produce a most violent explosion, which no precaution can prevent; and the employment of a larger proportion of spirit of wine would be unnecessary, as some of it seems to remain undecomposed in the residuum of the mixture from which the ether has been separated. When we add the acid to the spirit, we always hear a crackling noise, which M. Baumé thinks must arise from the action of the acid upon the portion of the spirit with which it comes into contact, and which it dilates to a very great degree. But what I chiefly wish to mark here, is the occurrence of a phenomenon which happened during the extreme cold of last winter (1783), in preparing nitrous ether after M. Baumé's method.

The phial in which the acid was added to the spirit was placed in a freezing mixture of snow and salt, and both the ingredients were taken immediately from another cold exposure.

After about half an ounce of acid was added, besides the crackling noise which M. Baumé describes, a number of spicular crystals were produced, which were collected, upon decanting the mixture, into another vessel.

So curious and unexpected an appearance induced me to repeat the experiment under the same circumstances; which I did as nearly as possible, though entirely without success, as no crystals were produced in the mixture. After this failure, I cooled both the acid and spirit separately, before mixture, down to 30 degrees below frost in Fareinheit's scale, and also brought the phial in which

they

The marine acid has no sensible action upon spirit of wine: this acid is only dulcified by simple mixture with this liquor, as the two others are, when mixed in small quantity with spirit of wine. M. Baumé, in his Dissertation upon Ether, says he obtained a little marine ether

they were to be mixed to the same temperature. Upon mixing, under those circumstances, the experiment succeeded to my wish; and exhibited a much larger number of spicular crystals than the first one. By some accident the vessel was broke in which those crystals were kept; and I have not since had a convenient opportunity to repeat the experiment. There is indeed something very inexplicable in this production of crystals, as the mixture of nitrous acid and good spirit of wine commonly produces a violent degree of heat, and causes the emission of vapours which cannot be repressed.

M. Baumé found the production of the elastic fluid so extremely troublesome and dangerous, that in order to avoid the inconvenience arising from the violent explosion, he contrived to fix an empty bladder to the mouth of the phial, to collect all the vapour which might be generated. This contrivance served the purpose in preventing the danger of explosions, though not without the loss of some ether, which was always collected in a smaller quantity when this apparatus was employed. But there is a process, very elegant in itself, and superior to M. Baumé's, from the absence of any troublesome collection of inconsiderable vapour, which has been practised for some years past by Dr Black professor of chemistry in the university of Edinburgh. Dr Black takes two ounces of strong nitrous acid, which he puts into a phial: he pours in, very slowly and gradually, a quantity of water nearly equal in bulk to the acid; so that by trickling down the sides of the phial, the water may float upon the surface of the acid without mixing: he next adds, in the same cautious and gradual manner, three ounces of highly-rectified spirit of wine; so that the spirit may in its turn float upon the surface of the water in the same way that the water floats upon the surface of the acid. By this attentive mode of proceeding, Dr Black keeps the three fluids separate by means of their different specific gravities, and has a stratum of water interposed between the acid and spirit. After this he sets the phial in a cool situation; the acid gradually rises up by attraction through the water, mixes with the spirit, and acts upon it; and in proportion to this forms a quantity of nitrous ether, without any chance of producing elastic vapours. The ether formed by this process is extremely pure, and very fragrant in the smell, and does not so necessarily require separation from the acid by rectification as that by M. Baumé's process.

ether by mixing the vapours of these two fluids. Ludolf and Pott employed butter of antimony. M. le Baron de Bornes prescribed dissolving flowers of zinc in marine acid, and distilling the salt concentrated by evaporation in close vessels with spirit of wine. By this process we very easily get marine ether. But no person has prosecuted the subject with so great ardour and success as M. le Marquis de Courtanvaux. . According to the process prescribed by this chemist, a pint of spirit of wine, with two pounds and a half of fuming liquor of Libavius, a very strong heat ensues, and a white suffocating vapour rises, which disappears upon agitating the mixture: an agreeable smell is disengaged, and the liquor assumes a citron colour. The retort is put into a hot sand-bath; two receivers are luted, the last of which is immersed into cold water. A dephlegmated spirit of wine soon comes over, and then the ether: it is perceived by its sweet smell and the striæ which it forms in the inner surface of the retort. As soon as this smell changes and becomes strong and suffocating, the receiver is changed, and the distillation continued: a clear acid liquor is obtained with some drops of a sweet oil; a yellow matter succeeds of a butyraceous consistence, a true butter of tin; and, lastly, a brown heavy liquor, which emits very copious white vapours. There remains in the retort a powdery grey matter, which is a calx of tin. The product of ether is poured upon some oil of tartar in a retort; a brisk effervescence and copious precipitate ensues from the tin, which the acid that had passed over with the ether kept dissolved. A little water is added, and the whole distilled with a gentle heat. About the half of this product is obtained as marine ether. All the liquors which pass after this marine ether are very copious in tin: they attract humidity from the air, they unite with water without precipitating. It is not well known to what this so rapid action of the marine acid, contained in the fuming liquor upon on ſpirit of wine, ought to be aſcribed, when this acid pure has not the ſmalleſt effect: but it appears from M. Scheele's experiments, that the marine acid is in its dephlogiſticated ſtate, or, according to the new doctrine, overcharged with air; and hence its action in this ſtate.

M. de la Planche, an apothecary, propoſed preparing marine ether by pouring upon decrepitated marine ſalt in a tubulated retort ſome oil of vitriol and ſpirit of wine. The marine gas, diſengaged by the vitriolic acid, meets in the globe the ſpirit of wine in vapour, with which it combines. An acid ether reſults, which is rectified with fixed alkali by diſtillation. Marine ether is very tranſparent, very volatile: it has nearly the ſame ſmell as vitriolic ether has; it burns like it, and yields a ſmoke ſimilar to it. But it differs from it in two properties; the one is, its exhaling, in combuſtion, an odour as pungent and as active as that of ſulphureous acid; the other is, its having a ſtyptic taſte, reſembling that of alum. Theſe two phenomena intimate, that this ether is different, and perhaps leſs perfect than the other two: no doubt, upon a farther examination of its other properties, more ſingular differences will yet be found *.

After

* To form pure marine ether was long conſidered as one of the moſt difficult problems in chemiſtry; and after all the perfection to which the proceſs has been brought by the labours of the Marquis de Courtanvaux, we cannot eaſily obtain marine ether, from the ſmall diſpoſition which the marine acid has to unite with ſpirit of wine. All chemiſts who have wrought upon the ſubject, know the amount of the difficulty from their own experience; and they have all been inclined to aſcribe the repugnance between the two fluids to the ſmall degree of attraction which ſubſiſts between this acid and phlogiſton. In the proceſs of making vitriolic ether, the vitriolic acid is known to act upon the principle of inflammability of the ſpirit of wine, ſo far at leaſt as to become highly phlogiſticated itſelf, and to be converted into the volatile ſulphureous acid. If, therefore, the muriatic acid could by any means be rendered capable of attracting phlogiſton, it might alſo acquire the power of decompoſing ſpirit of wine, and of forming marine ether. Struck with the force

After having given an account of the action of the three mineral acids upon ſpirit of wine, we muſt reſume the hiſtory of this fluid. The action of the other acids upon ſpirit of wine has been but very ſlightly examined. It is known that it eaſily unites with the acid of borax: this ſalt communicates to its flame a green colour; that the ſpirit of wine abſorbs more than its own volume of aërial acid. As to the neutral ſalts, M. Macquer has determined, that the vitriolic ſalts diſſolve but with difficulty in it, that the nitrous and marine ſalts unite much better, and that, in general, it diſſolves theſe ſubſtances ſo much the more eaſily that their acid is leſs adherent. Spirit of wine, boiled upon vitriolated tartar and Glauber's ſalt, diſſolved none of them. Aërated tartar and ſoda unite with it in ſmall quantity: the moſt part of the ammoniacal ſalts unite with it. The deliqueſcent earthy ſalts, ſuch as the calcareous nitrous and marine ſalts, and thoſe with baſe of magneſia, diſſolve very well in it. Some metallic ſalts alſo are very ſoluble in it; ſuch as martial vitriol in the ſtate of mother-water, nitre of copper, iron and copper in the marine acid, corroſive ſublimate: all the copper ſalts give a very beautiful green colour to its flame.

Spirit of wine does not diſſolve ſulphur in ſticks nor in

force of this analogy, I endeavoured to bring the conjecture to the teſt of experiment, by mixing dephlogiſticated marine acid with highly-rectified ſpirit of wine. For this purpoſe, I put ſome marine acid into a retort along with ſome black mangane ſe; placed the retort upon a ſand-heat, raiſed the fire, and directed the acid vapour into a receiver which contained ſpirit of wine. After a ſhort time, the marine acid vapour was perceived to make a ſenſible impreſſion upon the ardent ſpirit in the receiver; it formed an intimate union with the whole, and converted a portion into perfect ether, which was ſeparated from the reſt of the ſpirit by rectification. This etherial liquor, when purified, was found, upon examination, to be completely prepared, and in all its properties to correſpond with, if not to excel, the beſt marine ether made in a very tedious and complex manner. Its ſmell is highly fragrant, inclining to that of citrous, and its mobility nearly equal to that of either the vitriolic or nitrous: This proceſs is both eaſy and expeditious, and will afford a convenient method of obtaining marine ether to thoſe who wiſh to ſtudy the ſubject with greater accuracy.

in flowers; but it unites with it when these two bodies are in the state of vapour, and come in contact, according to the discovery of M. le Count de Lauraguais. His process consists in putting flowers of sulphur into a glass cucurbit, and in placing in the same vessel, and upon the flowers of sulphur, a glass-tumbler full of spirit of wine, heating the cucurbit in a sand-bath, and adapting a head and recipient. The sulphur is volatilised at the same time with the spirit of wine: these two substances combine; and the fluid which runs into the receiver is a little troubled, and emits a fetid odour. It contains about one grain of sulphur in the gros of spirit of wine.

It has no action upon the metals nor their calces. It in part dissolves some bitumens, such as succinum and ambergris; it does not touch those that are black and as if charry: it is observed, that when it is distilled from above salt of tartar it unites better with these bitumens; and that this salt, mixed with these last, rendered them much more dissoluble, undoubtedly by converting them into a soapy state.

There are few vegetable substances upon which spirit of wine has not a more or less remarkable action: the extracts lose their colouring part, and frequently all their substance, when they are of the nature of the extracto-resinous or resino-extractive bodies: the saccharine and soapy juices unite with it. M. Margraaf extracted by its means an essential saccharine salt from skirwort, parsnips, &c. But the substances with which it combines most easily are, the essential oils, the spiritus rector, camphor, balsams, and resins. Charged with the spiritus rector, it is called a *spirituous distilled water*. To prepare these fluids, spirit of wine is distilled with the odorous plants in a balneum mariæ. The spirit attracts the principle of odour, and is volatilised along with it: it dissolves also a part of their essential oil, which makes it whiten upon mixture with distilled water; but it is separated from this foreign principle by rectifying it

in

ıariæ with a very gentle heat; and care
w over only three-fourths of the ſpirit of
, in order to have only the ſpiritus rec-
rituous diſtilled waters acquire an odour
as they become old; and it appears that
nciple combines more and more intimate-
it of wine. Spiritus rector has ſo great
ſpirit of wine, that this laſt is capable
ſf from the eſſential oils and water. In
ng ſpirit of wine from the eſſential oils
rged with the odour of a plant, the ſpi-
dorous principle, and leaves the oil and
ıdour. It is obſerved, that ſpirit of wine
the heavy and thick eſſential oils, than
more fluid and light. Water diſunites
: it precipitates the oil in the form of
que globules; but the ſpiritus rector
: ſpirit of wine. Spirit of wine eaſily
or in the cold, but better by means of
ution, very copious, as two gros in the
of wine, mixed with water which has
dually and by drops, yields a cryſtalline
erved by M. Romieu: it is a perpendi-
upon which needles are implanted which
ament under an angle of ſixty degrees.
ıt ſucceeds but rarely; and it requires
the quantity of water, cooling, &c.
tinctures, elixirs, balſams, quinteſſence,
to the compounds of oily or reſinous
it of wine, which is ſufficiently charged
tances, to have a great colour, and to
recipitated by water. They are, as the
either ſimple, when they hold only one
lution; or compound, when they con-
ěſe medicines are prepared in general by
ice in powder, or the plant dry, whoſe
eſin is to be diſſolved, to the action of
ıe, which is aſſiſted by agitation and the

gentle heat of the sun or of a sand-bath. When the resins of several plants or any vegetable substances are to be extracted at a time, these substances must first be digested in the spirit which are most difficultly acted upon; and those must be successively exposed to its action which are the most soluble: when this menstruum is as fully charged as can be, it is filtrated. Sometimes a compound tincture is at once made, by mixing several simple tinctures: in this way the elixir proprietatis is prepared, by mixing the tincture of myrrh and that of saffron and aloes together. Resins and balsams may be separated from spirit of wine by pouring some water into the tinctures, or by distillation. But in these two cases the spirit retains the odorous principle of these two substances. Water cannot decompose the tinctures formed of the resino-extractive or extracto-resinous substances; as those of rhubarb, saffron, opium, gum-ammoniac, &c. because these two substances are equally soluble in these two menstrua.

Spirit of wine and aquavitæ are of very extensive and various uses in life. The latter is drunk to support oppressed spirits; but its excess is dangerous, as it dries the fibres, and produces tremors, palsies, obstructions, dropsies. Pure spirit of wine, or united with camphor, is employed as an external remedy in stopping the progress of gangrenes.

Distilled spirituous waters are administered in medicine as tonic, cordial, antispasmodic, stomachic, medicines. They are exhibited diluted with water or sweetened with syrup.

Drinks are made with these waters and sugar called *ratafias*. These, well prepared and taken in small quantity, may be very useful; but in general they agree with few persons, and they may be hurtful to many. Excess in these kinds of liquors bodes the greatest danger; and instead of strengthening the body and stomach, as is very commonly believed, they produce for the most part an entirely opposite effect; those which are

are the leaſt hurtful, when they are drunk rarely and moderately, ought to be prepared in the cold with one part of ſpirit of wine diſtilled upon the aromatic ſubſtance whoſe ſmell we wiſh to communicate to it, two parts of water, and one part of pure ſugar.

The tinctures have nearly the ſame virtues with the diſtilled ſpirituous waters; but their action is much ſtronger; for which reaſon they are employed in a much ſmaller doſe, or in the form of pills, or with wine, or alſo in aqueous liquors. The precipitate which they form in this liquor is likewiſe ſuſpended in the mixture; and beſides, the colouring part remains diſſolved in the ſpirit of wine.

Laſtly, ſpirit of wine, united with reſin-copal, the oil of ſpike or lavender, and that of turpentine, forms varniſhes called *dry varniſhes;* becauſe, by applying a layer of this compound to the bodies we want to varniſh, the ſpirit flies off readily, and leaves a tranſparent reſinous lamina. The eſſential oils which are mixed with them hinder them from drying too readily, and they prevent them from being brittle by the unctuoſity which they communicate.

LECTURE LV.

Of Tartar.

TARTAR is an eſſential acid ſalt, united with a portion of vegetable alkali and oil, which is depoſited upon the ſides of the barrels during the inſenſible fermentation of the wine. It is not a product of the ſpirituous fermentation, as ſome chemiſts have ſuppoſed, ſince M. Rouelle junior found it ready formed in muſt and verjuice.

It is in the form of irregular plates, diſpoſed in layers, often filled with brilliant cryſtals, of an acid and vinous taſte. We diſtinguiſh white and red tartar; which do not differ but in the red's containing more extractive colouring part. Crude tartar, exposed to the fire in close vessels, gives over an acid reddish phlegm, an oil at first light, then heavy, coloured, and empyreumatic; a little volatile alkali; and a great quantity of aërial acid, which Hales, Boerhaave, and several other chemists, mistook for air. A coal remains which contains much fixed alkali, and which incinerates easily. By the combustion and incineration of the tartar, a very pure vegetable fixed alkali is extracted. Some tartar in powder is put into cornets of paper, which are then steeped in water; the cornets are arranged in a furnace between two beds of charcoal, which is kindled: when the fire is out, the cornets, which preserve their form, are taken out; they are washed with cold distilled water, to see what they contain: the ley is filtrated, and evaporated to a pellicle; it is left to cool, to separate from it the vitriolated tartar: the water is poured off from this salt, evaporated, and crystallised afresh, till it yields no more vitriolated tartar: then it is evaporated to dryness; and by this means a vegetable fixed alkali is obtained, united with a portion of fixed air, and always containing a little vitriolated tartar.

Tartar dissolves in water with difficulty, since an ounce of this fluid, at the temperature of ten degrees above ice, takes up only four grains. As it contains much oily and colouring matter, it is purified by solution and crystallisation at Ancane and Calvisson, in the neighbourhood of Montpelier. We owe the details of this purification to M. Figes. He has added them in a memoir, printed among those of the Academy 1725. The tartar is made to boil in water; the solution is filtrated boiling: it turns turbid upon cooling, and deposites irregular crystals, which form a paste: this paste is made to boil in copper vessels with water, previously mixed with an argillaceous earth brought from the village of Merviel, two leagues from Montpelier: it throws off scums, which are carefully taken off, and then a saline pellicle is formed: the fire is lowered; the pellicle

licle is broken, which mixes with the cryſtals which are precipitated from the ſolution: the cryſtals are waſhed with water to carry off the earth, which renders them impure; and they are ſent to market under the name of *cream* or *cryſtals of tartar;* which differ only in the cream being depoſited at the ſurface, while the cryſtals are depoſited at the bottom of the liquor. It ſeems that the white clay ſeems to deprive the tartar of its oily matter and extractive part.

At Venice, tartar is purified in a manner a little different, according to M. Deſmaretz: the ſalt in powder is diſſolved in boiling water; the impure matters which it contains are left to depoſite, and they are carefully taken away: the liquor yields cryſtals by repoſe and cooling. Theſe cryſtals are rediſſolved in water, which is heated ſlowly: when this new ſolution is boiling, bruiſed whites of eggs are thrown in, and aſhes paſſed through a ſieve: the aſhes are mixed fourteen or fifteen times: we take off the ſcum which the effervefcence occaſions, and leave the liquor to ſettle. Very ſoon a pellicle, and very white ſaline cryſtals, are formed; the water is poured off, and the ſalt dried: this method alters the nature of tartar, and changes a part into vegetable ſalt. It is the cream of tartar, or tartar purified about Montpelier, whoſe properties we are going to examine.

Very pure cream of tartar is cryſtalliſed, but in a very irregular manner. It has a ſour and leſs vinous taſte than the crude tartar. When it is put upon a burning coal, it emits much ſmoke, which has a pungent empyreumatic ſmell: it becomes black and charry. Diſtilled in a glaſs retort in a pneumato-chemical apparatus, by conducting the fire gradually, a phlegm, at firſt ſlightly coloured and ſlightly acid, is obtained: then a ſtronger and deeper-coloured acid paſſes; an oil, which aſſumes gradually colour and conſiſtence, whoſe ſmell is empyreumatic; laſtly, ſome concrete volatile alkali, and a great quantity of aërial acid. A very copious coal remains in the retort, which, waſhed without incinera-

 tion,

tion, yields abundance of fixed alkali. All theſe products may be rectified by diſtillation with a gentle fire. The phlegm paſſes over almoſt without colour; the oil becomes very white and very volatile in this rectification; the volatile alkali is in part combined with the acid, and is obtained in a ſeparate and pure ſtate only by diſtilling the laſt portions of phlegm with the addition of fixed alkali. As for the coal, the fixed vegetable alkali which it contains is not produced in the operation, as ſome chemiſts have imagined, who were not fully acquainted with the nature of cream of tartar; but it is ready prepared in this ſubſtance. It is to this fixed alkaline ſalt that the production of volatile alkali is owing, formed by the reaction of the former upon the oil: the quantity of this volatile ſalt may alſo be augmented by diſtilling the oil obtained from the cream of tartar from above the coal which is left in its analyſis in the retort.

Cream of tartar ſuffers no alteration from the air.

It diſſolves in twenty-eight times its weight of boiling water; and it cryſtalliſes by cooling, but in a very remarkable manner. A certain quantity of earth ſeparates from the ſolution of this ſalt, ariſing undoubtedly from that employed in this purification. This ſolution reddens the tincture of turnſol, and has an acid taſte.

It is uncertain what is the action of the quartzy, argillaceous, and ponderous earths upon cream of tartar. Meſſrs the Chemiſts of the Academy of Dijon have obſerved, that magneſia, formed with cream of tartar a ſoluble ſalt, which the fixed alkali decompoſed, and which evaporated in the free air, gave ſmall priſmatic cryſtals, diſpoſed like the rays of the ſun. This tartarous ſalt, expoſed to the fire, bubbles up, and is converted into a light coal.

Several chemiſts have very fully deſcribed the action of lime and chalk. When ſome chalk is thrown into a ſolution of cream of tartar, an effervescence is produced from the diſengagement of the aërial acid, and a very

very copious precipitate is formed: this precipitate is the combination of the tartarous acid and lime: the supernatant liquor contains a neutral ſalt fully formed in the cream of tartar, and compoſed of its acid united with the vegetable fixed alkali: this ſalt, as we ſhall ſee more fully afterwards, is known by the name of *ſoluble tartar.*

To M. Rouelle junior this excellent analyſis of cream of tartar is owing: it proves, 1. That this ſubſtance is compoſed of a ſuperabundant oily acid, and of a certain quantity of this acid, united with the vegetable fixed alkali in the ſtate of a neutral ſalt. 2. That the combination of the tartarous acid with lime forms a neutral ſalt having very little ſolubility. M. Prouſt diſcovered, that the calcareous tartarous ſalt, diſtilled in a retort, leaves a reſiduum which kindles in the air like pyrophyrus. M. Bergman, in his Diſſertation upon the Elective Attractions, gives a proceſs for ſeparating the tartarous acid from this ſalt. He preſcribes waſhing with diſtilled water the precipitate formed by chalk thrown into a ſolution of cream of tartar, putting this calx tartariſata into a phial, and pouring above it eight times its weight of vitriolic acid, formed of one part of oil of vitriol and eight parts of water. This mixture is left in digeſtion for twelve hours, and frequently agitated with a wooden ſpatula; the clear liquor above the depoſition is poured off; this is waſhed with water till it has no taſte, and this ley mixed with the firſt liquor. This is the tartarous acid. It is evident, that in this experiment the vitriolic acid has decompoſed the calx tartariſata, and formed ſelenite by diſengaging the tartarous acid which the water diſſolved. The acid thus obtained, contains almoſt always a little vitriolic acid; it is purified by adding a little calx tartariſata, whoſe acid is diſengaged by the vitriolic and left pure. M. Bergman adds, that the ſolution of this acid, evaporated to the conſiſtence of a clear ſyrup, yields laminated cryſtals, very ſeparate from one another; that theſe cryſtals

 blacken

blacken on the fire, yield to the retort an acidulated phlegm and a little oil; and that the coal which they leave is neither acid nor alkaline. From these details it appears, that the tartarous acid contains oil like all vegetable acids.

Cream of tartar unites very well with the different alkalis. Some powdered cream of tartar is thrown into a solution of fixed salt of tartar; a brisk effervescence arises, produced by the disengagement of the aërial acid; cream of tartar is added to the point of saturation; the liquor is boiled half an hour and filtrated; it is evaporated and left to cool slowly; long square crystals form, terminated by two-sided figures. This salt is called *vegetable salt*, *soluble tartar*, *tartar tartarisata*: it has a bitter taste; it becomes coaly when it is greatly heated; it is decomposed in a retort, and yields an acid phlegm, oil, and a great quantity of fixed air; it attracts in a small degree humidity from the air; it dissolves in four parts of distilled water; it is decomposed by the mineral acids, and likewise by most of the metallic solutions.

Cream of tartar, combined with the mineral alkali, forms the *sal de Seignette*, so called from an apothecary of Rochelle, who was the first who made it known. It is prepared by putting twenty ounces of cream of tartar into four pounds of boiling water; very pure alkali of soda is added to the point of saturation, which is known when no more effervescence ensues upon addition of the alkali. This combination renders cream of tartar soluble. The liquor is evaporated to almost the consistence of a syrup, and it yields by cooling very beautiful regular crystals, often of a considerable size. They are prisms with six, eight, or ten unequal sides, truncated at right angles at their extremities. The prisms are for the most part cut into two in their length, and the larger side or base upon which they rest is marked with two diagonal lines which cross in the middle, and divide this base into four triangles. Sal de Seig-

Seignette, which was ſold at firſt as a ſecret, and diſcovered at the ſame time by Meſſrs Boulduc and Geoffroy 1731, has a bitter taſte; it is decompoſed in the fire as the vegetable ſalt; it effloreſces in the air, becauſe it contains much water of cryſtalliſation; it is almoſt as ſoluble as the vegetable ſalt, and decompoſable like it by the mineral acids and metallic ſolutions. The mother-water of this ſalt contains the portion of vegetable ſalt, which made part of the cream of tartar.

The volatile alkali forms with cream of tartar a tartarous ammoniacal ſalt, which cryſtalliſes very well by evaporation and cooling. M. Bucquet ſays, that its cryſtals are rhomboidal pyramids. M. Macquer obſerved ſome of them in large priſms with four, five, or ſix ſides; others ſwelled in the middle, and terminated by very acute points. And Meſſrs the Dijon Academicians, obtained ſome of it in parallelopipeds with two alternate two ſided figures. This ſalt has a cool taſte, and is decompoſed by the fire; it effloreſces in the air; it is more ſoluble in hot than in cold water, and cryſtalliſes by cooling; lime and the fixed alkalis diſengage its volatile alkali; it is decompoſed by the mineral acids and metallic ſolutions.

Meſſrs Pott and Margraaf heated cream of tartar with the mineral acids, and the latter extracted from it neutral ſalts reſembling thoſe which each of theſe acids forms with the vegetable fixed alkali; from whence he concluded, that this alkali is ready formed in the cream of tartar. M. Rouelle junior, who has made the moſt numerous and accurate experiments upon cream of tartar, obtained the ſame reſults. By putting a pound of very finely-powdered cream of tartar upon a pound of oil of vitriol, the mixture turns hot; their reciprocal action is favoured by the heat of a balneum mariæ and agitation with a glaſs-rod; this heat is continued for ten or twelve hours; the mixture becomes thick like milk, two or three ounces of boiling diſtilled water are poured upon it, which communicate fluidity; it is left in a bal-

balneum mariæ about two hours; then it is taken from the fire, and three pints of boiling diſtilled water are added. This ſolution is coloured and opaque, contains free vitriolic acid, a portion of cream of tartar undecompoſed, and vitriolated tartar. The exceſs of acid is ſaturated by chalk; the ſelenite precipitates with a little cream of tartar; the mixture is filtrated and evaporated; it yields a little cream of tartar and ſelenite till it be reduced to eighteen or twenty ounces; then it is decanted and evaporated a-new; it yields by repoſe cryſtals of true vitriolated tartar; all of which may be thus obtained by repeated evaporations and cryſtalliſations. This ſalt is mixed always with a little cream of tartar, and it conſumes upon a red iron; but by ablution with a proper quantity of diſtilled water it is diſſolved, and the cream of tartar remains at the bottom of the veſſel. Such is the proceſs deſcribed and repeated with ſucceſs by M. Berniard after M. Rouelle, *Journ. de Phyſique*, tom. 17. page 183, 184.

The nitrous and muriatic acids, treated in the ſame way with cream of tartar, yield nitre and febrifugal ſalt; which proves without any dubiety the preſence of fixed alkali in this ſubſtance.

Cream of tartar acquires ſolubility by the union of borax and ſedative ſalt, according to the experiments of M. de Laſſone. One part of this laſt ſalt may render ſoluble even four parts of cream of tartar. This mixed ſolution when evaporated yields a gummy, greeniſh, and ſtrongly acid ſalt.

Cream of tartar ſeems capable of uniting with moſt of the metallic ſubſtances, as demonſtrated by M. Monnet and the Dijon Academicians: But as all theſe combinations have been but ſlightly examined, we ſhall confine ourſelves to thoſe of antimony, mercury, lead, and iron, becauſe theſe compounds are moſt known, and moſt employed in medicine.

The combination of antimony with cream of tartar is called *ſtibiated* or *antimoniated tartar*. As it is one of of the moſt important remedies which chemiſtry affords to medicine, its properties muſt be carefully examined. Since the time of Adrien de Mynſicht, who firſt made it known 1631, different preparations of it have been offered. The pharmacopœias and chemical works all differ, either in the antimonial ſubſtances which ſhould be employed, or in their quantity, and likewiſe in that of the water and cream of tartar; or, laſtly, in the method of making it. In M. Bergman's diſſertation upon this medicine, we have a very good ſketch of the different proceſſes offered expreſsly for its preparation. It has been ſucceſſively ordered to uſe the crocus metallorum, liver, glaſs, and flowers of antimony: ſome have directed to boil theſe with the cream of tartar and a greater or leſs quantity of water for ten or twelve hours: others require only an ebullition of half an hour: and, laſtly, ſome authors wiſh the filtrated ley evaporated to dryneſs; and others require it to be cryſtalliſed, and only the cryſtals to be employed in medicine. From theſe different methods, it muſt happen that the antimoniated tartar is not always the ſame, poſſeſſing different degrees of ſtrength; ſo that there cannot be any certainty in its effects. Thus M. Geoffroy, who examined ſeveral ſtibiated tartars of different degrees of ſtrength, found by analyſis, that the weakeſt contain in the ounce from thirty grains to a gros and eighteen grains of regulus; thoſe of a middle emetic quality a gros and a half; and the moſt powerful two gros ten grains. The glaſs of antimony has been preferred to the other antimoniated ſubſtances, becauſe it is one of the moſt ſoluble by cream of tartar: but this glaſs may be more or leſs calcined; and theſe different degrees of calcination ought to influence its emetic quality. However, by taking very tranſparent and porphyriſed glaſs of antimony, boiling it in water with an equal part of cream of tartar till this laſt be ſaturated, filtrating and evaporating this ſolution with a gentle heat, cryſtals of ſtibiated tartar are obtained by cooling and repoſe,

pose, whose degrees of strength seem to be very constant. The liquor is decanted and evaporated, and furnishes, by different successive evaporations, new crystals. The mother-water contains some sulphur, vegetable alkali, and a certain quantity of liver of sulphur. When we filtre the mixture of cream of tartar, glass of antimony and water, a kind of yellow or brown jelly remains upon the filtre, which M. Rouelle has described. This jelly distilled yields a very inflammable pyrophorus discovered by M. Proust.

M. Macquer has proposed to substitute to the glass the powder of algaroth *, which by itself is a violent emetic;

* This precipitate is commonly made by the addition of a large quantity of water, which attracts the acid, and occasions the deposition of the metallic part. The Edinburgh college, in adopting this process, employ a solution of alkaline salt, which entirely frees the precipitate from any muriatic acid. It is evident, that the basis of which tartar emetic is made in this process, is always of a similar nature, as the metals lose phlogiston before they are dissolved in acids, and constantly fall equally calcined when the precipitation is not occasioned by a body which can furnish phlogiston. Neither water nor alkali contains phlogiston; so that by the use of either precipitant the powder of algaroth is exactly the same. In the common way of combining muriatic acid with antimony, this acid must first be united with mercury, as it does not directly attack the regulus; and the composition is completed by a decomposition during distillation. But as this process is extremely tedious and laborious, a set of experiments was lately made here, to accomplish the end in an easier and more expeditious manner. It was observed, in the note upon manganese, that the marine acid undergoes great changes, and becomes vastly more active when distilled from the black calx of this metal. A trial was made how far the acid, in this state, could dissolve antimony directly, without the intervention of corrosive sublimate. The marine acid and manganese were put into the retort, and the reguline antimony, as advantageously exposed to the incondensible steems of the acid as possible, into the receiver with a small quantity of water. When the acid came over, it was found to attack the antimony, and to dissolve it completely; so that when the process was finished, there was a perfect solution of antimony in muriatic acid. After the success of this experiment, a trial was farther made of the action of this dephlogisticated acid upon crude antimony. Here it was found to attack the antimony very

emetic; becauſe this powder, precipitated from the butter of antimony by water, is always of the ſame degree of ſtrength. M. Bergman adopted the opinion of M. Macquer; and ſince that time an emetic tartar is prepared in the laboratory of the Dijon academy, according to the method of this chemiſt and of M. de Laſſone. This medicine has been employed with the greateſt ſucceſs: it operates in the doſe of three grains without fatiguing the ſtomach or inteſtines. Stibiated tartar is cryſtalliſed in three-ſided pyramids: it is very tranſparent: it is decompoſed by the fire, and becomes charry;

very powerfully, and at the ſame time the ſulphur. However, the power of dephlogiſticated marine acid vapours to decompoſe the ſulphur, does not impede the ſolution of the antimony which is firſt attacked by them. All this increaſed activity of the marine acid ariſes from the loſs of ſome phlogiſton, of which it had been robbed by the mangaueſe. The ſolution formed in this way is equally perfect with any other, and furniſhes a precipitate of equal quality. This proceſs is vaſtly ſimpler than the common one, and at the ſame time leſs expenſive. Tartar emetic, prepared from this precipitate, is at preſent preferred above every other kind; at the ſame time it may be queſtioned, how far it is practicable to make tartar emetic of uniform ſtrength, by uſing glaſs of antimony, made according to M. Bergman's directions. In the ordinary proceſs, the degree of calcination which the metal ſuffers before fuſion is not determined with ſufficient accuracy. Therefore M. Bergman propoſes to ſubſtitute diaphoretic antimony, which is a perfect calx, with the addition of as much phlogiſton as is requiſite to vitrefy the mixture into a perfect glaſs. By this method a glaſs may be made which ſhall be almoſt conſtantly the ſame. Some ſulphur neceſſarily enters into the compoſition of the glaſs; without it no degree of calcination will produce any thing but an opaque white enamel. This is known by mixing regulus of antimony and diaphoretic antimony in various proportions, and expoſing the mixture to a melting heat, when we find the ſides of the crucible covered with a whitiſh glaſs, which does not bear the ſmalleſt reſemblance to true glaſs of antimony. Whether the powder of algaroth or glaſs of antimony be employed, we digeſt it in a ſufficient quantity of water along with the cream of tartar till the whole is diſſolved, and afterwards we evaporate the ſolution. There is ſome attention requiſite in collecting the cryſtals, as ſome of them ſhoot which contain no metal. After they are formed, they undergo a little change of colour from the diſſipation of ſome phlogiſton.

charry; it is efflorefcent, and becomes of a dirty and farinaceous white: it diffolves in fixty parts of cold water, and in much lefs of boiling water: it cryftallifes by cooling: the alkalis and lime decompofe it; calcareous earth and pure water in a great dofe are capable of decompofing it: whence it follows, that it fhould be adminiftered only in diftilled water. Liver of fulphur precipitates it in a red powder or kind of fulphur auratum, and may be ufed as a teft to difcover this falt in any water. Iron feizes upon its tartarous acid, and feparates the calx of antimony: on this account emetic tartar fhould not be made in iron veffels. M. Durande, a celebrated phyfician of Dijon, propofed making this medicine publicly, and by a uniform procefs, as is ufually done to the theriaca. We are of opinion, that it would be very ufeful only in procuring a uniform tartar emetic, upon whofe effects the phyfician might always depend.

The tartarous acid may be combined with mercury in two ways. One which M. Monnet mentioned, confifts in diffolving in boiling water fix parts of cream of tartar with one part of mercury, precipitated from the nitrous acid by fixed alkali. This liquor, filtrated and evaporated, yielded him cryftals decompofable by pure water. The fecond means of uniting mercury with tartarous acid, is to pour the nitrous folution of this metal into one of vegetable falt, or falt de Seignette: a precipitate is obtained, formed by the union of the tartarous acid and mercury, and ordinary or rhomboidal nitre remains in folution.

Cream of tartar acts in a fenfible manner upon the calces of lead. M. Rouelle junior fays, that the faturnine falt which is formed in this operation does not remain in folution in the liquor; and that this laft, when evaporated, furnifhes only fome pure vegetable falt, which was ready formed in the cream of tartar. This is one of the proceffes which was ufed to demonftrate the prefence of fixed alkali in the tartar.

Copper and its calces are very eaſily attacked by the tartarous acid; a ſalt of a beautiful green reſults ſuſceptible of cryſtalliſation, but as yet not fully examined.

Iron is one of the metals upon which cream of tartar acts moſt efficaciouſly. A medicine is prepared, called *tartarus chalybeatus*, by boiling in twelve pounds of water four ounces of porphyriſed iron-filings and a pound of white tartar. When the tartar is diſſolved, the liquor is filtrated, it depoſites cryſtals: more are obtained by the evaporation of the mother-water. The tinctura martis tartariſata is prepared by making a paſte with ſix ounces of iron-filings, a pound of white tartar in powder, and ſufficiency of water: the mixture is left in repoſe for twenty-four hours; then it is diluted with twelve pounds of water, and the whole boiled for two hours, taking care to add water to replace the evaporated part: the liquor is decanted, filtrated, and inſpiſſated to the conſiſtence of ſyrup, and an ounce of ſpirit of wine added to it. M. Rouelle ſays, that the vegetable fixed alkali is free in this tincture; and that in treating it by the acids neutral ſalts are obtained, which ſhow the preſence of this alkali. There are ſtill two medicines formed by the combination of the tartarous acid and iron; the one is the martial ſoluble tartar, which is a mixture of one pound of tartariſed tincture of mars and four ounces of vegetable ſalt evaporated to dryneſs; the other is known by the name of *martial balls*. They are made by putting one part of ſteel-filings and two parts of powdered white tartar into a glaſs veſſel, with a certain quantity of aquavitæ. When this liquor is evaporated, the maſs is pulveriſed, and ſome more aquavitæ added, which is again allowed to evaporate like the firſt. This proceſs is repeated till the mixture be thick and tenacious; then it is formed into balls.

Crude tartar is very uſeful in dyeing; it is uſed alſo by the hatters.

The

The different preparations of cream of tartar juſt now enumerated, are employed moſtly in medicine. Pure cream of tartar is conſidered as a refrigerant and antiſeptic: in the doſe of half an ounce or an ounce, it is a gentle, not nauſeating, laxative. The vegetable ſalt and that of Seignette are in frequent uſe, as aſſiſtant purges, in the doſe of a few gros. Stibiated tarter is one of the moſt uſeful and powerful remedies for which medicine is indebted to chemiſtry. This ſalt is emetic, purgative, diuretic, diaphoretic, according to the doſes and proceſſes employed for its adminiſtration. Many a time it produces all theſe effects at once. It ſhould be regarded alſo as a powerful alterative, and a proper remedy in overcoming conſtipations and obſtructions of the viſcera, when it is given in very ſmall quantity frequently repeated. It is adminiſtered in the doſe of a grain to four grains, diſſolved in a proper vehicle, as an emetic. It is mixed in the doſe of a grain with other purgatives, which it aſſiſts; and in that of half a grain, largely diluted, as a good alterative. M. de Laſſone found, that tartar emetic is rendered very ſoluble in water by the addition of ſal ammoniac, and that a mixt ſalt reſults reſembling ſal alembroth.

This new compound ſhould be reckoned capable of producing very powerful effects upon the animal œconomy. Tartarus chalybeatus, martial ſoluble tartar, tartariſed tincture of Mars, are employed as tonic and aperient.

LECTURE LVI.

Of the Acid Fermentation *and of* Vinegar.

MANY vegetable ſubſtances are ſuſceptible of undergoing the acid fermentation: ſuch are, the gums, the amylaceous feculæ, diſſolved in boiling water: but this property is very remarkable, in particular, in the fermented and ſpirituous liquors. All theſe liquors

liquors expoſed to heat, and in contact with the air, paſs to the acid fermentation, and yield the fluid called *vinegar*. It is eſpecially the wine of grapes which is employed for its preparation, although very good vinegar may be made from cyder, perry, &c.

There are three conditions requiſite in the acetous fermentation: 1. A heat of twenty or twenty five degrees of Reaumur's thermometer: 2. A viſcous and at the ſame time acid body, ſuch as a mucilage and tartar: 3. The contact of air. The change of wines which paſs to the ſtate of vinegar, is to be aſcribed only to the inteſtine motion excited in theſe fluids by the preſence of a certain quantity of a mucous body not altered, and capable of undergoing a new fermentation. The preſence of an acid matter, ſuch as tartar, is neceſſary to determine the acid fermentation. Laſtly, the contact of the air is indeſpenſable; and it appears that there is a portion of it abſorbed during this fermentation, as M. l'Abbé Rozier has proved.

All wines are equally proper to form vinegar. The worſt of them are preferred, becauſe they are cheaper; but the experiments of Beccher and Cartheuſer demonſtrate, that the generous wines, and thoſe abounding in ardent ſpirit, yield in general the beſt vinegars.

Boerhaave, in his Elements of Chemiſtry, has deſcribed a very good proceſs for the formation of vinegar. Two barrels are taken, and a hurdle of twigs placed at ſome diſtance from their bottom, upon which are extended vine branches and ſtalks; ſome wine is poured upon them, ſo that one of the barrels is full and the other half empty. The fermentation commences in this laſt: when it is ſet agoing, this barrel is filled with the wine contained in the firſt. By this means the fermentation ſtops ſhort in the full veſſel, and is well forwarded in that which is half empty. When it has advanced to a conſiderable degree, this laſt barrel is filled with the liquor of that which fermented the first time; so that the fermentation recommences in the first and stops short in the second. Thus the two barrels are alternately filled and emptied till the vinegar be entirely formed; which happens generally in twelve or fifteen days.

Upon observing what passes in this fermentation, much bubbling and hissing is perceived; the liquor is heated and troubled; it exhibits a great number of filaments, which run through it from all sides; it emits a very acid smell, not at all dangerous; it absorbs a great quantity of air: the fermentation must be stopped twelve of the twenty-four hours. These phenomena gradually abate, the heat diminishes, the motion stops, the liquor becomes clear; it lets fall a sediment in reddish slimy flocci, which stick to the sides of the barrels. Numerous experiments have taught us, that the smaller the quantity of wine the fuller is the contact of air, and the more quickly it passes to the state of vinegar. Care is taken to draw off the vinegar clear when it is made, in order to separate it from above its lee; which without this precaution would make it pass to the putrid fermentation. Vinegar does not deposite tartar as wine does. This salt is dissolved and combined with the ardent spirit and water during the fermentation; it is even probable that it is the presence of this salt which contributes to the taste and the other acid properties of vinegar. This fluid has more or less colour, according to the wine employed in its preparation: but in general, the least coloured vinegars are much more so than the white ones; because they hold dissolved the colouring matter of the tartar, which has been also evolved in the fermentation *.

Vine-

* When the acetous fermentation is completely finished, and all the wine converted into vinegar, we are no longer able to discover any vestige of the ardent spirit which the wine contained, or to extract a single drop of it from old vinegar by distillation; the inflammability, along with all the other properties of the spirit of wine,

Vinegar prepared as juft now defcribed, is very fluid, of an acid and fpirituous odour, of a more or lefs ftrong bitter tafte: it reddens the blue colours of vegetables. Expofed to a gentle heat in veffels flightly fhut, it is altered, lofes its fpirituous part, depofites a great quantity of mucous tufts and filaments, and affumes a putrid fmell and tafte. It appears to be a compound of water, ardent fpirit, a particular acid, a little tartar, and a colouring extractive matter. Thefe matters

wine, being totally loft or deftroyed by the progrefs of the fermentation, and never to be recovered again by the employment of any procefs with which we are acquainted: But the vinegar cannot be reftored to the ftate of wine; it has a natural tendency to proceed towards the putrefactive fermentation, and to lofe all its diftinguifhing qualities, and to be no longer acid. This difpofition to change proves frequently troublefome to vinegar-brewers. The progrefs of the fermentation may however be ftopped by quenching a red hot iron in the liquor; and the acidity may be again imparted by the addition of fome fpirit of wine. So that the ftrength of the vinegar feems in fome meafure to depend upon the quantity of ardent fpirit which the wine contains. In confirmation of this, M. Cartheufer obferves, that vinegar becomes ftronger when fome fpirit is put into the fermenting mixture: and Becher remarks, that by corking up a bottle clofe, and ufing a greater degree of heat, he produced a ftronger and better vinegar than what is to be made in the common way. But the ftrongeft of all vinegars is procured by decompofing fome neutral falt by means of heat, as Mr Fourcroy has explained. There is, however, another method of concentration practifed in convenient fituations. This method is to expofe the acid to a cold mixture, when the aqueous part congeals, and leaves the acid ftill fluid. By expofing this concentrated acid to a ftill greater degree of cold, fome more of the water is feparated; and by repeating this mode of procedure fufficiently often, the vinegar may at laft be brought to about feven times its original ftrength, according to the experiments of M. Geoffroy. Vinegar in this ftate is perhaps in fome refpects nearly as ftrong as fome of the weaker mineral acids, efpecially the marine. But however much concentrated, it differs from all of them in the extreme degree of volatility; for in place of attracting humidity from the air, it flies off by expofure, and is entirely diffipated: and when united to a metallic, earthy, or faline bafis, the falt may be decompofed, and the vinegar expelled, by means of heat alone; which is not the cafe with neutral falts formed by the mineral acids.

may be separated by the action of the fire, as wine is done.

In distilling vinegar in a naked fire in a cucurbit of free-stone covered with a head, or in a glass retort placed upon a sand bath, at first a phlegm of a brisk agreeable odour passes over, but very slightly acid; to it very soon a very white acid liquor succeeds, having a fragrant smell. This is distilled vinegar; what distils after that has less smell and more acidity: it becomes the more acid as the distillation advances. All these products may be separated, and distilled vinegars obtained, differing from one another in acidity and smell. About two thirds of liquor, which constitutes the purest vinegar, are sufficient to be drawn over in this process. The portion which comes over after that, is more acid, but has an empyreumatic smell, which may be dissipated by exposition to the air: it assumes also a small colour. This operation shows, that the acetous acid is heavier than water; the remaining vinegar is thick, of a deep and dirty red colour: it deposites a certain quantity of tartar; it is considerably acid. If it is evaporated in an open fire, it assumes the form of an extract: and if, when it is dry, it is distilled in a retort, it affords a reddish acid phlegm; an oil, at first light and coloured, and after that weighty; and a little volatile alkali: the coal which is left contains much fixed alkali.

Vinegar may be concentrated by exposition to a congealing heat. It is then observed to be very acid, and copious in colour, but very easily altered.

The acid of vinegar, separated from tartar and from its colouring part by distillation, is susceptible of uniting with a great number of bodies.

It combines but imperfectly with argillaceous earth, and forms with it small needled crystals.

It unites easily with magnesia, and yields a salt very soluble in water, uncrystallisable, but which furnishes by evaporation a viscous deliquescent mass. This salt

salt is decomposed by heat and the mineral acids: it is very soluble in spirit of wine.

The acid of vinegar combines with lime, and it decomposes chalk, dissipating its acid. This salt which it forms with lime, is susceptible of crystallising into very fine needled prisms like satin. Calcareous acetous salt is bitter, sour, efflorescent, and decomposable by fire and the alkalis.

The combination of vinegar with the vegetable alkali is called *terra foliata tartari*. It is made by pouring upon white salt of tartar very pure distilled vinegar; the mixture is agitated, and vinegar added, to the point of saturation and solution of the salt; even an excess of acid should be added: the liquor is filtrated, and evaporated with a very gentle fire in a porcelain or pure silver vessel; when it becomes thick, the evaporation is continued in a sand-bath till the salt be very dry. By this means a very white terra foliata is obtained. If it is too much heated, it acquires a grey or brown colour, because a portion of the vinegar is burnt. Some chemists say, that this salt may be obtained in a regular form, by leaving the solution to cool when evaporated to a thick pellicle. The terra foliata tartari has a pungent, acid, and urinous taste. It is decomposable by the action of the fire, and yields in a retort an acid phlegm, an empyreumatic oil, some volatile alkali, and a great quantity of a fragrant gas, formed of aërial acid, and inflammable gas. The remaining coal contains much uncombined fixed alkali. This salt strongly attracts the air's humidity; it is very soluble in water. Vitriolic acid decomposes it. This decomposition is effected by pouring upon two parts of terra foliata, introduced into a tubulated glass retort, to which a recipient is adapted, one part of oil of vitriol: a vapoury fluid is instantly disengaged with a brisk effervescence, of a penetrating smell, which condenses in the receiver in form of a liquor called *radical vinegar*. This vinegar is very concentrated, of a very strong acidity; but it is not pure, and is always mixed with a certain quantity of sulphureous acid, distinguishable by its smell. Cream of tartar likewise decomposes it, and forms with its base vegetable salt.

Vinegar unites perfectly with marine alkali, and forms with it a salt called *acetous mineral salt.* This salt differs from terra foliata only in being more susceptible of crystallising in striated prisms, very similar to Glauber's salt, and in being not deliquescent.

In order to obtain it well crystallised, its solution must be evaporated till a pellicle appear on the surface, and then set in a cool place. The acetous mineral salt is decomposable by the fire and the mineral acids, like the terra foliata. To these details we shall add, that when a strong fire is added in distilling the acetous calcareous and alkaline salts, the residua of these salts are so many pyrophori, and burn when exposed to the air. M. Proust, to whom these discoveries are owing, is of opinion, that to produce a pyrophorus, only a charry residuum is required, divided by an earth or a metallic calx.

Acid of vinegar forms with volatile alkali a salt called *spirit of Mindererus.* This salt cannot be evaporated without losing the greatest part of it on account of its volatility; however, by a long evaporation, needled crystals are obtained, whose taste is hot and pungent, and which attract very readily the air's humidity. This acetous ammoniacal salt is decomposed by the action of heat, by lime and the alkalis, which disengage its volatile alkali; and by the mineral acids, which dislodge its acid.

Vinegar acts upon almost all the metallic substances; and in these combinations presents very important phenomena.

It does not seem to dissolve immediately calx of arsenic: but this substance, distilled with equal parts of terra foliata, yielded to M. Cadet and Messieurs the Dijon Academicians a red fuming liquor, of a very stink-

ſtinking ſmell, very tenacious, and of a very ſingular nature. M. Cadet has already obſerved, *Acad. des Sciences, Savans Etrangers, tom.* 2. *p.* 633. that this liquor was capable of kindling fat lute. Meſſieurs the Dijon Academicians, deſirous to examine a yellowiſh matter of an oily conſiſtence collected at the bottom of the veſſel, which contained the arſenico-acetous fuming liquor, decanted a portion of this ſupernatant liquor, and poured the reſt upon filtrating paper. Scarce had a few drops paſſed through when a very diſagreeable ſmoke inſtantly aroſe, which formed a column from the veſſel to the ceiling; a kind of boiling was excited at on the ſides of the matter, and a beautiful roſy flame iſſued out, which continued for a few inſtants.

In the third volume of the Dijon Elements of Chemiſtry, p. 41—47. the detail may be ſeen of the beautiful experiments which theſe learned academicians made upon the ſubject. They compare this liquor to a liquid phoſphorus: we think it is a kind of pyrophorus, like theſe, which we ſhall deſcribe more fully afterwards. The reſiduum of the terra foliata with the calx of arſenic, is in great meaſure vegetable fixed alkali.

Vinegar diſſolves the calx of cobalt, and forms a pale roſe-coloured ſolution.

It has no action upon biſmuth or its calx.

It directly diſſolves nickel, according to M. Avidſon. This ſolution yields green cryſtals formed in ſlices.

This acid does not act upon regulus of antimony; but it ſeems to diſſolve the glaſs of this ſemimetal, as Angelus Sala prepared an emetic ſalt with theſe two ſubſtances.

Zinc, and likewiſe its calx, diſſolves very well in diſtilled vinegar. M. Monnet obtained from this ſolution, when evaporated, flat laminated cryſtals. The acetous ſalt of zinc fulminates upon coals, and emits a ſmall blueiſh flame. To diſtillation it yields an inflammable liquor, a yellow oily fluid, which becomes very ſoon of a deep green, and a white sublimate, which burns in the light of a candle with a beautiful blue flame. The residuum is in the state of a pyrophorus which is not very combustible.

Acid of vinegar does not dissolve mercury in its metallic state. However, this combination is effected by minutely dividing the metal by means of mills, as Keyser did. Calx of mercury easily unites with vinegar. This acid is boiled upon precipitate *per se*, upon turbith, or upon the mercury precipitated from the nitrous solution by the fixed alkali. The liquor becomes white, and when it boils clear it is filtrated; it precipitates by cooling spangled argentine crystals, resembling sedative salt. This salt has been called *terra foliata mercurialis*. It is at once prepared by pouring a nitrous solution of mercury into one of terra foliata tartari: the nitrous acid unites with the fixed alkali of this last, with which it forms nitre, which remains in solution in the liquor; and the calx of mercury combined with the vinegar is precipitated in the form of brilliant spangles: the mixture is filtrated, the terra foliata mercurialis remains upon the filtre. This salt is decomposed by the action of the fire: its residuum yields a kind of pyrophorus. It is easily altered by combustible vapours.

Tin is but very little altered by vinegar. This acid dissolves but a very small quantity of it; and the solution evaporated yielded M. Monnet a yellowish pellicle, resembling gum, and of a fetid odour.

Lead is one of the metals upon which it has the most action. This acid dissolves it with the greatest facility. This metal in laminæ, exposed to the action of hot vinegar, is covered with a white powder called *cerufs*, which is only a calx of lead This calx bruised with a third of chalk, forms the white-lead of the painters. In order to saturate vinegar with as much lead as it can dissolve, this acid is poured upon the cerufs in a matrass; the mixture is set to digest in a sand-bath; the liquor is filtrated after several hours digestion; it is evaporated

porated to a pellicle; it furnishes by cooling and repose white crystals, forming either shapeless needles, if the liquor has been too much evaporated, or flattened parallelopipeds terminated by two surfaces, disposed like the top of a house, when the operation has been properly conducted. They are called *salt* or *sugar of lead*, on account of its saccharine taste; this taste is at the same time styptic. A similar salt is prepared with litharge and vinegar: equal parts of these two substances are boiled together till they be saturated; the liquor is evaporated to the consistence of a clear syrup: in this state we have Goulard's extract of lead, known long before him by the name of *vinegar of lead*. Saccharum saturni is decomposed by heat; it furnishes an acid, reddish, very fetid liquor, very different from radical vinegar. The residuum is a very fine pyrophorus. This salt is decomposed by distilled water, lime, alkalis, and mineral acids. The extract of lead, diluted with water and mixed with a little aquavitæ, forms Goulard's vegeto-mineral water.

Vinegar dissolves iron with activity. The effervescence which happens in this solution is owing to the disengagement of an inflammable gas, whose properties have not been examined. The liquor assumes a red or brown colour. By evaporation it yields only a gelatinous magma, mixed with a few brown elongated crystals. This martial acetous salt has a styptic sweetish taste; it is decomposed by the fire, and lets go its acid; it attracts the air's humidity; it is decomposed in distilled water. If it is heated till it emit no more smell of vinegar; it leaves a yellowish calx attractible by the magnet.

The acetous solution of iron yields a very black ink with the gall-nut; and it might be employed with success in the tincture. The Prussian alkali precipitates from it a very pretty Prussian blue. Martial æthiops, the precipitates of iron, the croci of Martis, the sparry ore of iron, yield with vinegar solutions of a very beautiful red.

Copper diſſolves with great facility in diſtilled vinegar. This ſolution, aided by heat, gradually aſſumes a green colour; but it is more eaſily effected with this metal already altered and calcined by vinegar. Copper thus prepared is verdegris. It is prepared about Montpelier, by putting laminæ of this metal into earthen veſſels with grape ſtalks, which have been firſt moiſtened and fermented. The ſurface of theſe laminæ is very ſoon covered with a green ruſt, which is alſo augmented by heaping them together; then the copper is ſcraped, and the verdigris ſhut up in leather ſacks and ſent to market. M Monnet, an apothecary at Montpelier, has fully deſcribed this manipulation in two papers printed among thoſe of the Academy of Sciences 1750-3. Verdegris diſſolves with readineſs in vinegar. This ſolution, which is of a beautiful green colour, furniſhes by evaporation and cooling green cryſtals in quadrangular truncated pyramids, which he calls *verdigis*, or *cryſtals of Venus*.

Thoſe which are prepared in commerce, and called *diſtilled verdigris*, becauſe they are prepared with diſtilled vinegar, are in the form of a beautiful pyramid: its cryſtals preſent this arrangement, becauſe they are depoſited upon a rod cleft into four, whoſe branches have been ſeparated by a piece of cork.

Verdigris has a very ſtrong taſte, and is a violent poiſon. It is decompoſed by the action of the fire; it is effloreſcent, and is covered with a powder whoſe green colour is much more pale than that which diſtinguiſhes this ſalt when not altered: it diſſolves completely in water without being decompoſed. Lime-water and the alkalis precipitate this ſolution.

When this ſalt, in powder, is diſtilled in a glaſs or earthen retort with a recipient, a fluid, at firſt white and a little acid, is obtained, but which very ſoon acquires a conſiderable acidity, and ſuch as equals the concentration of the mineral acids. The retort is changed to have ſeparately the phlegm and the acid. This is called *radical*

dical vinegar or *vinegar of Venus.* This acid is coloured green by a certain quantity of calx of copper, which carries over in the diſtillation. When no more paſſes and the retort is red, the reſiduum which it contains is in the form of a brown powder of the colour of copper, which frequently communicates the brilliancy of this metal to the ſides of the retort. This reſiduum is ſtrongly pyrophoric, as Meſſrs le Duc d'Ayen and Prouſt have obſerved. Vinegar of Venus is rectified by diſtilling it with a gentle heat; then it is perfectly white, provided the fire is not urged too far towards the end of the operation, and the portion of calx of copper remaining in the retort not too much dried.

Radical vinegar, thus rectified, has a ſmell ſo pungent and penetrating, that it is impoſſible to endure it any time: its cauſticity is ſo great, that when applied to the ſkin it reddens and corrodes it. This fluid is conſidered as the moſt pure acid of vinegar, the moſt concentrated and free of the phlegm which covered its properties. This acid is extremely volatile and inflammable; heated with the contact of air it kindles and burns ſo much the more rapidly the higher the ſtate of rectification. This experiment led chemiſts to believe that vinegar is an acid combined with ardent ſpirit; perhaps it might even be conſidered as a kind of natural ether. This idea agrees with the penetrating and agreeable odour which the firſt diſtilled portions of this acid emit. Radical vinegar evaporates entirely in the air; it unites with water, and produces great heat; it forms with the earths, alkalis, and metals, the ſame ſalts with ordinary vinegar; but it acts upon combuſtible bodies, generally much more remarkably than this laſt. M. le Marquis de Courtanvaux, *Acad Savans Etrangers*, vol. 5. p. 72. that it was only the laſt portion of acetous fluid, obtained in the diſtillation of verdigris, which was inflammable; and that it poſſeſſed alſo the property of congealing by cold. This laſt portion, rectified, cryſtalliſed in the recipient in large laminæ and needles; and returns to its fluid ſtate only

only when exposed to a heat of thirteen or fourteen degrees above the freezing point. This liquor is a kind of glacial vinegar.

The acid of vinegar, aided by heat, dissolves gold precipitated by the fixed alkali. This solution precipitated by the volatile alkali yields fulminating gold, as demonstrated by M. Bergman. The same thing happens to platina and gold: they are both dissolved. Radical vinegar has no action upon these metals in a metallic state, but dissolves them in the state of a calx.

Vinegar is susceptible of combining with several of the immediate principles of vegetables; it dissolves the extracts, mucilages, and essential salts; it unites with the spiritus rector; it has been considered as the proper solvent of gum-resins; it has through length of time or by distillation also a very remarkable action upon the fat oils, which it converts into a kind of soapy state: the combination of vinegar with vegetable substances has not yet been examined. This acid is made use of to extract some of the principles, and particularly the odour of these bodies; and vinegars of different natures simple or compound, are prepared in medicine. Acetum scilliticum, colchicum, afford an example of the first kind; theriacal vinegar belongs to the second. These medicines are prepared by continued maceration and digestion for several days. As this acid is volatile, it is distilled upon aromatic plants, whose odorous principle it attracts; such is the distilled vinegar of lavender, which is employed for the toilette. These liquors are in general less agreeable than the spirituous distilled waters.

Radical vinegar decomposes spirit of wine, and forms ether with as much facility as the mineral acids, as discovered by M. le Count de Lauraguais. For this purpose, radical vinegar is poured into a retort upon an equal part of spirit of wine; a considerable heat is produced; the retort is put in a hot sand-bath, and two recipients are adapted, the last of which is plunged into

cold

cold water or pounded ice; the mixture is made to boil quickly. At firſt a dephlegmated ſpirit of wine paſſes over, then the ether, and at laſt an acid, which turns ſtronger as the diſtillation advances; a brown maſs reſembling a reſin remains in the retort. Care is taken to change the recipient as ſoon as the etherial ſmell turns acrid and pungent, and the acid is collected apart. The ether is rectified with a gentle heat with cauſtic fixed alkali, and a great deal of it is loſt in this operation.

M. de la Planche an apothecary, prepared this ether by pouring upon ſaccharum ſaturni, introduced into a retort, oil of vitriol and ſpirit of wine. The theory and practice of this operation correſpond with thoſe made uſe of in preparing the marine, nitrous, and vitriolic ethers.

Acetous ether has an agreeable ſmell like the other ethers; but it is always mixed with that of vinegar, although it is not acid. It is very volatile and inflammable; it burns with a vivid flame, and leaves a charry mark after its combuſtion.

Vinegar is employed as a ſeaſoning to meat. Much uſe is made of it in medicine; it is refrigerant and antiſeptic; with ſugar a ſyrup is made of it, which is given with great ſucceſs in ardent and putrid fevers; applied externally it is aſtringent and reſolvent. All its combinations are uſed likewiſe as very good medicines.

Terra foliata tartari and acetous marine ſalt are powerful deobſtruents and aperients: they are adminiſtred in the doſe of half a gros, and even a gros.

Spirit of mindererus, given in a few drops in convenient vehicles, is aperient, diuretic, cordial, antiſeptic. It does ſervice frequently in leucophlegmaſias or ſwellings of the external parts of the body.

The terra foliata mercurialis is a very good antivenereal: it made the baſe of Keyſer's pills.

The extract of lead, acetum ſaturni, aqua-vegeto-mineralis, are employed externally as diſcutients: theſe me-

medicines being ſtrong repellents ought to be be adminiſtred with great prudence; particularly when applied to the ſkin upon uncovered ulcerated parts. Boerhaave met with ſeveral girls ſeized with phthiſis after the external uſe of the preparations of lead.

Ceruſs enters the compoſition of drying ointments and plaſters, and verdigris compoſes ſeveral collyria and ointments.

Radical vinegar is employed as a very powerful ſtimulent like the volatile alkali. It does ſervice in caſes of debility, when it is reſpired. To have it in a commodious form, a certain quantity of it is poured upon groſsly powdered vitriolated tartar, which is put into a veſſel well ſhut. This medicine is every where known by the name of *ſal aceti*.

Acetous ether has not yet been much uſed; and it is uncertain if it has any other virtues which the other ethereal liquors do not poſſeſs.

Of the Putrid Fermentation of Vegetables.

All vegetable bodies which have undergone the ſpirituous and acid fermentations, are ſtill ſuſceptible of a new inteſtine motion, which alters their nature. This motion is called the *putrid fermentation*. Stahl and ſeveral other chemiſts have ſuppoſed that this ſpecies of fermentation is only a continuation of the two former; or rather, that theſe three phenomena depend only upon a ſingle indvidual motion, which tends to deſtroy the texture of the ſolids and to change the nature of the fluids: and indeed it is obſerved, that if certain vegetable ſubſtances are left to themſelves, they ſuffer the three fermentations ſucceſſively and without interruption. For inſtance, all the ſaccharine matters, diluted with a large quantity of water, and expoſed to a degree of heat from twelve to twenty degrees, yield, firſt wine, then vinegar, and at length their acid character is very ſoon loſt; they are altered, putrify, loſe all their volatile principles, and finiſh with leaving only a dry, inſipid, earthy

earthy ſubſtance. However, it muſt be obſerved, tha a great number of vegetable ſubſtances do not undergo, at leaſt in a ſenſible manner, theſe three kinds of fermentations in the order mentioned The taſteleſs mucilages, the gums diſſolved in water, paſs to the acid ſtate without becoming manifeſtly ſpirituous: the glutinous matter ſeems to paſs directly to putrefaction without having been aceſcent. It appears, therefore, that though, in ſeveral principles of vegetables, theſe three fermentations follow and ſucceed each other, there is, however, a great number of others which are ſuſceptible of undergoing the two laſt and not the firſt, or even the putrefactive ſtate, without having manifeſted any ſigns of acidity.

The inteſtine motion, which changes the nature of vegetable ſubſtances, and reduces them into their elements, requires particular conditions, which it is neceſſary to know. Humidity, or the preſence of water, is one of the moſt neceſſary: the dry and ſolid vegetables, ſuch as wood, are not at all altered as long as they retain this ſtate; but if they are moiſtened, and their fibres divided, then the inteſtine motion very ſoon begins. Water, then, appears to be one of the cauſes of putrefaction. Heat is not leſs neceſſary. Cold, or the freezing point, not only oppoſes this ſpontaneous deſtruction, but it even retards its progreſs, and forces theſe ſubſtances which have been influenced by it to take as it were a retrograde motion. The degree of heat neceſſary to putrefaction is much leſs than that which the acid and ſpirituous fermentations ſupport, as this phenomenon begins at the temperature of five degrees; but a more conſiderable heat is better, one at leaſt ſo weak as not to diſſipate the moiſture and entirely dry the ſubſtance which is putrefying. The acceſs of air is alſo a condition which ſingularly favours putrefaction, as vegetable ſubſtances are kept very well *in vacuo*. However, this preſervation is limited; and the contact of air does not ſeem to be ſo indiſpenſable to the putrid fermentation as the two conditions mentioned before.

The putrefaction of vegetables has its particular phenomena. Vegetable fluids which are putrefying, are troubled, lose their colour, deposite different sediments; air-bubbles rise to their surface, and mouldy tufts form in its beginning. Vegetable substances simply moistened, and those which are soft, exhibit the same phenomena. The motion which is here exerted is never so considerable as that by the spirituous and acid fermentations. The volume of the substance which is putrefying does not appear to be augmented, nor its heat increased; but the most important phenomenon is the change of smell and the volatilisation of an acrid, pungent, urinous principle, resembling the volatile alkali, and which in fact is it: hence putrefaction has been called the *alkaline fermentation*, and the volatile alkali has been considered as its product. The pungent odour gradually exhales, a nauseous insipid odour succeeds, which it is difficult to describe. In this state the decomposition is at its height; the putrefied vegetable is very soft, like thick milk; it suffers a great number of successive modifications in the odorous principle, which it exhales: in the last place, it dries; its disagreeable odour gradually dissipates, and it leaves only a blackish, as if charry, residuum, in which only a few saline and earthy substances can be found.

Such is the order of the phenomena which are observed in the spontaneous decomposition of vegetables which are putrefying; but this decomposition, pushed to the point of reducing them to their earthy or saline state, is very tedious in its performance; and it should be also added, that it has not been properly observed by any person. Naturalists and chemists have been reproached for their neglect with respect to animal-substances; but they are more deservedly blamed for their neglect of vegetables. No intelligent person has yet undertaken to observe the complete putrefactio of these, although many

many have defcribed the phenomena which happen in that of animal-matters. Upon this account we think, that the hiftory of the fpontaneous and natural analyfis of vegetables fhould now be concluded ; adding only, 1. That the little which has been faid is fufficient to fhow, that vegetable putrefaction attenuates, volatilifes, and deftroys all the humours of thefe bodies, and reduces them to the earthy ftate. 2. That nothing yet is pofitively known of the phenomena and limits of this kind of putrefaction, properly to warrant its diftinction from that of animal-matters. 3. As this fermentation is much more remarkable, and has been more obferved in the humours and folids of animals, the details, which we fhall add at greater length in the examination of thefe fubftances, fhall complete the fketch which we have juft now given, and fhall terminate the hiftory of the facts known upon putrefaction, which interefts efpecially young phyficians, for whom this part of our work is particularly defigned.

ANIMAL KINGDOM.

LECTURE LVII.

Natural History of ANIMALS *.

ANIMALS are in general diſtinguiſhed from vegetables by loco-mobility and more perfect organiſation. However, there are ſome entire claſſes of theſe ſubſtances that are fixed to one place like vegetables; as the lithophytes, the zoophytes, known by the name of *polypi*, which are born and die upon the ſame ſoil: And on the other hand, ſome vegetables exert as much motion in their leaves and flowers as certain animals; for inſtance, the cruſtaceous worms. Organiſation appears even leſs perfect in the polypi than in the moſt part of plants. It hence follows, that it is very difficult to eſtabliſh an exact line of ſeparation between theſe two kingdoms; and naturaliſts have been neceſſarily obliged to confound them into one, known by the name of *Organic Kingdom*.

How-

* We propoſe to give here only a ſummary of the methods of the naturaliſts, to facilitate to beginners the ſtudy of natural hiſtory and the underſtanding of good authors. As to the general conſiderations about the nature of animals, the order which we have adopted does not permit us to enter into details upon this ſubject, eſpecially when treated in ſo excellent and philoſophical a manner by M. le Comte de Buffon and M. Bonnet.

However, upon confidering only perfect animals, great differences are found between thefe fubftances and vegetables. Numerous and very diftinct organs, a more complicated ftructure, more numerous and extenfive functions, are the characters to which thefe differences fhould be defcribed; notwithftanding, it is not the lefs difficult to give a good definition of thefe fubftances.

Confining ourfelves to the moft general characters, animals may be defined to be bodies endued with underftanding and motion, faculties neceffary for the prefervation of their life. All can be reproduced; fome, by the union of the two fexes, bring forth living young ones; others lay eggs, which require only heat to bring forth the young; there are fome which are multiplied without the affiftance of their mates; laftly, feveral are reproduced when they have been cut, as is done with the roots of plants.

It is very difficult for naturalifts to affign to animals the true character of their kind. The mixture of races produces varieties without number: tranfportation into different climates alfo occafions various changes in form, in colours, &c. Thofe only whofe form is conftant, and which are perpetuated by the reproduction of the individuals, fhould therefore be accounted as diftinct kinds. With regard to the alterations produced by the growth of the fpecies, the climate, and cohabitation, they ought to conftitute only varieties.

The number of animals which cover the furface of our globe being very confiderable, man never would have arrived at diftinguifhing and recognifing each of them, had not nature prefented, in the diverfified form of the external parts of their bodies, remarkable differences, by means of which it was eafy for him to eftablifh the diftinctions. Naturalifts have at all times felt the utility of thefe differences, and have made ufe of them with advantage in dividing animals into more or lefs numerous claffes, and forming what are called *methods*. Although it be demonſtrated, that theſe kinds of claſſifications do not exiſt in nature, and that all the individuals which nature produces form an uninterrupted chain without diviſion, it cannot, nevertheleſs, be denied, that they aid the memory, and are very proper to guide our ſtudy of natural hiſtory. The methods, then, ought to be conſidered as inſtruments appropriated to our weakneſs, which may be employed with vaſt ſucceſs in taking a ſuperficial view of the vaſt field of the riches of nature. Ariſtotle eſtabliſhed only general and ſimple diviſions; but his excellent reflections upon the external and internal organs of animals have formed a baſe upon which in a great meaſure have been founded the diviſions of the firſt methodical naturaliſts, ſuch as Geſner, Aldrovand, Johnſton, Charleton, Ray, &c. A great number of other naturaliſts ſucceeded theſe, who have completed the methods, and who have added to the knowledge acquired in this department of ſcience; but among theſe laſt, thoſe whoſe works it is neceſſary to be well acquainted with, and from whom we ſhall borrow what we are to ſay here, are Meſſrs Klein, Linnæus, Briſſon, Geoffroy, &c.

After man, whoſe organiſation and underſtanding require that he ſhould be ſet at the head of animal bodies, and which makes him alone one claſs apart, all the other animals may be divided into eight claſſes; which are, the quadrupeds, the cetaceous animals, birds, amphibious animals, fiſhes, inſects, worms, and polypi. Perhaps it might be poſſible to multiply theſe claſſes: but then, by augmenting the diviſions, the difficulties would be multiplied; and this is what muſt be avoided in the artificial method, whoſe ſimplicity and perſpicuity conſtitutes its only merit.

Claſs I. QUADRUPEDS. ZOOLOGY.

QUADRUPEDS are animals which have four feet, whoſe body is covered for the moſt part with hair: they

they respire by means of lungs, similar to those of man; they have a heart, with two ventricles: they are viviparous. These animals are those whose structure approaches most to man: there are some, as the ape and some others, which Linnæus thought might be confounded even in the same order with man. This naturalist gives the name of *mammalia* to this class of animals, in which he comprehends the cetacea; because all these have dugs, and suckle their young.

Although this class of animals seems to approach to man, they are, however, very different; but it is requisite to unite them here. The differences are the horizontal position of their bodies; the form of the extremities; thickness; the hardness of their skin, covered with hair or furnished with a hard shell, as if horny; the vertebral column, prolonged to form the tail; the anterior part of the cranium, flattened and horizontal; the ears, large and elongated; the bones of the nose and upper jaw, very long, and placed obliquely. Upon comparing this structure with that of man, whose body is erect and perpendicular, the radius of the arm is moveable, the fingers are fully separated, the thumb is opposed to the four others, the skin is sleeked and thin, it will soon be perceived how much this conformation exalts his sensibility, and renders him superior to the most perfect animals. The anatomy of his internal organs, and the history of his functions, also add much weight to these important considerations.

The ancient naturalists, at the head of whom may be placed Aristotle and Pliny, distinguished quadrupeds only by the places which they inhabited. So, for want of exact descriptions and sure characters, we are at a loss sometimes to discover what animals they are describing. The naturalists, who perceived the disadvantages of this method, adopted a very different manner of treating the subject. They made use of the external form of the most apparent parts of animals to render their characters easy to be understood, by means of which they might be furely diftinguifhed from one another. We fhall relate here only three artificial methods with refpect to quadrupeds; thofe of Linnæus, Klein, and Briffon.

Method of Linnæus.

Linnæus divides the animals with dugs, *Mammalia*, into feven orders. The firft, which comprehends thofe which he calls *primates*, has for characters, dentes incifivi in both jaws; two mammæ, fituated upon the breaft; the arms extended from the clavicles. This order contains four genera, viz. man, the ape, profimia, and vefpertilio. We cannot but fay, that this method is far from being natural, as it approximates bodies of very diftinct natures, as man and a bat.

Animals of the fecond order are called *bruta*. Their characters are, the abfence of the dentes incifivi, the feet armed with ftrong claws, flow walking. This order comprehends fix genera; which are, the elephant, trichecus, bradypus, myrmecophaga, manis, dafypus. The two firft genera are very far diftant from the four others.

In the third order, which the Swedifh naturalift defigns by the name of *feræ*, wild beafts, he comprehends all animals with mammæ, whofe dentes incifivi are conical; and for the moft part fix to both jaws, whofe canini are very long, and molares not flat; whofe feet are armed with fharp claws; and, laftly, which devour their prey, and live upon plunder. There are ten genera in this order; phoca, canis, felis, viverra, muftela, urfus, didelphis, talpa, forex, and erinaceus.

The fourth order, intitled *glires*, is diftinguifhed by the following characters: The animals which compofe it have two dentes incifivi to each jaw, no canini; their feet are armed with claws, and proper for jumping. They eat bark, roots, &c. This order comprehends fix genera; which are, the hiftrix, lepus, caftor, mus, fciurus, ſciurus, and the American dormouſe, called by Linnæus *noctilio*.

This naturaliſt united in the fifth order, by the name of *pecora*, the quadrupeds which have dentes inciſivi to the inferior jaw, and which have none to the upper; whoſe feet are forked, and which are ruminant. Camelus, moſchus, cervus, capra, ovis, bos, are theſix genera which compoſe this order.

The ſixth order comprehends, under the name of *belluæ*, the quadrupeds which have obtuſe dentes inciſivi, and feet with claws. The four genera which compoſe this order, viz. equus, hippopotamus, ſus, rhinoceros, are diſtinguiſhed very well from one another by the number of their teeth and form of their feet.

Laſtly, the ſeventh order, which comprehends the *cetacea*, *cete*, is diſtinguiſhed from all others by the form of the feet, which reſemble fins. But as we ſuppoſe, with ſeveral modern naturaliſts, a particular claſs of cetacea ſhould be made, we ſhall treat of them after the quadrupeds.

The method of Linnæus appears to be defective in ſome points; not only in its approximating bodies ſo diſtinct as man and the bat, and in its ſeparating animals ſo ſimilar as the rat and the mouſe; but ſtill more ſo in the diviſions being not numerous enough, and in their not being fit for diſtinguiſhing a quadruped eaſily: now this ought to be the ſole merit of a method, and its only advantage.

Method of M. Klein.

M. Klein divided the quadrupeds into two great orders. In the firſt he has compriſed thoſe which have feet with claws, *pedes ungulati*, *pedes cheliferi*; in the ſecond thoſe whoſe feet have toes, *pedes digitati*.

The firſt order is divided into five families, whoſe character is taken from the diviſion of the pedes ungulati into ſeveral pieces. The firſt family, called *monochela*, comprehends the genus of the horſe. The ſecond, whose individuals are called *dichela*, comprehends all those which have forked feet, *bisula*. Some have horns, as the bull, the ram, the goat, the stag, &c.; others have none, as the wild boar, sow, babyraussa. The *trichela*, or animals whose foot is divided into three parts, compose the third family, in which there is only the rhinoceros. The fourth family, whose characteristic is to have the foot separated into four pieces, *tetrachela*, contains only the hippopotamus. The fifth, which is distinguished by the feet divided into five pieces, *pentachela*, contains only the elephant.

The second order of quadrupeds, which comprehends those which are *digitati*, is likewise divided into five families. The first, designed for the animals which have two toes to the foot, *didactyla*, comprehends the camel, and the silenus, or the sloth of Ceylon. The second family, in which are comprised the animals with three toes to the feet, *tridactyla*, comprehends the sloth and the ants. In the third, M. Klein comprised under the name of *tetradactyla*, animals with four toes, the armadillos and the Guiney pigs, which seem to be kinds of rabbits. The fourth family, which has for characters five toes to the feet, *pentadactyla*: it is the most numerous of all; it contains the rabbit, squirrel, dormouse, rat, mouse, mole, bat, weasel, porcupine, dog, wolf, fox, coati, cat, tyger, lyon, bear, ape: the number of the kinds comprehended under these different genera is very considerable. It must be observed, that M. Klein, in all these characters taken from the form of the feet, considers only the fore-feet for the distinction of the families. Lastly, the fifth family of the *digitati*, is formed of the animals whose feet are irregular, *anomalopedia*, the otter, beaver, sea-cow, and calf.

The same objection might be made to M. Klein's method as to that of Linnæus. Although his first divisions are well adapted for families, the genera are not easy to be distinguished according to his method, particularly those of the fourth family of the digitati.

Me-

Method of BRISSON.

M. BRISSON has avoided the greateſt part of theſe inconveniences, by combining all the characters given by the naturaliſts who preceded him. The number of their teeth is made uſe of, their abſence, the form of the extremities, that of the tail, the nature of the appendices, as the horns, the ſcales, prickles. His combined method is without contradiction the moſt complete, and the moſt proper to diſtinguiſh a quadruped, and to refer it to the genus to which it belongs.

Claſs II. CETACEA.

THE cetacea are large animals which inhabit the ſea, and which, by the ſtructure of their lungs and of their blood-veſſels, may live in water, as we ſhall more fully relate in the hiſtory of reſpiration. They reſemble quadrupeds in the ſtructure of their mammæ, becauſe they bring forth their young alive, and in general in their internal organs: but they differ in the form of their extremities, which are formed into fins, and in two large openings placed at the top of their heads, by which they reject the water to a more or leſs conſiderable height. Naturaliſts call theſe conduits *ſpiracula*. The number of the genera of theſe animals is much leſs than that of quadrupeds. M. Briſſon diſtinguiſhed them, 1. Into cetacea which have no teeth; ſuch as the whale, balæna: 2. Into cetacea which have teeth only in the upper jaw; ſuch as the monodon or monoceros: 3. Into cetacea which have teeth only in the under jaw; ſuch as the phyſeter: 4. Into cetacea which have teeth in both jaws; ſuch as the dolphin, delphinus.

LEC-

L E C T R U E LVIII.

Clafs III. Birds. Ornithology.

THE birds are two-feeted animals, which move in the air by means of their wings, which are covered with feathers, and which have a bill of a horny fubftance. Thefe animals prefent a great number of interefting facts, relatively to the varied form of their bill, the ftructure of their feathers, the motions which they make, their manners, &c. From the abridgement of phyfiology, which we fhall offer in a more full manner afterwards, we fhall learn the moft important of thefe facts. We muft confine ourfelves to the external characters, which naturalifts have made ufe of to diftinguifh the birds, and to clafs them methodically.

The firft philofophers who treated of this part of natural hiftory, eftablifhed no other differences between birds except which nature prefented, relatively to the places frequented by thefe animals. Thus they diftinguifhed them into birds of the woods, of the plains, bufhes, feas, rivers, lakes, &c. Some others have diftinguifhed them, by their way of feeding, into birds of prey, granivorous. &c.

But the methodifts have followed another line to diftinguifh the birds. Linnæus has divided them, from the form of their bill, into fix orders, like the quadrupeds, with which he compared them. But thefe divifions feem to be not fufficiently detailed, efpecially in obferving, that the number of the fpecies is much more confiderable in birds than in quadrupeds, as M. le Compte de Buffon has brought the number of known quadrupeds up to 200, and the birds to 1500 or 2000. We fhall confine ourfelves to the methods of Meffieurs Klein and Briffon.

M. Klein divided the birds into eight families, from the form of their feet. The firft, under the name of *didactyli*, comprehends thofe which have two toes to the

the feet; the ostrich is the only one in this division. The second comprehends the tridactyli; as the casoar, the bustard, the lapwing, the plover. The third, the tridactyli; those which have two toes before and two behind; as the parrot, the wood-pecker, the cuckow, the king's-fisher. The fourth, the tetradactyli which have three of their toes before and one behind. This family is the most numerous of all; it comprehends the birds of prey, diurnal and nocturnal, the corbies, magpies, starlings, thrushes, blackbirds, larks, &c. The fifth family comprehends the tretradactyli which have three toes before, united by a membrane, and the posterior one is free. These birds are called *plain footed;* the geese, ducks, gulls, puffings. The sixth comprehends the tetradactyli whose four toes are united by one membrane: These are called in Latin *planci.* The pelican, the cormorant, booby, &c. are ranked by M. Klein in this class. The seventh is composed of those which have only three toes united by one membrane; these are the tridactyli palmipedes: the chaffinch, &c. belong to this class. Lastly, the eighth comprehends the tetradactyli whose toes are furnished with membranes, fringed or fimbriated. They are also called *dactylobi.* The pigeons and moor-hens compose this last family. Klein's method, though more detailed than that of Linnæus, is still pregnant with difficulties in distinguishing the genera, particularly those of the fourth family. For this reason, we prefer that of M. Brisson. It is true, that this last, in which the author has used all the characters united, as in the case of the quadrupeds, at first sight appears very complicated; but upon being reduced into a table, all the divisions are at once presented, and from these any bird may be at once distinguished.

Class IV. Amphibious Animals.

The amphibious animals are those which live upon land and water equally. They may remain in these a longer

longer or ſhorter time. The toes of thoſe of theſe animals which have any toes, are almoſt always united by membranes; their ſkin is ſleek and ſcaly: ſometimes they are covered with an oſſeous ſhell. Some are oviparous; however, there are others that are viviparous; ſome ſwim, others walk as quadrupeds; laſtly, ſome creep like the worms. From theſe differences, Linnæus divided the amphibia into three orders. The firſt comprehends the reptiles which have feet; the ſecond the ſerpents; and the third the kinds of ſwiming animals called *cartilaginous fiſhes*.

LECTURE LIX.

Claſs V. FISHES. ICHTHYOLOGY.

FISHES are very different animals from the preceding, whoſe internal organs have a ſtructure quite particular, as we ſhall notice in our abridgement of phyſiology. They are diſtinguiſhed from other animals by their want of feet, but they have fins which ſerve to move them in the water; and by their reſpiring water inſtead of air. The fiſhes are much more difficult to know than the other animals; therefore their natural hiſtory is in general much leſs advanced.

In order to underſtand the methodical diviſion which we ſhall propoſe, from Artedi, Linnæus, and Gouan, it is requiſite to take a curſory view of their external anatomy. The bodies of fiſhes may be divided into three parts, viz. the head, the trunk, and the fins.

The head of theſe animals has different ſhapes. It is flattened either horizontally or laterally, or it is round; bare, or ſcaly; ſleek, or full of prickles, tubercles, &c. It is remarked, that their mouth is furniſhed with fleſhy or oſſeous lips, of appendices ſoft and very moveable; the teeth adhering to the jaws, the palate, the tongue, the gullet; the eyes, two in number, immoveable, without lids; two noſtrils on each ſide; the

the opening of the gills, or the bronchiæ; the coverings, or the round, triangular, ſquare bones, deſigned to ſecure the opening of the bronchiæ; the bronchial membrane placed under the coverings, ſupported upon ſeveral props or bones in form of an arch, whoſe number varies from two to ten. This membrane is double under the coverings; and it is very important to examine its ſtructure and varieties, becauſe the characters of the genera are moſt frequently taken from the number or form of its fibres.

The trunk, like the head, differs in its form: it is either round, globular, long, flat, or angular. Its lateral line muſt be obſerved, which appears to divide each ſide of the body into two parts; the thorax, placed under the gills, at the beginning of the trunk, and filled by the heart and bronchiæ; the belly, whoſe ſides form the machinery, continued from the head to the tail, and which contains the ſtomach, inteſtines, liver, bladder, the parts of generation; the opening of the anus, which is common to the inteſtines, the bladder, and the parts of generation: laſtly, the tail, which ends the trunk, is of various form and extent.

The members or fins, pinnæ natatoriæ, are formed of membranes formed of ſmall fibres; ſome of which are hard, oſſeous, and terminated in a ſharp point, which conſtitutes the fiſhes called *acanthopterygious* by Artedi; others are flexible, ſoft, obtuſe, as if cartilaginous, which conſtitutes the fiſhes called *malacopterygious*. Five kinds of fins are diſtinguiſhed relatively to their ſituation; the dorſal, pectoral, abdominal, that of the anus, and that of the tail.

The dorſal fin is ſingle; it keeps the fiſh in equilibrium; it varies in ſituation, number, figure, proportion.

The thoracic fins are ſituated at the opening of the gills: they are two in number; they ſerve for arms, ſome-

ſometimes even for wings: they differ in the place of their inſertion, their extent, their figure, &c.

The fins of the belly are the moſt important to be known, becauſe Linnæus has made uſe of their ſituation to determine diſtinctly the claſſes of fiſhes. Theſe fins are placed at the inferior part of the body before the anus, and always lower than the pectoral. They are ſometimes wanting; and as Linnæus has compared them to the feet, he has called thoſe fiſhes *apodes*, or *without the feet*, which want theſe kinds of fins. They are preſent, however, in the greateſt number of fiſhes; but their inſertion varies: when they are placed above or below the opening of the gills and the pectoral fins, they are called *jugular;* and likewiſe the fiſhes receive that name which have their fins inſerted in that place: if they are attached to the thorax, and behind the opening of the gills, then they are called *thoracic;* and likewiſe the fiſhes with this ſtructure have received the ſame name from Linnæus: Laſtly, when they are placed under the belly, nearer the anus than the pectoral fins, they are called *abdominal;* which name the fiſhes likewiſe get when they are thus formed.

The fin of the anus is ſingle. It occupies all or part of the region ſituated between the anus and tail; it differs in form, extent, number, although it never has been found double, except in the golden fiſh of China.

The fin of the tail is placed vertically at the extremity of the body, and terminates the tail: This is the helm of the fiſh, the inſtrument by which it changes, at its pleaſure, its direction according to its various motions. It exhibits alſo ſeveral varieties, in form, adherence or connection, extent, &c.

From theſe details upon the external anatomy of fiſhes, we paſs to the methodical diviſions of the naturaliſts. Before the time of Artedi, no naturaliſt had yet attempted a methodical diſpoſition of the fiſhes, although methods for claſſing the other animals had been offered

offered. This philosopher is the first who proposed a system of ichthyology, from the nature of the bones of the fins, hard or soft, prickly or obtuse, or from the form of the gills. He next laboured to multiply the divisions from other parts; but a premature death cut off his design. Linnæus thought of establishing an ichthyological method, from the various situation of the fins of the belly: and M. Gouan, a celebrated professor at Montpelier, combined with great ingenuity the systems of Artedi and Linnæus. This naturalist first divides the fishes into those which have the gills complete, that is to say, formed of an opening and a bronchial membrane properly organised; and those which have the gills incomplete, that is, which want either the bronchial membrane or the opening, or both. The first, then, are distinguished by the form of their fins; in short, these parts are formed either of hard and acute bones, or of soft fibres, as if cartilaginous. These differences constitute three classes of fishes, viz. 1. the acanthopterygious; 2. the malacopterygious; 3. the branchiostigious. In each of these classes of fishes, the abdominal fins being found either absent, or placed at the neck, thorax, belly, M. Gouan has divided each class into four orders; that is, into apodes, jugular, thoracic, and abdominal.

As, from these first general divisions, there are no more characters, to divide the genera into sections and articles, as there are in the method of M. Brisson for quadrupeds and birds, we shall have no need of reducing the method of Gouan into a table; and it will suffice to offer the genera under the classes and orders to which they belong.

Class. 1. ACANTHOPTERIGIOUS FISHES.

Order 1. *Apodes.*

Gen. 1. Trichiurus,
2. Xiphias,
3. Ophidium.

Order 2. *Jugular.*

Gen. 1. Trachnius,
2. Uranoſcopus,
3. Callyonimus,
4. Blennius.

Order 3. *Thoracic.*

Gen. 1. Gobius,
2. Cepola,
3. Coryphæna,
4. Scomber,
5. Labrus,
6. Sparus,
7. Chœtodon,
8. Sciœna,
9. Perca,
10. Scorpæna,
11. Mullus,
12. Trigla,
13. Cottus,
14. Zeus,
15. Trachipterus,
16. Gaſteroſteus.

Order 4. *Abdominal.*

Gen. 1. Silurus,
2. Mugil,
3. Polynemus,
4. Theutis,
5. Elops.

Claſs 2. MALACOPTERYGENEOUS FISHES.

Order 1. *Apodes.*

Gen. 1. Muræna,
2. Gymnotus,
3. Anarhichas.

Gen. 4.

Gen. 4. Stromateus,
5. Ammodytes.

Order 2. *Jugulares.*

Gen. 1. Lepidogaſter,
2. Gadus.

Order 3. *Thoracic.*

Gen. 1. Pleuronectes,
2. Echeneis,
3. Lepidopus.

Order 4. *Abdominal.*

Gen. 1. Loricaria,
2. Atherina,
3. Salmo,
4. Fiſtularia,
5. Eſox,
6. Argentina,
7. Clupea,
8. Exocætus,
9. Cyprinus,
10. Cobites,
11. Amia,
12. Mormyrus.

Claſs 3. BRANCHIOSTEGIOUS.

Order 1. *Apodes.*

Gen. 1. Syngnathus,
2. Baliſtes,
3. Oſtracion,
4. Tetraodon,
5. Diodon.

Order 2. *Jugulares.*

Gen. 1. Lophius.

Order 3. *Thoracic.*
Gen. 1. Cyclopterus.

Order 4. *Abdominal.*
Gen. 1. Centrifcus,
2. Pegafus.

The characters of the genera are taken from the fhape of the body, of the head, mouth, bronchial membrane, and particularly from the number of fibres which fupport this membrane.

Clafs VI. INSECTS. ENTOMOLOGY.

INSECTS are animals which are known by the form of their bodies, as divided by rings, and by the prefence of two moveable horns which they have above the head, and which are called *antennæ*. Infects compofe one of the moft numerous claffes of animals, owing, undoubtedly, to their fmallnefs; as it has been obferved that the fmaller thefe bodies are, the more their reproduction is multiplied. The hiftory of thefe animals is one of the moft agreeable, the moft amufing, and perhaps not the leaft ufeful; as properties belonging to them may be found ufeful to medicine and the arts.

Infects prefent in their claffes an example of almoft all the other animals, relatively to their manners, their form, their habitation.

Some walk like quadrupeds; others fly like birds; fome fwim and live in waters, like fifhes; laftly, fome leap or crawl, like certain reptiles. This analogy may be pufhed even farther upon examing in detail the ftructure of their extremities, of their mouth, their internal organs, &c.

Infects, confidered externally, are compofed of three parts; head, breaft, and belly.

The head differs in form, extent, and pofition. It is fometimes very large, comparatively to the bulk of the infect, and fometimes very fmall; it is either round, fquare, ſquare, long, ſleek, or rough, or full of tubercles, or covered with hair in certain places. It is obſerved, 1. The antennæ placed in the neighbourhood of the eyes, formed of different pieces, articulated and moveable, reſembling a thread, terminated in a point or by a maſs. The form of theſe organs is eſſential to be di-diſtinguiſhed, becauſe it is almoſt always the character uſed to diſtinguiſh the genera. 2. The eyes, which are of two kinds, in facets or a net-work, ſleek and ſmall: theſe organs are ſometimes very large, at other times very ſmall; their number varies. There are ſome inſects which have only one, as the monocolus; others two, as the ſpider. 3. The mouth, which is formed either of ſtrong horny jaws, placed and moved laterally, or of a trunk more or leſs long, dilated, ſpiral, &c. or only of one ſlit. This part is frequently adorned with ſmall moveable appendices, called *antennulæ*, two or four in number.

The breaſt of inſects: it is placed between the head and the belly; it is ſometimes round, ſometimes triangular, cylindrical, broad, narrow. It ſhould be conſidered as compoſed of ſix ſides, and likewiſe a kind of cube, whoſe form it aſſumes ſometimes. The face, or the anterior extremity, is hollowed to receive the head: this articulation is made ſometimes only by a thread, as in the fly. The occiput is generally round, and articulated with the firſt ring of the belly. Sometimes it is joined with this part only by a thread. The upper ſide is often flat and ſleek, ſometimes round, prominent, full of appendices, tubercles, terminated by a kind of projecting edge; which conſtitutes the thorax marginatus. It is to the hind part of this ſide to which the wings are attached. It is known that the greateſt part of inſects is provided with theſe organs, but they differ ſingularly from one another; and as it is upon theſe differences upon which are founded the principal diviſions of the claſſes adopted by the methodiſts, it is of importance to conſider them. Theſe wings are either in the number of of two or four. Among thoſe which have two tranſparent ones, as the fly, midge, &c. theſe wings are always accompanied towards their inſertion and below it with a minute filament terminated by a round button, which is called *halter*, and which is covered with a concave membranous appendix called the *ſpoon*.

In a great number of inſects theſe two wings are very ſtrong, fold back, and are platted, under hard, horny, moveable caſes, called *elytra*. Theſe caſes differ in form: ſome cover all the belly; others are as if cut tranſverſely, and cover only a part of the belly; ſome are hard, others ſoft; moſt are accompanied to the height of the ſuture or line, by which they approach with a ſmall triangular piece ſoldered to the breaſt, which is called *ſcutellum*; this piece is wanting in ſome: laſtly, with inſects, in ſome caſes, the elytra are ſoldered, as if formed of a ſingle piece and immoveable.

The wings are often four in number; then they are either membranous and tranſparent, as in the beetles, waſps, &c. or they have upon each of their ſides a coloured powder, which with the microſcope preſents ſcales laid upon the wings like tiles upon a houſe, imbricatim.

The inferior part of the breaſt is irregular, formed of ſeveral pieces connected to one another, and in it is inſerted a part of the legs. The number of theſe laſt varies in inſects: many have ſix, others eight, as the ſpiders; in ſome there are ten, as in the crab: and ſome inſects have a much greater number. Sixty are to be counted in the wood-looſe; and ſome kinds of ſcolopendras have even ſeventy, and a hundred and twenty on each ſide. In thoſe which have only ſix, eight, or ten, they are all attached to the breaſt, according to Geoffroy: in thoſe which have a greater number, a part of the legs is inſerted in the rings of the belly.

The leg of an inſect is compoſed always of three parts; the thigh, leg, and tarſus. There is often beſides theſe an intermediate piece between the body and

and the thigh. The tarſus is formed of ſeveral pieces or rings articulated to one another: the number of theſe rings varies, and extends from two to five. There are even inſects whoſe tarſus of the fore feet is more conſiderable than the hind ones: which eſtabliſhes an analogy between the ſtructure of theſe ſmall animals and that of a great number of quadrupeds, whoſe fore-feet before have a greater number of toes than the poſterior feet. M. Geoffroy has taken part of this character for his diviſion; as we ſhall notice more afterwards. The tarſus is terminated by two, four, or ſix ſmall claws or hooks, and furniſhed below frequently with bruſhes or ſpongy balls, which ſupport and connect the inſect upon the moſt poliſhed bodies, as glaſſes, &c.

Upon each ſide of the breaſt one or two oblong oval openings are obſerved, which are called *ſtigmata*, and by which the inſect reſpires.

The third part of inſects is the belly. For the moſt part it is compoſed of rings, or of horny half rings, which are inchaſed into one another. Sometimes no rings at all are obſerved, and the belly appears formed only of one ſingle piece. It is generally larger in females than in males. It has the parts of generation fixed to its extremity. In its ſides a ſtigma is obſerved upon each ring, except upon the two laſt. At the poſterior part of the belly ſeveral inſects carry their ſtings; ſome of which are very acute and ſharp, others ſerrated, and others like a wimble. They ſerve either as inſtruments of defence, or to pierce the places where the inſect is to depoſite its eggs.

The moſt ſingular phenomenon in inſects, and that in which they differ entirely from moſt of the other animals, is the changes of ſtate through which they paſs, or the metamorphoſes which they undergo before they become perfect inſects. There are ſome inſects, and almoſt all thoſe of the apteræ, which ſuffer no ſuch changes; but the greateſt number are ſubmitted to them. No inſect comes out of its egg with the mother's form; but it appears in the form of a worm, with or without legs, and whose structure of the head and wings greatly varies. This first state is called *larva:* under this kind of mask the insect eats, turns larger, moves, and changes its skin several times. When it has come to its full growth, it changes skin for the last time, but it is no longer in the form of a worm or larva, but in quite a different one, called *nympha*, *chrysalis*, *aurelia*.

M. Geoffroy distinguishes four kinds of nymphæ. The first is that which does not resemble an animal; some rings are observed only on the under part, and the upper presents only indistinct impressions of antennæ, legs, and wings. The skin of this kind is hard, cartilaginous, and it has only a few motions in its rings, as in the butterfly, &c.

The second kind of chrysalis discovers the parts of the perfect animal inveloped in a very minute and soft skin. It is immoveable like the first. Insects in cases, those with four bare wings, and those with two, furnish examples. The third kind is that whose wings are fully evolved and apparent, and which move. Such are the midges, and the insects which pass the two first states of their life in water.

Lastly, the fourth kind comprehends those which resemble the perfect insect, in the form of the body, the presence of the antennæ, and paws. These nymphæ move and eat. They differ from perfect insects only in the absence of the wings, and in their being unfit for generation. The nymphæ of the beetles, bugs, locusts, &c. are of this kind.

The history of insects laboured under the same difficulties as that of other animals. The ancient naturalists had distinguished them only by the places which they inhabit. Before Linnæus no person had undertaken to dispose them methodically, and to offer characters by which they might be distinguished. This naturalist has the merit of the first systematic division of these animals. Geoffroy afterwards undertook to class them in a more

exact

exact manner. His division of sections and genera is a master-piece of accuracy, precision, and perspecuity. We shall adopt his division.

Geoffroy divided the insects into six genera, from the absence, structure, and number of the wings. The first section comprehends the coleopteræ, or insects whose wings are covered with a case. Their mouth, of two lateral horny jaws, also forms a second general character of this section. The May-bug presents these two characters.

The second section comprehends the hemipteræ; the upper parts of whose wings are either somewhat thick and coloured, or half hard and opaque: but the character of the wings, which is not exact in this section, is replaced by that of the mouth, which is constant. This mouth is a long acute trunk, folded back below between the paws. The wood-bug and cicada belong to this section.

The third section is composed of *tetrapteræ* with farinaceous wings, whose four wings are coloured by a scaly powder, and which have a trunk more or less long, frequently twisted spiral-ways, as the butterfly. Linnæus calls these *lepidopteræ*.

In the fourth section, the *tetrapteræ* are comprehended, with bare wings. Their four wings are membranous; they have hard jaws: of this sort is the wasp. Linnæus made two orders of these insects: the *neuropteræ*, whose anus has no sting, and whose wings are marked with nerves; and the *hymenopteræ*, which have the anus armed with a sting, and the membranous wings without any very apparent nerves.

The fifth section contains the *dipteræ*, or those with two wings; their mouth is most frequently formed of a trunk, and they have balancers and spoons under the origin of their wings.

In the sixth and last section, the *apteræ*, or those without wings, are ranked; as the spider, louse.

Besides these first divisions, Geoffroy has established others to facilitate the distinction of insects.

LECTURE LX.

Class VII. Vermes.

WORMS are soft animals, of a very different form from that of the insects, with which they have been confounded by several naturalists. They have no bones, properly so called; and their members are not conformed as those of insects; they are not obliged like them to pass through different states. It is the want of feet, properly formed like those of insects, and the constant and immutable form, which they keep all their life, which make their proper characters, and which distinguish them from other animals.

The class of worms is the most numerous and least known of all animals. There are few dead or organised substances in which some worms are not found, which find their nourishment there. The most part of naturalists, and Linnæus himself, have put in the same class the vermes and polypi; but their internal structure and functions are entirely distinguished. A heart and vessels are found in most worms; and nothing similar is found in the polypi.

We must fully distinguish the worms of which we are now treating, from the animals which are only larvæ of insects, and which have for that reason received the name of *worms* on account of their form. Their head, furnished with a jaw, and the legs, which they have more or less, most commonly six, afford characters by which they may be easily distinguished.

The worms have a great deal of motion: they love and seek for humidity: some have no very distinct head: the most part are hermaphrodite. Those which have a head have it armed with two moveable retractible horns, called *tentacula*. It seems, that almost all the worms, of which we shall give an abridged account, have

have the property of being reproduced when they are cut through; which indicates a ſimple organiſation, and that their nature approximates to that of the polypi.

The worms may be divided into three ſections. The firſt ſhall contain the bare worms, whoſe organiſation is beſt known, and which approximate in this character to the other animals. In the ſeeond, we ſhall rank the worms inveloped in a teſtaceous covering, or the ſhell-worms: their organs are leſs known than thoſe of the firſt: however, the excellent reſearches of M. Adanſon prove, that their ſtructure approximates to the bare worms. The third ſection will comprehend the worms with a cruſtaceous covering: their organiſation is not ſo well known as that of the preceding, and their external form and ſtructure of the mouth alone has as yet been examined.

Sect. I. *Vermes Nudi.*

It is eaſy to figure the difference which exiſts between the worms compriſed in this ſection and thoſe of the two others: they are bare, and without any calcareous or cruſtaceous covering. There are ſix different genera in this ſection.

Genus 1. *Gordius.*

This worm, according to Linnæus, has a body like a round wire, the mouth forked, the jaws horizontal and obtuſe: its body is pale, and its two extremities are black. Geſner and Aldrovande have called it *vitulus aquaticus*. The vena medinenſis belongs to this genus.

Genus 2. *Lumbricus.*

This worm's body is formed of rings: its anterior extremity is pointed; its genital parts are at the neck: it is hermaphrodite and oviparous. The earth-worm, that of the inteſtines of men, and the ſea-worm, are the three kinds well known.

Genus

Genus 3. *Aſcaris.*

Its body is ſleek, and its extremities are acute: it is to be ſometimes found in the rectum of children.

Genus 4. *Hirudo.*

Its body appears to be ſwelled in the middle: its extremities dilate into a roundiſh and flat body: its head is armed with three ſmall tuberoſities, which make a wound with three angles upon an animal's ſkin, to which it adheres. Its mouth is the pump, and a ſort of fleſhy tuberoſity, placed at the bottom of this cavity, is the piſton. Such is the mechaniſm by which the leech ſucks the blood.

Genus 5. *Limax.*

Its body is oblong, covered with a ſhield or fleſhy mantle above, and formed of a muſcular ſheath below. There are ſeveral kinds of ſnails, which differ in ſize, colour, ſhades.

Genus 6. *Tænia ſolium, Vermis ſolitaris.*

It has received this name, becauſe it has been ſuppoſed to be found alone in the inteſtines of animals: but it is an error; ſeveral have been found in man. There are ſometimes dozens in horſes, dogs, fiſhes, &c. The character of this animal is of being flat like a riband, and formed of diſtinct articulated rings, having a fine elongated extremity, which has been ſuppoſed its head. It is thought that each ring is a particular animal, which lives and is nouriſhed ſeparately. It is of a very conſiderable length, frequently ſeveral yards long. It is reproduced when it is cut; ſo that animals are not cured unleſs it be taken out entire. It lives in the inteſtines of almoſt all animals. The different kinds of tænia have not yet been well examined. Two are diſtinguiſhed in man; the one with ſhort rings, the other with long. The nature of this worm is not yet

pro-

properly known, notwithſtanding the labours of Meſſrs Andry, Tyſon, Herrenſchvands, Bonnet, Butini. It ſeems to approach much to the polypi. Linnæus has deſcribed a round kind of them articulated, of a ſhining white, ſaid to be found in marſhes. He ſpeaks alſo of a flat white worm, which he calls *faſciola*, very different from the tænia, in its not having articulations. A very ſmall kind is found near rivulets, and likewiſe in the liver of ſheep; there is another much longer, and is found in the inteſtines of fiſhes.

SECT. II. *Vermes Teſtacei.*

THESE are covered with a ſhell of the nature of chalk. Moſt naturaliſts have paid attention only to this covering, and have founded methods upon the form of the ſhells. M. Adanſon is one of the firſt who has undertaken to deſcribe the inhabitants of the ſhells, and to diſtinguiſh theſe animals from their internal form. This work is very little advanced and very difficult. We ſhall point out the road of the naturaliſts in the diſtribution of the ſhells. M. Dargenville is the author who has given the moſt clear and moſt complete method. He divides the ſhells into three orders. The firſt comprehends the ſhells of a ſingle piece, *univalvæ*, *cochleæ*; the ſecond, the ſhells of two equal and ſymmetrical pieces, *bivalvæ*, *conchæ*; the third, the *multivalvæ* or *polyvalvæ*, which are formed of more than two pieces united.

Order 1. *Univalvæ, Cochleæ.*

Familiæ ſive genera 1. Patella,
2. Haliotis,
3. Tubulus, dentalium,
4. Nautilus,
5. Cochlea,
6. Nerites,
7. Trochus,
8. Cylindrus,

Fa-

Familiæ ſive genera 9. Voluta,
10. Strombris,
11. Buccinum,
12. Murex,
13. Purpura,
14. Porcellana,
15. Globus.

Order 2. *Bivalvæ, Conchæ.*

Familiæ ſive genera 1. Oſtrea,
2. Chama,
3. Concha cordis,
4. Pecten,
5. Mytulus,
6. Solen.

Order 3. *Multivalvæ vel polyvalvæ.*

Familiæ ſive genera 1. Pholas,
2. Balanus,
3. Concha anatifera.

SECT. III. *Vermes Cruſtacei.*

THIS ſection is the leaſt numerous and leaſt known ; comprehends two genera well characteriſed.

Genus 1. *Aſterias.*

ITS mouth is in the middle of the body ; its cruſtaceous covering is divided into filaments, which have the form of a net-work.

Genus 2. *Echinus.*

ITS mouth is at the baſe of a cruſtaceous round and vaulted covering, ſet full of prickles. M. Dargenville has placed the urchins in the order of the multivalvæ ; M. Klein has made a complete work upon theſe animals, intitled, *Naturalis diſpoſitio echinodermatum.* The ſituation of the anus appeared to him ſuſceptible of furniſhing a proper method for diſtinguiſhing them. He has

has divided them into three claſſes. The firſt comprehends the anocyſti, whoſe anus is placed at the top or higheſt place of the vault: the ſecond contains the catocyſti, which have the anus placed below. In the third claſs, he places the pleurocyſti, whoſe anus opens neither at top nor bottom, but at the ſurface or the ſide.

The characters of the ſections and genera are taken from the form of the arch, and from their points or adpendices.

Claſs VIII. Polypi.

The polypi conſtitute the laſt claſs of animals. They are the leaſt perfect, and ſome of them ſo reſemble vegetables, that naturaliſts have for a long time compriſed them in the number of theſe laſt.

The ancients gave the name of *polypi* to animals more or leſs voluminous, having a great number of feet, and which are uſed as a delicate diſh.

By this name we underſtand all moving animals, whoſe organiſation ſeems ſimple, which are not known to have a heart, and which are formed only of a membrane, turned round into ſacs, furniſhed with a great number of feet or horns. The polypi have alſo a property which diſtinguiſhes them, which is, they have the power of contracting and entirely loſing their form. They live for the moſt part, united in a conſiderable number, in the ſame habitation. They are reproduced by kinds of grains which are ſeparated from their ſurface, and adhere to bodies; or they grow like vegetables. If they are cut, they ſhoot and are renewed like a ſlip of a tree. We divide them into four ſections.

Sect. I. *Polypi Nudi.*

These polypi are not covered with any hard ſubſtance or caſe: they fix to rocks or any ſolid body. Two genera are diſtinguiſhed.

Genus

Genus 1. *Polypi aquæ dulcis; Hydra Linnæi.*

M. Trembly has deſcribed theſe ſmall animals with great accuracy. They are formed of a body or long tube, at the extremity of which is a dilitation, which is the mouth. The ſide of this cavity is ſurrounded with a great number of filaments, or contractile moveable horns, which ſerve the animal for catching and devouring its prey. This author has diſtinguiſhed ſeveral kinds from their ſize, colour, form: Such is the polypus with arms, polypus of the colours of the peacock, &c.

Genus 2. *Sea-Nettle.*

It repreſents a truncated conical body, flabby, contractile, whoſe mouth is armed with horns or feet. Linnæus, and moſt part of the naturaliſts, have ranked them among the zoophytes, or animals which reſemble plants *.

Sect. II. *Polypi in cellulis corneis, ſive quaſi ligneis.*

The character of the polypi of this ſection is, they adhere to the branches of ſubſtances, either horny, or flexible and fibrous, as young twigs of trees and plants. Theſe ſubſtances were compriſed alſo among vegetables, down to Peyſonel, Ellis, Donati; who have proved that they really belong to animals. They have been called *zoophytes* on account of their form, and *lythophytes* on account of their hardneſs.

Genus 1. *Keratophytes*, or *Lithophytes*, *properly ſo called.*

It is a polypus frequently elaſtic, formed by a matter of

* It has been uſual to rank the cuttle-fiſh here; but their more perfect organiſation, the preſence of a heart and a ventricle, obſerved and deſcribed by Swammerdam, make them approach the fiſhes. M. le Cat calls the cuttle fiſh a fiſh-inſect. It is a ſubſtance, indeed, which ſeems to hold the middle-rank, and to ſerve as a paſſage between theſe two claſſes of animals.

of different colours, which cuts and fuſes like horn, poliſhes like it, and burns with its ſmell. Black and articulated coral are of this ſort.

Genus 2. *Sertularia ſive corallina.*

NOTHING more reſembles plants than this production of polypi. M. Ellis has given a deſcription of a great number of kinds of this ſubſtance. He diſtinguiſhes them into four families; to wit, the veſicular, the tubular, the cellular, and articulated. The two laſt families ſeem to be not of the ſame horny nature as the two firſt.

SECT. III. *Polypi in Cellulis Cretaceis.*

THIS kind of polypi conſtructs and inhabits ſolid ramifications of a various form, but they are all of the nature of chalk. They are in general more voluminous than the former, and particularly the corallina. The genera which compoſe this ſection are diſtinguiſhed, like thoſe of the preceding, by their habitation; the animals which build them reſemble, for the moſt part, thoſe of the preceding, viz. they are ſmall contractile facs, furniſhed with a greater or leſs number of feet at their extremity, and which differ in ſize, colour, form, &c.

Genus 1. *Corallium.*

THE coral is a habitation of polypi, hard and calcareous, which has the form of a plant. This habitation is formed of layers; it is ſolid only in the middle; it is covered with a kind of bark, pierced with a great number of holes, in which the polypi are implanted. Three or four kinds of coral are diſtinguiſhed in the ſhops; the red, the roſe, the white, &c.

Genus 2. *Madrepora.*

THE madrepore differs from the coral in its having a more tender texture and more ſtony aſpect. Beſides, its

its ſurface is pierced with a great number of holes, which penetrate into the heart of the calcareous ſubſtance, whilſt theſe holes are found only upon the bark of the coral, and the bark is wanting in madrepores. The ramified form, like vegetation, or in maſſes ſimilar to muſhroms, a cap, a hand, has given different names to madrepores. The beſt method of diſtinguiſhing them, conſiſts, in general, in the characters taken from the form of the holes. Dr Pallas, in a very good work upon the zoophytes, intitled, *Elenchus Zoophytorum*, has given the beſt diviſions of marine animal productions whoſe form imitates that of vegetables.

SECT IV. *Polypi in cellulis mollibus et ſpongioſis.*

It is eaſy to diſtinguiſh the polypi of this fourth ſection from thoſe of the two preceding, by the ſoft, flexible, and cellular nature of their habitations. We are acquainted with three kinds of habitations which ſeem to belong to this ſection.

Genus 1. *Eſcarra.*

This matter is known by its ſofter, ſometimes friable, frequently pliant nature, and by its being formed of very fine celullæ, which imitate the holes and the texture of a wide wrought web: in theſe cells the polypi lodge.

Genus 2. *Spongia.*

The texture of the ſpongia is as if parenchymatous, A great number of kinds has been diſtinguiſhed by the form of the holes, the texture. Their figure has made them get different names, as Pan's flute, wax-candle, Neptune's cup, ſea agaric, &c. All the cells of the ſpongia communicate, and they ſerve as the habitation of a great number of polypi.

Genus 3. *Alcyonium.*

The name of *alcyonium* is given to the ſpongy, fleſhy, or gelatinous bodies, irregularly rounded, which lodge the

the polypi in their ſubſtance. This ſingular marine production has not yet been rightly examined: it has got names relatively to its form only; ſuch as the ſea-beet, ſea-waſp. They ſeem to be the laſt link of the chain of animals.

LECTURES LXI. LXII.

Of the FUNCTIONS *of* ANIMALS, *conſidered from Man down to the Polypi.*

THE characters peculiar to living and organic bodies are, as has been ſeveral times obſerved, the different functions performed by their different organs. We have conſidered them in vegetables: the order which we have adopted likewiſe requires that we conſider them in animals.

The part of medicine which treats of the functions of animals is Phyſiology. This excellent ſcience ought not to be limited to man alone: it ſhould be extended to all animals: and under this point of view we ſhall curſorily conſider it.

The functions of animals may be reduced to the following:

1. Circulation; 2. Secretion; 3. Reſpiration; 4. Digeſtion; 5. Nutrition; 6. Generation; 7. Irritability; 8. Senſibility. Theſe different functions are found in man, quadrupeds, cetacea, birds, fiſhes, reptiles, inſects: the worms and polypi poſſeſs none of theſe at all, and the claſſes before theſe two laſt do not poſſeſs them in the ſame degree.

1. Circulation is one of the firſt functions; it is the ſupport of life: when it ceaſes, the animal inſtantly dies: the organs which preſide over it are, the heart, arteries, and veins.

The heart is a conical muſcle which has at its bottom two cavities, called *ventricles*. At its baſe are two other facs, called *auricles*. From the left ventricle a very large artery goes out, called *aorta*, which diftributes the blood to all the body: from the right ventricle another artery alfo goes out of an equal fize, called *pulmonary artery*, becaufe it is ramified in the lungs. The right auricle receives the blood in its return from the whole body by the two venæ cavæ: this fluid paffes into the right auricle, from thence into the right ventricle; from this laft it is poured into the lungs through the pulmonary artery, and it returns by the pulmonary veins into the left auricle; from thence it paffes into the left ventricle, which fends it through the whole body by the aorta. This motion, thus going on in man, conftitutes two kinds of circulation, that through all the body, and that through the lungs: this laft was known before the other; the general circulation was difcovered by the celebrated Dr Harvey an Englifh phyfician.

Among the quadrupeds, the cetacea, and birds, this function is the fame way performed as in the man. Among the fifhes the heart has only one ventricle, and the lungs or gills receive no blood by a particular cavity from the heart. In reptiles it is performed as in fifhes. Infects and worms have a heart formed of a fet of knots, which contract one after the other; their veffels are very fmall; their blood is cold and colourlefs. The polypi have neither heart nor veffels; they are lefs perfect than vegetables with refpect to this kind of function.

2. Secretion is a function by which juices are feparated from the blood in the different organs deftined for particular ufes, as the bile in the liver, &c. This function is one of the moft extenfive in all animals: it is found in all animals; but it is impoffible to defcribe it, without entering into very extenfive details. It fhall therefore fuffice to obferve, that in all animals in which there is a true circulation, the fecretion follows the fame laws as in man; and it alfo appears to be performed in the moft part of animals which have no heart. Befides the

ana-

analogy which ſubſiſts neceſſarily between man and the animals which poſſeſs the ſame organs with him relatively to this function, each claſs of animals preſents particular ſecretions which are not found in man; ſuch as muſk and civet in quadrupeds; ſpermaceti in the cetacea; the oily juice deſtined to cover the feathers of birds; the poiſonous humour of the viper; the gluey fluid of the ſcales of fiſhes; the acrid juices of ants and waſps; the viſcous mucilage of ſnails; the colouring juices of purſlain, and a great number of others which the natural hiſtory of each animal makes known.

3. Reſpiration, conſidered in all animals, is a function deſtined to put the blood into contact with the fluid which they inhabit: man and quadrupeds have for this purpoſe an organ, called *lungs*. This viſcus is a maſs of hollow veſicles, which are only expanſions of a membranous and cartilaginous canal, called *trachea arteria;* and of blood-veſſels, which ſpread out, forming a great number of little holes at the ſurface of the bronchial veſicles. Theſe veſicles and veſſels are ſupported by a cellular texture, looſe and ſpongy, which forms the parenchyma of the lungs. The air diſtends theſe veſicles in reſpiration: the pure and vital portion of this fluid is abſorbed by the blood, which it renews, and to which it gives colour and concreſcibility. The air which is thrown out by reſpiration is impure; it troubles lime-water; it reddens the tincture of tournſol; extinguiſhes candles, and can no longer ſerve another reſpiration.

Among the cetacea this function is likewiſe performed; only as there is an immediate communication between the auricles, theſe animals can remain ſome time without reſpiring.

Though the reſpiration of birds be analogous to that of the preceding animals, this function ſeems to be much more extenſive in them. Indeed, anatomiſts have diſcovered in the belly of birds ſpongy veſicular organs which communicate with their lungs, and theſe laſt open juſt into the bones of the wings, which are hollow and without marrow, by a canal placed above the breaſt, which opens into the upper part and ſwell of the os humeri. This pretty diſcovery, made by M. Camper, ſhows us, that the air paſſes from the lungs of birds into the bones of the wings; and that this fluid, rarefied by the heat of their bodies, renders them very light, and ſingularly promotes their flight.

Fiſhes have gills, or bronchiæ inſtead of lungs: theſe organs are formed of membranous fringes, placed upon a bony arch, and full of blood-veſſels. The water enters by the mouth of the fiſhes; it paſſes acroſs the fringes, which ſeparate from one another; it ſqueezes and agitates the blood; and it goes out again by openings placed at the lateral and poſterior parts of the head, upon which two mvoeable bony pumps are placed, called *lids*, and ſupported by the bronchial membrane. Duverney thought that the bronchiæ ſeparated the air contained in the water. M. Vicq d'Azir, who ſo ſucceſsfully ſtudied comparative anatomy, and in particular that of fiſhes, ſuppoſes that the water does the office of air in the bronchiæ of theſe animals.

Inſects and worms have no lungs: they have canals or tracheæ placed all along the back; upon which, on each ſide, other ſmaller canals border, which terminate at the lateral part of each ring by a ſmall fibre called *ſtigma*. The ſtigmata are deſigned for reſpiring the air and expiring it. When they are covered with oil, the inſect ſuffers greatly: it turns convulſed, and dies. Worms have a ſtill more perfect organiſation. No kind of reſpiration is ſenſible in the polypi, which are leſs perfect in this function than vegetables, in which we have found tracheæ.

4. Digeſtion is the ſeparation of the nouriſhing matter contained in the aliments; and its abſorption by particular veſſels called *chyliferous;* it goes on in a canal continued from the mouth to the anus, and this canal is ſwelled out near about the top of the abdomen. This ſwelling is called the *ſtomach*. The alimentary

ca-

canal then follows: it is twisted to different sides, and is called *intestines*. This long tube, which is formed of muscles and membranes, is designed to retard the aliments, so as to extract all the nourishing substance which they contain. There are, besides, in the neighbourhood of the stomach, other glandular organs, whose office is to prepare fluids proper for stimulating the stomach and intestines, and for extracting the nourishing part of the food; these are, the liver, the spleen, and the pancreas: the bile and pancreatic juice flow into the first intestine, called *duodenum*, and are mixed with the aliments, to which they communicate an animal character, which assimilates them to the humours.

All the passage of the first intestines is filled with vascular mouths, destined to forward the chyle. These vessels carry it into the lumbar reservoir, into the thoracic canal, and the chyliform fluid is carried into the left subclavian vein, in which it mixes with the blood. Such are, in few words, the mechanism and phenomena of digestion in man.

Quadrupeds differ greatly in the form of their teeth, stomach, and intestines. There are some of these animals which have no teeth at all; as the ant, which eats only soft aliments: others have only dentes molares, as the tatou: some, as the elephant and sea-cow, have molares and canini: lastly, most of them have three kinds of teeth, molares, canini, and incisivi; but their number, position, and strength, vary prodigiously. What is most astonishing in this different structure of the teeth is, that there is a constant relation between the number and position of the teeth and the form of the stomach, according to the remark of Aristotle and Galen. Indeed, all quadrupeds which have dentes incisivi in both jaws, as the horse, the ape, the squirrel, the dog, the cat, have only a membranous ventricle, like man. Anatomists call these animals *monogastrici;* digestion is performed entirely in the same manner in them as in man. The quadrupeds which have dentes incisivi only in the lower jaw are the polygaſtric and ruminant; the camel, camel-opardalis, ram, goat, ox, ſtag, and kid. These quadrupeds are generally *biſulci*, and armed with horns. They all have four ſtomachs. The firſt is called, in the ox, the *paunch*, the receiver of the herbs: it is the greateſt, and divided into four other ſacs: it receives the aliment at the ſame time with the ſecond, which opens into the paunch by a large orifice: the herbaceous aliments contained in theſe organs dilate, the air is rarefied in them: they ſtimulate the nerves of theſe viſcera, and produce an antiperiſtaltic motion, which drives them up into the œſophagus and mouth, where they are again chewed by the dentes molares: reduced into a kind of ſoſt paſte by this operation, they are, as a fluid, carried, by a new deglutition, into the third ſtomach, the manyplies, the omaſus, by means of a half-hollowed canal from the œſophagus to this ventricle: laſtly, they ſoon paſs from hence into the fourth ſtomach, where they undergo true digeſtion. The inteſtines of ruminating animals are alſo far more extenſive than thoſe of monogaſtric quadrupeds. The cetacea reſemble entirely theſe laſt in the mechaniſm of this function *.

The

* The muſcular ſtructure of the œſophagus of herbivorous animals is very evident, which we find to terminate juſt where two ſacs or ſtomachs unite, the largeſt of which is on the left ſide: it has a very curious ſtructure; we diſcover in it a number of round papillæ, which increaſe the ſurface of it greatly. We call it the *venter, ventriculus*, or *ingluvies*, which may be compared to the croop of birds. The ſecond ſtomach has ſtill a more curious ſtructure, reſembling very much a ſection of a honeycomb, which is perhaps a very good name for it. This leads us to the third ſtomach; in which we find a number of beautiful proceſſes, and on every one of them a number of papillæ; or we find a number of plicæ, and hence the beſt name *manyplies*. Now, the laſt ſtomach reſembles our own much more nearly than any of the reſt does; it has a number of valvulæ conniventes, diſpoſed longitudinally, and which are more remarkable here than in man: then this ends in the pylorus.

Every perſon may recollect what probably they may have ſeen, that a cow or ſheep, after feeding a certain time, commonly lies down, and appears ſtill to be eating or chewing; and a ball is obſerved

The birds differ from one another in the ſtructure o
their ſtomach; ſome are membranous, others fleſhy o
muſcular. The firſt, called *hymenogaſtric*, are carnivof
rous. The ſecond, which deſerves the name of *myo
gaſtric*, live only on grains: their ſtomach is formed o

ſerved to aſcend and to deſcend alternately in the throat. The creature makes an effort; the ball riſes; then chewing a certain time, it ſwallows the ball; and the motion is ſo rapid, that the eye ſcarcely follows the ball.

The animal then takes in a ſufficient proviſion, receives it into the firſt ſtomach, which, after having lain there a certain time, it is brought from thence up into the mouth; we muſt ſuppoſe it returns, not into the ſame, but into the ſecond ſtomach, from the ſecond into the third, and from this into the fourth. When drink is taken, we do not obſerve that the animal receives the drink into the firſt ſtomach, brings it up to the mouth, and ſo back again; but taking a large quantity, it fills the firſt ſtomach, and makes it paſs by degrees into the reſt; and a calf does the ſame with milk. Nay, it may run directly from the œſophagus along the guttter into the fourth ſtomach.

Now we ſee at once the intention of nature in beſtowing ſo many ſtomachs; it is to enable the animal to convert the moſt groſs parts of vegetables into aliment, that nothing may be loſt. The dung of a horſe, which does not ruminate, is a much richer manure than the dung of a cow, from which the nutritious part is more thoroughly extracted. But we apply this farther in phyſiology. We have the cleareſt proof of a living principle, ſuperior to reaſon and experience, operating upon the active parts of animals. We ſee the ſtomach acting in various ways, according to the utility, without the animal knowing more of its inward make than we do. It is evident that every ſtomach poſſeſſes a different organiſation; and the concluſion to be hence inferred is, that the liquors from the mouth and ſtomach ought not to be conceived as ſerving merely for dilution, but that they act as menſtrua; for if a greater extent of ſurface alone was neceſſary, the ſame ſtructure would ſerve in all; but in every ſtomach a certain operation is performed by its liquor, as a menſtruum. Pieces of bones ſwallowed by animals are digeſted, and ſcarcely any veſtige of them is to be afterwards found: hence we may imagine, that the action of the ſtomach is much more complex than it has hitherto been ſuppoſed.

We obſerve, that the alimentary canal of the herbivorous animal is much longer than that of the carnivorous; ſo further changes need to be made, to render the food like to the conſtitution of the animal, and ſafe to be received; and the changes happening in the

laſt

a quadrigastric muscle, invested with a hard and thick membrane proper for trituration. These birds also have a double cæcum.

Fishes have a membranous stomach, long, furnished with several appendices: their intestines are in general short. We find a liver, but no pancreas. Reptiles present the same structure: their stomach is distended in a wonderful manner. Serpents are observed to swallow animals which are often larger than themselves.

Insects have a stomach and intestines fully organised. Swammerdam and Perrault say, that the mole has four stomachs: it is a stomach swelled and divided into four sacks, as we may be convinced by dissection. Worms have a very irregular stomach; small intestines are also found. The polypus seems to be only a stomach, for it digests very quickly: the same opening serves for a mouth and an anus.

5. Nutrition is a continuation of digestion and circulation: the solids always losing by the motion which they perform, must be repaired; and they are so by nutrition.

last part of the canal are necessary for the crude nature of the aliment in the upper end; and we are yet far from knowing the power of the intestinum cæcum. In the human body, it may be considered as an organ, chiefly for preparing a slimy matter; but in many animals, especially those living solely upon herbs, we find it of a very great size, nearly equal to the diameter of the stomach itself: in the bat, it is to the feel larger than the stomach; so a very great deal, in every part of digestion, still remains to be discovered by experiment.

The nourishment is absorbed here as in man, and conducted through the lacteals into the mesenteric glands: and the only circumstance here is, that the several glands are collected near the root of the mesentery, which is more transparent than in us, the fat following the course of the principal vessels.

After the lacteals have passed through these glands, they proceed in the usual manner upwards, till the nourishment gets into the circulating mass; though, after it gets there, we can still see it floating upon the surface of the blood for some time: so that authors who have supposed that the lungs and other organs contribute much to the mixture of the chyle with the blood, though proceeding on conjecture, have really proposed true opinions.

trition. In the firſt age of life, they acquire ſize, and the animal grows larger. Generally the cellular texture is conſidered as the organ of this function, and the lymph as the humour adapted to repair the ſolids. However, it appears, that each organ is nouriſhed with a proper and particular matter, which it ſeparates, either from the blood, the lymph, or any other fluid which moiſtens it. For example, the muſcles are nouriſhed from the fibrous matter which they ſeparate from the blood: the bones extract a calcareous phoſphoric ſalt and a lymphatic matter: the pure lymph is dried into plates in the cellular texture; the concreſcible oil is depoſited in theſe plates, to give riſe to the fat: each viſcus, then, has its particular manner of being nouriſhed, and the nutrition of each of them is a true ſecretion. Quadrupeds and the cetacea reſemble man perfectly in this function: the ſame thing happens alſo among birds: in fiſhes it is performed leſs quickly; on that account theſe animals live a very long time, and even the age of ſome is not known; in general, the ſlower the nutrition and growth, the longer the life.

Inſects have nothing particular in this function; only they grow ſolely under the form of larvæ, and not in the form of chryſalides and perfect inſects. Swammerdam and Malpighi have demonſtrated, that the larva contains, under ſeveral ſkins, the perfect inſect ready formed: the caterpillar alſo contains the butterfly, whoſe wings and feet are folded back.

In the worms and polypi, nutrition is performed in the cellular texture; it is done alſo in the ſame way in vegetables, by means of the reticular and veſicular textures.

6. Generation, conſidered in all animals, is performed in many different ways: the moſt require coition, and poſſeſs two diſtinct ſexes; ſuch are man, quadrupeds, and cetacea.

Female quadrupeds have a matrix, ſeparated into two cavities, *uterus bicornis*, and dugs in a greater number: they

they are not ſubjected to the menſtrual flux; the moſt of them bring forth ſeveral at a birth; their geſtation continues a ſhorter time: ſeveral have a particular membrane, deſtined to receive the urine of the fœtus; this membrane is called the *alantoides*.

The generation of birds is very different: the males have a very ſmall genital organ without a cavity; it is often double. Among the females, the vagina is placed behind the anus: theſe are ovaria without matrices, and a canal, deſtined to convey the egg from the ovarium into the inteſtine; this is called *oviductus*. The egg of the chicken, fecundated and not fecundated, has preſented facts unattended to by phyſiologiſts who have examined the phenomena of incubation. Malpighi and Haller are the only perſons who have made important diſcoveries. The laſt found the chick ready formed in eggs not fecundated.

Among fiſhes there is no decided proof of coition: the female depoſites its eggs upon the ſand; the male paſſes over them, and ejects upon them his ſeminal liquor, undoubtedly requiſite to fecundate them. Theſe eggs then hatch in a certain time.

The moſt part of male reptiles have a double or forked organ: theſe animals are oviparous, except the viper.

Inſects themſelves alone preſent all the varieties to be found among other animals. Some have two ſeparate ſexes in two ſeparate individuals; it is even the greateſt number. Others produce with or without coition, as in the vine-fretter: one of theſe inſects, contained alone under a glaſs, produces a great number of others. M. Bonnet has fully proved this fact by experiments repeated with the greateſt attention. The organ of the males is contained in the belly; it is made to come out by preſſing ſlightly upon the extremity of this part: it is generally furniſhed with two crotchets, deſigned to ſeize upon the female. The place of this organ is very various: in ſome, it is at the top of the belly

belly and before the breaſt, as in the female of the libellula; at other times, it is at the end of the antenna, as in the male ſpider. Inſects multiply prodigiouſly: they are almoſt all oviparous, except the wood-louſe.

Inſects are produced like men: each individual has two ſexes; and the coition is double, as is obſerved in the earth-worm and ſnail.

M. Adanſon adds, that the bivalves, animals in cockles or ſhells, have no organs of generation, and reproduce their young without coition: theſe worms are viviparous. The univalves, or ſnails, are oviparous; the young having come out either from the belly of the mother or from the eggs, have their ſhell ready formed.

The polypi are the moſt ſingular animals in the act of generation; they produce by ſlips: from every polypus in full vigour a button is ſeparated, which attaches itſelf to any neighbouring body, and grows upon it: polypi are alſo formed at their ſurface, as branches which produce trunks of trees.

In generation, only the phenomena are known; and all the ſyſtems which have been invented to explain the myſtery, always preſent unſurmountable difficulties: they are found collected in Haller's Phyſiology, *la Vénus Phyſique de Maupertuis*, Buffon's Natural Hiſtory. M. Bonnet is one of the naturaliſts who has ſaid moſt upon the ſubject, in his Conſiderations upon Organiſed Bodies. M. le Count de Buffon has offered an ingenious ſyſtem; and his work ought to be conſulted.

7. Irritability is the property which certain organs, called *muſcles*, have of contracting, that is, of ſhortening by the impreſſion of a ſtimulus. M. Haller has treated excellently of this doctrine. The muſcles of man, of quadrupeds, cetacea, and birds, reſemble one another: they are all equally red, formed of fibres united in bundles of different forms, covered and furniſhed with argentine membranes called *aponeuroſes*, terminated by flat roundiſh ſtrings called *tendons*.

Among

Among the fishes the muscles are white, and much more irritable than those that are red. In reptiles, irritability is still stronger and more tenacious; it continues a long time after the death of the animal; which appears to be common to all animals whose blood is cold, whilst the hot-blooded lose this property as the blood cools.

Insects have their muscles placed in the heart of thin bones, which are hollow, and of the nature of horn. This structure may be very well observed in the large green locust, or in the lobster.

The muscles of worms are very pale and irritable; they are even very strong, particularly in the worms with coverings, which have a weighty shell to move.

The polypi are very irritable; they contract into one point; they move their arms with a singular agility; they fold them back very readily. Their structure, however, does not appear to be muscular. It is irritability which gives animals the power of transporting themselves from one place to another, and of performing a great number of motions, in order to select the hurtful things, and to procure for themselves those that are useful. In the history of this function, then, these motions ought to be considered; standing and walking, jumping, flight, the pace of reptiles, swimming, are so many combined actions or results of muscular contractions proper to each class of animals. A full explanation of the cause of standing would require the examination of the musculi extensores of the human thigh. For leaping, we require the examination of the extremities, the form of the body, the long and acute face, and the thorax of quadrupeds, laterally compressed; for flying, the structure of the feathers of the sternum, pectoral muscles, the bill, and the tail, and the inner texture of the bones of birds. For this purpose, it would be necessary to consider in detail the muscular rings, the scales or tubercles, which serve reptiles instead of feet; the form of the body, the structure of the fins, that of the air-bladder, and its communication with the stomach,

ſtomach, in fiſhes. In inſects, the ſtructure, the number and poſition of the feet, the appendices of the tarſi, the form, poſition, and nature of the wings, of the balancers. At preſent it is ſufficient to have ſhown the importance of theſe conſiderations, and thoſe which merit in particular the attention of phyſiologiſts.

Laſtly, there is a reflection which appears to me to have been made improperly; it is, that a muſcle may be conſidered as a ſecretory organ, deſtined for the ſeparation of the fibrous and irritable matter, which we ſhall take notice of again in another place, and that the vitiated ſtate of this ſort of ſecretion ſhould be obſerved moſt attentively by phyſicians. We ſhall reſume the conſideration of this ſubject when we come to the blood.

8. Senſibility is a function, by means of which animals ſuffer the ſenſations of pleaſure and pain, according to the nature of the bodies which are in contact with theſe organs. The ſenſes depend upon the brain, the medulla oblongata, or ſpinal marrow, and the nervous cords, or pairs of nerves, which divide into a great number from theſe three foci. Without theſe organs they can have no ſenſibility.

In order to extend the mechaniſm of this function, theſe organs may be divided into three regions, which are continued, and ſeem to be only one, which phyſiologiſts have called the *ſenſible man.* Theſe three regions are, the focus, compriſed in the brain, cerebellum, and medulla oblongata; the middle part, orcommun ication which points out the nervous cords; and the ſenſitive expanſion, or the dilated expanſion of the nerves. This extremity or expanſion preſents a various form in different organs; ſometimes it is cellular and reticular, as the ſtomach and inteſtines; ſometimes it is ſoft and pulpy, as the bottom of the eye and the labyrinth of the internal ear; then it preſents the form of papillæ, as under the ſkin, upon the tongue, the glans penis; there

there it is expanded into long, soft, and flat filaments, as upon the nasal membrane of Schneider.

The brain of man is the most voluminous and most organised: hence the cause of its intelligence. Among quadrupeds it is much smaller: in recompense, the nerves are more sensible, and the sensations more acute, particularly that of smell, whose organ is very dilated, and as if multiplied by the number of the ethmoidal laminæ. The skin being thick and covered with hair, carries off the sensibility, and destroys the touch. The taste is very nice among animals. The ear presents the same apparatus as that of man. The cetacea have almost no brain relatively to the mass of their bodies: this organ is encircled with an oily thick fluid: their sensations are obtuse.

The brain of birds has not the same structure and the same apparatus of folds, eminences, and concavities, with that of man and quadrupeds. The beautiful structure of the eyes of these animals, their largeness, the thick and cartilaginous tunica sclerotica, the inner eye-lid, membrana nictitans, moved by particular muscles, the crystalline and vitreous humours, the bag of black matter contained in the extremity of the optic nerve, the brilliant covering of the choroid; all these announce a complete organisation; a care taken by nature to render the sight of birds piercing, and withal to enable them to discern at a distance their prey, and to shun the dangers which the rapidity of their flight might expose them to; in a word, to favour their agility and mobility, which seem to form the division of these animals. The sense of hearing is less perfect than their sight: they seem to show only a small sensibility to the smell and taste of aliments: the situation of the nostrils, and the hard membrane which covers the bill, fully explain these phenomena.

Among reptiles sensibility is very small. The brain is very inconsiderable; the nerves have no ganglions; the senses appear in general less active, although the eye and internal ear preſented a very beautiful organiſation to Meſſrs Klein, Geoffroy, and Vicq d'Azir.

Fiſhes have a very ſmall brain, and their cranium is filled with an oily maſs: their ſenſes, and particularly their ſight and hearing, are very delicate. The laſt of theſe organs is very well conformed, as Meſſrs Geoffroy, Klein, Camper, and Vicq d'Azir, have obſerved. The naturaliſts who imagined that fiſhes were deaf, are therefore much miſtaken.

Inſects have no brain, but a medulla oblongata, cylindrical, and full of knots, which pervade the whole of their body.

From this medulla ſhoot nervous filaments, which accompany the diviſion of the trachea; only the eyes of inſects are known. Swammerdam has deſcribed an optic nerve, which is divided under the cornea of the eyes like a net-work, into as many filaments as there are facets in this membrane. It is uncertain if they have any organ of hearing.

Almoſt no more traces of the ſenſible organ are to be found in the worm. Swammerdam has found a brain with two lobes, and moveable in the ſnail; two eyes, placed either at the baſe or at the point of the tentacula, and the contractile optic nerve, as well as theſe kinds of horns. M. Adanſon ſays, that in worms the eyes are ſometimes awanting, or that they are covered with an opaque ſkin.

As to the polypi, they have no organ of ſenſe, tho' they ſeem to ſeek for the light.

Senſibility, then, is the function which is imparted to man in a much greater degree than to all the other animals. It is that which diſtinguiſhes and places him at their head. This function ought to be underſtood at large, by the legiſlator, philoſopher, and phyſician.

LECTURE LXIII.

Of the CHEMICAL ANALYSIS *of* ANIMAL SUBSTANCES.

THE analyſis of animal ſubſtances is the moſt difficult and leaſt advanced part of chemiſtry: Ancient chemiſts were content with diſtilling theſe ſubſtances in a naked fire; but we now know that this operation alters, and diveſts entirely of their nature, the ſo compounded bodies as the fluid or ſolid animal ſubſtances are: only a few of the humours in man, and of certain quadrupeds, have been analyſed.

Many reaſons have been advanced againſt the progreſs of this branch of chemiſtry: the difficulty and diſagreeableneſs of theſe labours; the few reſources which ſcience affords, in treating of the animal ſubſtances, without their undergoing conſiderable changes; the impoſſibility of effecting the moſt diſtant ſyntheſis, in endeavouring to reproduce theſe ſubſtances; and in particular the ſmall intereſt which the moſt chemiſts, when not phyſicians, have hitherto had in the knowledge which this analyſis might furniſh; theſe make the principal motives which have ſtopt the progreſs of this ſubject. Nevertheleſs, the reſearches of ſome moderns, particularly Meſſrs Rouelle, Macquer, Bucquet, Poulletier de la Salle, Berthollet, Prouſt, Scheele, and Bergman, have diſcovered a new ſcene, and announce, that the healing art may reap conſiderable advantages from this kind of inveſtigation.

The bodies of the principal animals, ſuch as man and quadrupeds, which we attend to in particular, are formed of fluids and ſolids. The humours of animals are diſtinguiſhed into three claſſes, relatively to their uſe. The firſt comprehends the recrementitial humours, deſtined for the nouriſhment of ſome organs: the ſecond comprehends the excrementitial humours, which are ejected from the body by certain emunctories,

ies, as if useless, and susceptible even of hurting, f too long retained. In the third, the humours are ranked, which contain the two preceding, and of which a part is recrementitial and a part excrementitial. The first are, the blood, lymph, gelatinous, fibrous, or glutinous parts, the fat, marrow, matter of internal perspiration, and the osseous juice. The second comprehend the fluid of transpiration, sweat, the mucus of the nostrils, the wax of the ears, the wax of he eyes, the tears, the urine, and the excrements. The last are, the saliva, the tears, the bile, the pancreatic juice, the gastric and intestinal juice, the milk and seminal liquor. As there is great need that all hese fluids be known, we shall here treat of those only which the chemists have examined.

Of the Blood.

Among the excrementitial humours, the most important, the most compounded, and the most incomprehensible, is the blood. It shall be the first, because, according to the doctrines of the best physicians, it is he source and focus of all the other animal fluids. Several physicians, and in particular M. Bordeu, considered it as a kind of liquid flesh, and as a compound of all the animal humours. This opinion, however, although it be very probable, is not yet entirely demontrated.

The blood is a fluid of a fine red colour; of an uncuous fat consistence, as if soapy; of an insipid, somewhat saline, taste; which is contained in the heart, areries, and veins. This fluid greatly differs, according o the regions which it runs through; and it is not the ame, for example, in the arteries and veins, in the breast and in the region of the liver, in the muscles, and in the glands. This is a fact not sufficiently atended to by chemists who have treated of it in their works.

Upon considering the blood through the whole animal

kingdom, it is obſerved, that it varies ſingularly in different animals, in the colour, conſiſtence, odour, and particularly the temperature. This laſt property is the moſt important, and ſeems to depend upon circulation and reſpiration. Man, quadrupeds, and birds, have their blood hotter than the medium in which they live; for that reaſon they are called animals with hot blood. Among the fiſhes and reptiles, it is of a temperature equal to the ſurrounding medium; they are called therefore animals with cold blood. It is probable it might be the ſame caſe with the other properties of this fluid, and particularly the chemical qualities or characters, if the blood of all animals were diſtinctly known.

Human blood, which is here particularly meant, varies according to the age, the ſex, the temperament, and health. In infancy, among women and pituitous perſons, it is more pale and leſs conſiſtent; in robuſt ſtout men, it is of a deep red, almoſt black, and of a much more ſaline taſte. Phyſicians know alſo, that in perſons affected with the lues and ſcrophula, the tetters and gout, the blood is acrid, and deſtitute of taſte and ſweetiſhneſs, which are peculiar to it in the ſtate of health.

Before entering upon the analyſis of the blood, its phyſical properties muſt be known; its colour, heat, taſte, ſmell, and particular conſiſtence already mentioned. The microſcope diſcovers in it a great number of globules; which, when they are paſſing through the ſmall ducts, loſe, according to Leeuwenhoek and Boerhaave, their red colour, become yellow, and laſt of all white: ſo that, according to the Leyden phyſician, a red globule is an aſſemblage of ſeveral ſmaller white globules, and owes its colour only to the aggregation. The blood preſents alſo a ſingular phyſical property. While it is warm and in motion, it remains conſtantly fluid and red: when it is cool and in repoſe, it turns ſolid, which gradually ſeparates of itſelf into two parts; the one red, which ſwims above, of a deep colour,

colour, remaining concrete till it be altered; it is called the *craſſamentum*: the other, which occupies the bottom of the veſſel, is of a greeniſh yellow, gluey, called *ſerum* or *lymph*. This ſpontaneous coagulation and ſeparation of the two parts of the blood takes place in the laſt inſtants of animal life, and gives riſe to theſe concrete ſubſtances which are found after death in the heart and large veſſels, and which have been conſidered falſely as polypi.

Blood, expoſed to a gentle heat long continued, paſſes to the putrid fermentation. If it is diſtilled in a balneum mariæ; it yields a phlegm of an inſipid ſmell, which is neither acid nor alkaline, but which eaſily paſſes to the putrid ſtate by means of an animal ſubſtance which is diſſolved in it. Blood, very ſtrongly heated, coagulates *, and dries gradually, as M. De-

haen

* The coagulation of blood, with all the circumſtances which attend this appearance, bears but little analogy to the conſolidation of animal jellies or of mucilaginous ſubſtances by expoſure to cold. A quantity of blood, recently drawn from a vein, coagulates in the ſame time in any degree of heat between 56° and 104° of Farenheit's ſcale: however, the coagulum formed in the higheſt degree of heat is ſomewhat firmer; ſo that cold does not ſeem in the leaſt to promote the coagulation of blood. Neither will reſt alone produce it, as appears from the experiments of Mr Hewſon, who applied two ligatures upon the jugular vein of a living animal; and upon opening the vein after a ſtagnation of two hours, he found the blood contained between the ligatures ſtill poſſeſſed fluidity. But if the air of the atmoſphere gain admiſſion to a vein, the blood inſtantly coagulates; and this change ſeems to be produced in conſequence of ſome chemical attraction, ſince all kinds of air will not produce this effect: on the contrary, the admiſſion of phlogiſticated air prevents, or at leaſt retards, the progreſs of coagulation, and diſpoſes the blood to aſſume a deeper and more purple tint of colour. This deepneſs of colour diſtinguiſhes arterial blood from venous blood: and accordingly Dr Hamilton found, that arterial blood acquired all the properties of venous blood by mixing ſome dephlogiſticated air with it.

The lymph of blood contained in the umbilical cord is made to coagulate by expoſure to 118° degrees of heat, the ſerum requires about 160°; and ſo far is cold from aiding the coagulation of

blood,

haen has discovered: it loses seven-eighths of its weight, and makes an effervescence with acids. It may be hardened sufficiently by a well-conducted fire, so as to form a kind of horny substance. If some dried blood is exposed to the air, it slightly attracts humidity; and in some months a saline effervescence is formed, which M. Rouelle took for mineral alkali. Distilled in a naked fire, it yields an alkaline, not acid phlegm, as some chemists have pretended, and particularly Vieussens: then a light oil passes over; next an oil coloured and weighty, some concrete volatile alkali, or aërated ammoniacal salt, made impure by some gross oil: a spongy coal remains in the retort, very difficult of incineration, in which we find some marine salt, mild mineral alkali, some iron, and an earthy matter, whose nature is not yet known.

Blood united to the alkalis becomes more fluid by repose. The acids instantly coagulate it, and alter its colour; then the neutral salts, formed with the mineral alkali, and any of the acids indiscriminately employed, are extracted by filtration, evaporation, exsiccation with a gentle fire, and then by washing the remaining dried matter. Spirit of wine coagulates blood.

Experiments made upon recent blood do not point out the nature of the substances of which this fluid is composed; but the spontaneous decomposition of the blood, and the separation of its two parts, the crassamentum and serum, affords a method of acquiring this knowledge, by examining each of these matters in particular. It is but a few years since the chemical analysis of the blood was no farther advanced than what has been just now delivered; but the labours of Mess. Menghini, Rouelle junior, and Bucquet, have investigated this humour in a very different manner; particularly the two last have proved in their works, to what perfection

blood, that the blood contained in a portion of a vessel may be frozen, and afterwards thawed without congealing, but will speedily coagulate upon the admission of air.

fection the analyfis of animal fubftances might be brought by following their footfteps. The refearches of thefe philofophers fhall direct our following confiderations of the properties of each of the fubftances which compofe the blood.

The ferum is very far from being pure water: it is a particular matter of important confideration, to which we have given the name of *lymph*. It is of a yellowifh white, inclining a little to green; its confiftence is unctuous and gluey. Expofed to heat, it coagulates and hardens a long time before it comes to the boiling point: it turns the fyrup of violets green. Diftilled in a balneum mariæ, it yields a phlegm of a mild infipid tafte, neither acid nor alkaline, but which putrefies readily: after that it is hard, dry, and tranfparent like horn: it is incapable of folution in water: diftilled in a retort, it yields an alkaline phlegm, a great quantity of concrete volatile alkali, and a very fetid thick oil. All thefe products have in general a very fetid particular odour. The coal of the lymph, diftilled in a naked fire, fills the retort almoft entirely. It is fo difficult of incineration, that it muft be kept burning for feveral hours, and a great furface muft be prefented to the air, before it be reduced to afhes. Thefe are of a blackifh grey, containing marine falt, mineral alkali, often a little iron, and an earthy matter, which has not yet been properly examined. Might not this earthy matter be perhaps a kind of neutral phofphoric falt like the bafe of the bones?

The lymph, expofed for fome time to a warm temperature in an open veffel, paffes eafily to putrefaction, and then gives a great quantity of concrete volatile alkali of an infupportable fmell. It putrefies fo rapidly, that M. Bucquet could not afcertain if it paffed to the acid fermentation before it became alkaline. This liquor unites with water in all proportions; then it lofes its confiftence, tafte, and greenifh colour: the mixture muft be agitated to favour the combination, as the different denſity of theſe two fluids is an obſtacle to their union. The lymph, poured into boiling water, coagulates in great meaſure, and inſtantly. A portion of this fluid forms with the water a kind of an opaque and milky white liquor; which has, according to M. Bucquet, all the characters of milk, viz. which grows ſour like this fluid, coagulates by heat, the acids, &c.

The alkalis united with the lymph render it more fluid, occaſioning a kind of ſolution. The acids alter it in an oppoſite manner; they give it conſiſtence, and coagulate it. Upon filtrating this mixture, and evaporating the fluid obtained by this filtration, we obtain that neutral ſalt which the acid employed ſhould form with the mineral alkali; which proves that this laſt ſalt exiſts uncombined, and poſſeſſing all its properties in the lymph. The coagulum, formed in this liquor by the addition of an acid, diſſolves very readily in the volatile alkali, which is the true ſolvent of the lymph; but it does not diſſolve at all in pure water: an acid ſeparates the lymph united with the volatile alkali. The ſame coagulum, diſtilled in an open fire, yields the ſame products as pure dried lymph; and its coal contains a great deal of mineral alkali: which proves, according to M. Bucquet, that there is a portion of this ſalt intimately combined in the lymph, which the acid employed to coagulate it does not ſaturate.

Lymph does not decompoſe the argillaceous and calcareous neutral ſalts; but it decompoſes very well the metallic ſalts. It is coagulable by ſpirit of wine: this coagulum differs greatly from that which is formed by the acids in its ſolubility in water, according to the diſcovery of M. Bucquet. Therefore, from theſe details, the lymph appears to be an animal mucilage, compoſed of water, oil, marine ſalt, mild mineral alkali, and an inſoluble matter, which has been conſidered as a particular earth, although it is perhaps an earthy ſalt. The moſt ſingular property of this mucilage, and one which merits

merits the attention of physicians, is that of becoming concrete by the action of heat.

The crassamentum, exposed to heat in a balneum mariæ, yields an insipid water; it dries and becomes brittle. In a retort it yields an alkaline phlegm, a thick oil of a fetid and empyreumatic smell, and much concrete volatile alkali. Its residuum is a spongy coal, of a brilliant and metallic aspect, difficult of incineration, and which, when heated with the vitriolic acid, yields Glauber's salt and martial vitriol. After this operation, it yields an earth unknown. The crassamentum putrefies very readily in warm air. When it is washed with water, this fluid separates into two very distinct matters; the one, which it dissolves, communicates to it a red colour. This solution, treated by different menstrua, presents all the characters of the lymph; but it contains a much greater quantity of iron. This metal is extracted by incineration, and by washing the incinerated coal, to separate the saline matters. The residuum of this ley is in the state of crocus martis, of a beautiful colour: it is generally attractible by the magnet. The colour of the blood is attributed to this metal. Iron in very great quantity has been extracted from this fluid by Messrs Menghini, Rouelle, and Bucquet.

The crassamentum, after having been washed and deprived of all the red lymph it contained, is in the state of a fibrous white matter, which remains to be examined.

The fibrous part of the blood is white and colourless when it is well washed; it has only an insipid taste. By distillation in a balneum mariæ, an insipid phlegm, susceptible of putrefaction, passes over. The most gentle heat singularly hardens the fibrous matter. When it is exposed to a brisk heat, it contracts like parchment: distilled in a retort, it yields an alkaline phlegm, a heavy, thick, and very fetid oil, much concrete volatile alkali, tainted by a portion of oil. Its coal is not large, but compact, weighty, less difficult to incinerate than that of the lymph. It cinder is very white; it contains no ſaline matter, which has been carried off by the waſhing of the craſſamentum, nor iron; it is a kind of reſiduum, whoſe aſpect is earthy, but whoſe nature has not been examined.

The fibrous part putrefies very quickly, and with great facility. When it is expoſed to a warm and moiſt air, it ſwells, and yields much volatlle alkali. It is inſoluble in water. When it is boiled with this fluid, it hardens and acquires a grey colour. The alkalis do not diſſolve it; but even the weakeſt acids combine with it. The nitrous acid, concentrated, diſſolves it with efferveſcence and the diſengagement of nitrous gas; it forms with it a yellowiſh mucilage. With ſpirit of ſalt it yields a kind of green gelly. The acid of vingar diſſolves it by means of heat: water, and particularly the alkalis, precipitate the fibrous part united with the acids. This animal matter is decompoſed in theſe combinations; and when it is ſeparated from the acids by any means, it does not preſent the ſame properties. The neutral ſalts, and other mineral ſubſtances, have no action upon it. It unites with the lymph, particularly that which is coloured, to form the craſſamentum. This laſt is ſoluble entirely in the acids, as the fibrous part is, on account of the combination of this matter with the red lymph. Hence it is evident, that the fibrous part is widely different from the *lymph*, properly ſo called. This is a ſubſtance more animalyſed than this laſt; a ſort of animal gluten, which has great relation with that of the farina, and particularly which has the very remarkable property of becoming concrete by cooling and repoſe. It cannot be doubted, that this ſubſtance, which has not yet been ſufficiently diſtinguiſhed from the lymph by phyſiological and pathological authors, is of particular conſequence in the animal-œconomy. Might not this be what is depoſited in the muſcles, and that which conſtitutes the fibrous part of theſe organs, and the ſubſtance irritable from its excellence? If this aſſertion

were

were as fully demonſtrated, as it is probable, ſhould it not be important to pay more attention to this ſubſtance than has yet been done, and to regard it as capable, by its abundance or deficiency, of inducing particular diſeaſes? I ſhall have occaſion ſome other time to reſume this important ſubject.

Notwithſtanding all theſe reſearches about the blood's properties, all its chemical qualities are very far from being known. It is uncertain what is the difference between the lymph and the fibrous part. The blood has not been examined in all its ſtates; and eſpecially in the different diſeaſes, in which this fluid ſuffers conſiderable alterations, viz. in ſtrong inflammations, in the chloroſis, &c. Phyſicians are more acquainted with it in this point of view than chemiſts. M. Rouelle has examined the blood of ſome quadrupeds, as the ox, the horſe, the calf, the ſheep, the ſow, the aſs, and goat. They all yield the ſame products as that of man, but in different proportions.

LECTURE LXIV.

Of the Milk.

MILK is a recrementitial humour, deſtined for the nouriſhment of young animals in the firſt ſtages of life. It is of a dirty white; of a ſaccharine ſweet taſte; of a ſlightly aromatic ſmell. It is ſeparated immediately from the blood in the female mammæ of animals: it is conveyed principally by the mammary arteries. Man, quadrupeds, and the cetacea, are the only animals which give milk. All other animals want the organs which are deſtined for the ſecretion of this humour.

Milk differs greatly in the different kinds of animals: in women it is very ſaccharine; that of cows is ſweet, and its principles are well united: thoſe of the goat and aſs have a particular virtue; they are often ſlightly

ſlightly aſtringent. As to its other properties, its differences depend generally upon the aliments which nouriſh the animals.

Cow milk, which we ſhall make our example in this analyſis, becauſe it is eaſily procured, is a compound of three different ſubſtances; ſerum, which is fluid and tranſparent; the butter and cheeſe, both of which have a greater conſiſtence. Theſe three parts are mixed and ſuſpended, ſo that they form a kind of animal emulſion.

Milk, expoſed to the action of heat, and to a balneum mariæ, yields a taſteleſs phlegm, of a weak ſmell, and ſuſceptible of putrefaction. With a little ſtronger heat it coagulates like the blood, according to the obſervations of M. Bucquet. By agitation and gradual exſiccation, it forms a ſort of ſaccharine extract. This extract diſſolved in water conſtitutes the weak milk of Hoffman. Diſtilled in a naked fire, this extract yields ſome acid, fluid oil, concrete oil, and volatile alkali. Its coal contains very little fixed vegetable alkali, febrifugal ſalt, and a earth little known.

Milk, expoſed to a warm temperature, is ſuſceptible of the ſpirituous fermentation, and of forming a kind of wine; but it muſt be in great quantity. The Tartars prepare a ſpirituous liquor with the milk of mares. Milk paſſes to the acid ſtate readily, and then it coagulates. The cheeſy part forms into a maſs, and the ſerum ſeparates.

The acids inſtantly produce the ſame effect upon milk: they coagulate it: the alkalis, and eſpecially the volatile, rediſſolve the coagulum. Boerhaave ſays, that in boiling milk with oil of tartar, this fluid becomes yellow, then red, and of the colour of blood. He thinks alſo, that it is a ſimilar combination which converts the milk to the ſtate of true blood in the human body.

Weak milk is prepared by heating ordinary milk along with runnet. This ſubſtance formed by the milk, rendered ſour in the ſtomach of calves, is a ferment which coagulates the cheeſy part. When this coagulum

is

is formed, the milk is paſſed through a cloth. The herb gallium and thiſtles act as runnet does upon milk.

Serum prepared in this manner is turbid; it is clarified by means of the white of an egg and cream of tartar; it has a ſweet taſte; it contains a ſaccharine eſſential ſalt, a gelatinous matter little known, a colouring extractive ſubſtance, and ſome febrifugal ſalt diſſolved in a great quantity of water. In order to ſeparate theſe principles, the water is volatiliſed by evaporation to the conſiſtence of ſyrup; it is left to cool; a ſalt at firſt a little red cryſtalliſes, which is the ſugar of milk, and is purified by new ſolutions and cryſtalliſations. This ſalt cryſtaliſes nearly like vitriolated tartar: it has an earthy and a moderately ſaccharine taſte; it diſſolves in one part and a half of hot water; it yields to diſtillation an acid phlegm, a light oil, a ſpongy coal, which contains very little fixed alkali; it is very difficult of incineration. It does not ſeem ſuſceptible of paſſing *per ſe* to the ſpirituous fermentation; but when it is combined with the cheeſy matter and the gelatinous mucilage, it may form a kind of wine, ſince M. Spielman has proved by fair experiments, that milk fermented in a great maſs yielded ardent ſpirit by diſtillation.

From ſerum a new quantity of ſaccharine matter may be obtained by a ſecond, and even a third evaporation. The mother-water which remains is gluey; it turns frequently into a gelly by cooling, as M. Rouelle has obſerved; it contains a gelatinous mucilage, an extractive matter, and febrifugal ſalt, which is extracted by a new ſolution and evaporation. This ſalt may be detected alſo by oil of vitriol.

The ſugar of milk, expoſed to an open fire, fuſes, bubbles up, and burns like ſugar. It emits a ſmell of caramel; its coal is difficult of incineration; it yields very little aſhes, as a pound of the ſugar of milk furniſhed M. Rouelle only about twenty-four or thirty grains. This cinder contains febrifugal ſalt and a little vegetable fixed alkali, which this chemiſt aſcribed to the extractive matter. M. Vulgamoz found, according to M. Pœrner, the same analogy as M. Rouelle, between the salt of milk and sugar.

The cheese forms into a mass, and separates from the other parts of the milk by the action of heat, by the acid fermentation which this liquor is capable of undergoing, and by the mixture of acids. This matter, when well washed, is white, solid, as if fibrous: the action of a gentle fire hardens it. Distillation in a balneum mariæ extracts from it an insipid phlegm, which putrefies.

Dried cheese, distilled in a retort, yields an alkaline liquor, a heavy oil, and much concrete volatile alkali. Its coal is dense, very difficult of incineration, and yields no fixed alkali.

Cheese putrefies in a warm temperature: it bubbles up, emits a disagreeable smell, puts on a semifluidity, becomes covered with a scum, owing to the disengagement of a strongly-smelling and very mephitic gas, which escapes with difficulty from this viscous matter.

Cheese is insoluble in cold water; hot hardens it.

The alkalis dissolve it; and particularly the volatile, which, poured into milk coagulated by an acid, in the dose of a few drops, very soon resolves the coagulum.

The acids, concentrated, also dissolve it.

The neutral salts, and particularly marine salt, retard its putrefaction.

It appears from all these details, that cheese is a substance akin to the lymph; but as from its nature it is not soluble in water, it is by means of the gelatinous substance, and of the extractive and saccharine matters contained in the serum, that it, as well as the oily part, is kept in solution.

Butter partly separates from the milk by repose; it collects at its surface; but as it is mixed with much serum and cheesy matter, it is completely separated from them by a rapid motion: this constitutes the art of beating butter. The serum, which swims above the beat-

butter,

butter, retains a portion of this oily ſubſtance; it is yellow, ſour, and greaſy; it is called *butter-milk*. Cream is a mixture of cheeſe and butter taken from the top of the milk. It is of much more difficult digeſtion than milk. This ſubſtance is ſuſceptible of lathering by a great agitation. In this ſtate it is the whipt cream.

Pure butter is concrete and ſoft, of a golden yellow colour, of a ſweet agreeable taſte. It is a fat oil, rendered concrete by an acid: it fuſes with a gentle heat, and becomes ſolid by cooling. Diſtilled in a balneum mariæ, it yields an almoſt inſipid phlegm. In a retort, it gives over an acid of a very pungent ſtrong odour; an oil at firſt fluid; then a concrete coloured oil, of the ſame pungent odour as the acid. By rectifying theſe products, the oil is rendered fluid and as volatile as the eſſential oils. Only a ſmall quantity of coal remains.

Butter eaſily becomes acid and rancid in a hot air. Its acid is then evolved, and it has a diſagreeable taſte: Water and ſpirit of wine bring it back to its former ſtate, by diſſolving the acid. The fixed alkali diſſolves butter, and forms with it a kind of ſoap little examined.

From theſe details it is evident, that butter is an oily matter, of the nature of the vegetable concrete fat oils.

Freſh butter is ſweet, relaxing, and cooling: but it eaſily ſours, and in general agrees ill with moſt ſtomachs. Red butter, whoſe acid is evolved, is one of the moſt unhealthful aliments, and of the moſt difficult digeſtion.

Milk is an agreeable and uſeful aliment in many caſes. It is alſo one of the moſt valuable medicines which medicine poſſeſſes. It ſweetens the acrid humours in diſeaſes of the ſkin and joints; ſuch as the rheumatiſm, gout, &c. It cicatriſes ſome ulcers of a good nature. It may be impregnated with ſome aromatic parts; and is then an excellent medicine in the phthiſis pulmonalis. All ſtomachs do not digeſt milk. When acidities exiſt in in the firſt paſſages, it is generally hurtful. It generally requires great prudence in its adminiſtration. We uſe, ſometimes with ſucceſs, milk rendered medicinal by certain ſubſtances which we give to the animal which furniſhes it.

The milk of different animals has ſome particular virtues. That of women is mild, very ſaccharine, and agrees well in the maraſmus. Aſs-milk is ſucceſsfully employed in the phthiſis pulmonalis and gout: generally it is relaxing. Mare's milk approaches to aſs's milk. Goat-milk is ſerous and ſlightly aſtringent. Cow-milk is the thickeſt, fatteſt, and moſt nouriſhing: it is alſo the moſt difficult of digeſtion; and it is often neceſſary to take a little water along with it, or ſome aromatic infuſion, particularly if it is nauſeous and provokes vomiting.

Milk is employed alſo externally, as ſoftening and emollient. It readily calms pains; it ripens collections of matter and abſceſſes, and accelerates their ſuppuration. It is applied hot, and contained in a bladder, upon the affected parts.

LECTURE LXV.

Of the FAT.

THE fat is an oily concrete matter contained in the cellular texture of animals; it is white or yellowiſh, generally of an inſipid ſmell and taſte: in all animals it differs in ſolidity, colour, taſte, &c. Age alſo increaſes theſe differences: in infancy it is white, inſipid, and not very ſolid; in the adult it is firm and yellowiſh; in old age its colour is deeper, its conſiſtence very various, and its taſte in general more ſtrong.

That of man and quadrupeds is conſiſtent, white or yellow; that of birds is more fine, ſweet, unctuous, and in general leſs ſolid; among the cetacea and fiſhes it is almoſt fluid, and often placed in particular reſervoirs, as as in the cavity of the cranium. It is found in reptiles, insects, and worms; but in these animals it accompanies only the viscera of the lower belly, upon which it is laid in knots, only a small quantity is found upon the muscles and under the skin.

It has been observed, that the fat of frugivorous and herbivorous animals is firm and solid, whilst that of the carnivorous is more or less fluid. It must be, however, remarked here, that the fat is always less solid and concrete in a living and warm animal than it appears to be in a dead one, when cold and submitted to dissection.

The fat varies also, according to the different places of the body of the animal which conceal it: it is solid about the reins and under the skin; it is less so between the muscular fibres, or in the neighbourhood of the moveable viscera, such as the heart, stomach, intestines: it is more copious in the winter than summer: it seems to be destined to support the heat in the regions where it is placed; as many facts collected by physiologists demonstrate: it appears even to contribute to the nourishment of animals, as is observed in the bears, monkeys, rats; and in general all animals subjected to long abstinence, among which the fat melts, and is gradually destroyed.

When we want grease for pharmaceutical purposes, or for the examination of its chemical properties, it must be cut into pieces; the membranes and vessels which run through it must be separated, then washed with a great quantity of water: it is melted in a new earthen vessel, water being gradually added: when this fluid is dissipated, and no more boiling is excited, it is put into an earthen vessel, where it fixes.

All the chemical properties of fat have not yet been investigated. The action of heat, air, and two or three menstrua, only are known. However, it is one of the animal-matters which is the most necessary to be known, to enable us to judge of its uses, of which we know nothing

thing certain, and particularly of the alterations which it is susceptible of undergoing in living bodies.

The fat of any animal whatever, exposed to a gentle fire, liquefies, and then congeals by cooling. If it is strongly heated, and with the contact of air, it emits a fume of a pungent odour, exciting tears and coughing, and it enflames when it is hot enough to be volatilised: it yields but a small quantity of coal. If it is distilled in a balneum mariæ, a vapid water is got over, of a slight animal odour, which is neither acid nor alkaline, but which very soon acquires a putrid smell, and which deposites as if mucilaginous filaments. This phenomenon, which takes place in the water obtained by the distillation of all animal-substances in a balneum mariæ, proves, that this fluid carries along with it some mucous principle, which is the cause of its alteration. Fat distilled in a retort gives over a phlegm, at first aqueous, then strongly acid; an oil, partly liquid and partly concrete: there remains a very small quantity of coal, very difficult of incineration. These products have an acid, pungent, penetrating smell, as strong as that of the sulphureous acid; the acid is of a particular nature, and has not yet been examined. The concrete oil may be rectified by repeated distillation, so as to be very fluid, volatile, and penetrating; in a word, to exhibit all the marks of a true essential oil.

Fat, exposed to a hot air, is very readily altered: however sweet and destitute of smell when it is fresh, it becomes strong and pungent; it turns rancid: this alteration seems to be a real fermentation, which evolves the acid and sets it free. It is not the oily part of the fat which occasions this sort of change, but a particular animal-mucilage, which a farther analysis will show. Rancid fat may be corrected in two ways: water alone is capable of carrying off the acid which it contains, as M. Pœrner has observed; spirit of wine aso has the same property, according to M. de Machy. This proves, that the acid of rancid fat puts this fat matter into a kind

of

of ſoapy ſtate, and thus renders it ſoluble in water and ſpirit of wine. Theſe two fluids, then, may be ſucceſsfully uſed to deprive fat of its rancidity.

When fat is waſhed with a great quantity of diſtilled water, this fluid diſſolves a gelatinous matter, which may be demonſtrated by evaporation; but the fat always retains a certain portion of this matter, which is intimately combined with it, and upon which depends its fermenting property. The action of water upon this animal-ſubſtance has not been determined.

We are ignorant how lime and magneſia act upon greaſe; we know the alkalis combine with it, and form ſoap. The acids alter and burn it, as they do the fluid oils; they are capable even of converting it into the ſtate of an acid ſoap, ſoluble in water.

Sulphur unites very well with fat, and forms a combination which has not yet been fully examined.

Fat is ſuſceptible of diſſolving certain metals: it is alloyed with mercury in the preparation called *unguentum mercuriale*. The metal is triturated with hog's-lard for a ſufficiently long time: the mercury is divided, attenuated, and united ſo intimately with the fat, that it communicates to it a ſlaty colour, and it appears no longer in a metallic form. However, this union ſeems to be only an extreme diviſion; or, at moſt, there is only a ſmall portion of the mercury diſſolved by the acid of the fat; ſince, by means of a microſcope, globules of mercury are always perceived in the beſt prepared ointment.

Lead, copper, and iron, are the three metals which are the moſt alterable by fat; their calces likewiſe combine very eaſily with it: therefore it is dangerous to keep aliments prepared with greaſe in copper veſſels, and even in earthen ones whoſe covering contains glaſs of lead.

The moſt part of vegetable matters are capable of uniting with fat: the extracts and mucilages give it a kind of ſolubility, or at leaſt favour its ſuſpenſion in water. It combines in all proportions with oils, and communicates to them a part of its consistence.

Such are the known chemical properties of fat: they show us, that this substance is very similar to butter; that is, it is a kind of fat oil rendered concrete by a portion of acid.

As to its uses in the animal-œconomy, besides the heat which it supports in the parts which are adjacent; besides the round, flexible, and agreeable forms, and the whiteness which it gives the skin; it seems also, according to M. Macquer, to serve the purpose of absorbing the superabundant acids which may be found in the bodies of living animals, and it is as if the reservoir of these salts. It is known, however, that a very great quantity of acid, introduced into the body of an animal, dissolves and melts the grease, which is owing surely to its conversion into a kind of soap, and to its rendering it more soluble.

The excessive abundance, and particularly the alterations of the fat, produce in the animal-œconomy fatal diseases, whose symptoms and effects have not yet been fully examined. M. Lorry is particularly busy about this subject; and he has established between this substance and the bile a striking analogy. He has given several memoirs upon this subject, which will be printed among those of the Royal Society of Medicine.

Fat is also used as a seasoner to our food: it is nourishing to persons who have a good stomach. In medicine it is used externally as an emollient and anodyne: it enters the composition of several ointments and plasters.

Of the Bile.

THE bile or gall is a fluid, of a more or less yellowish green colour; of a very bitter taste; of an insipid disagreeable odour, which is separated from the blood in a glandular viscus, known to every body by the name of *liver*. In the greatest number of animals, except insects and

and worms, it is collected in a membranous reſervoir adjoining to the liver, called *gall-bladder*. Human bile has not been accurately examined, from the difficulty of procuring a great quantity: it is ox-bile which has ſerved for all our chemical experiments.

This liquor is of a conſiſtence almoſt gelatinous or glairy; it is ropy, like a moderately clear ſyrup: by agitation, it lathers like ſoapy water.

If it is diſtilled in a balneum mariæ, it yields a phlegm which is neither acid nor alkaline, but which is capable of turning putrid in a certain time. This phlegm often preſents a ſingular character; that of exhaling a very remarkable ſweet ſmell, very analagous to that of muſk or amber. This phenomenon has been ſhown in my own courſes; and ſeveral perſons have ſince confirmed it. When all the water which the bile contains has been extracted in a balneum mariæ, it is found in the ſtate of an extract more or leſs dry, of a dirty green and brown colour. This bilious extract attracts humidity from the air; it is very tenacious and pitchy, and it is entirely ſoluble in water. By diſtillation in a retort, it yields ſome volatile alkali, an empyreumatic animal oil, and much concrete volatile alkali. After this operation, a very voluminous coal remains, leſs difficult of incineration than thoſe hitherto ſpoken of. According to M. Cadet, who has given a very excellent paper upon the analyſis of the bile, in the Memoirs of the Academy, 1767, this coal contains mineral alkali, a ſalt which he imagines is of the ſame nature with the ſugar of milk, an animal-earth, and a ſmall portion of iron. It muſt be obſerved, that the diſtillation muſt be conducted ſlowly, becauſe this ſubſtance bubbles up conſiderably.

Bile, expoſed to a warm temperature, is very readily altered: its ſmell at firſt becomes more and more diſagreeable and nauſeous; its colour is deſtroyed; it precipitates whitiſh mucilaginous flocci; it loſes its viſcidity, and very ſoon aſſumes a fetid and pungent odour.

 When

When its putrefaction is far advanced, its smell becomes sweet like amber.

Bile dissolves very well in water.

The acids decompose it like soaps: they occasion a coagulum. If this mixture is filtrated, and the filtrated liquor evaporated, a neutral salt is obtained, formed by the acid employed and the mineral alkali; that is, Glauber's salt with vitriolic acid, rhomboidal nitre with the nitrous acid, and marine salt with muriatic acid. This excellent experiment, due to M. Cadet, demonstrates the presence of the mineral alkali in the bile. The substance remaining upon the filtre in these experiments is thick, viscous, very bitter, and inflammable; its colour and consistence vary, according to the nature and the degree of concentration of the acid employed in its separation. I have observed, that the vitriolic acid gives it a deep green colour, the nitrous a brilliant yellow, and the muriatic a beautiful clear green. This precipitate is a true resin: it bubbles up, melts, and kindles upon burning coals; it totally dissolves in spirit of wine; and water precipitates it as it does the resinous juices. The action of the acids upon bile, then, demonstrates, that this humour is a true soap, formed by an oil of the nature of resins, united with the mineral fixed alkali.

The neutral salts, mixed with bile, prevent its passing to the putrefactive state.

The metallic solutions are decomposed by the bile, which they decompose at the same time: the fixed alkali of this humour unites with the acid of the solution; and the oil of the bile precipitates with the metalic calx.

Bile easily unites with the oils, and takes them out of stuffs as soap does.

It dissolves in spirit of wine: this solution is of a greenish yellow; it deposites a gelatinous matter, which made a part of the bile, and which is insoluble in ardent spirit. The tincture of bile is not decomposed by water;

water; which demonſtrates, that this ſubſtance is a true animal-ſoap, equally ſoluble in aqueous and ſpirituous menſtrua. Ether alſo diſſolves it very readily.

Vinegar decompoſes bile like all the acids; upon evaporating the filtrated ſolution, a mineral acetous ſalt, well cryſtalliſed, is obtained.

From theſe different experiments it follows, that the bile is a compound of a great quantity of water, of an aromatic ſpiritus rector, a gelatinous mucilage, of an oil of the nature of the reſins, and of mild mineral alkali. M. Cadet found in it a ſalt of the nature of ſugar of milk. Beſides that, M. Poulletier de la Salle, to whom anatomy and animal chemiſtry are much indebted, found in human bile, or rather in biliary calculi, which are only bile thickened and concrete, a ſalt of a particular nature, which we ſhall ſpeak of preſently.

Bile, conſidered in the animal-œconomy, is a fluid which ſeems to ſerve the office of digeſtion. Its ſoapy quality renders it capable of uniting the oily matters with water. Its bitter taſte ſhows, that it ſtimulates the inteſtines, and promotes their action upon the aliments. M. Roux, a celebrated phyſician and chemiſt of the Faculty of Medicine of Paris, who to the loſs of theſe two ſciences died too ſoon, imagined, that the bile had ſtill a principal uſe, that of ejecting from the body the colouring part of the blood. Perhaps this humour itſelf is decompoſed in the duodenum by the acids, which are almoſt always evolved in digeſtion; it is certain at leaſt that it is very much altered, particularly in its colour, as it makes a portion of the excrements, which it colours. On this account, phyſicians draw very uſeful inductions from the inſpection of the fæces, to know what is the ſtate of the bile, and that of the liver which ſeparates it. The extract of ox-bile and of ſeveral other animal-matters is employed as a very good ſtomachic. It ſupplies the defect and inertia of the bile; it gives tone to the ſtomach, and re-eſtabliſhes the functions of this weakened viſcus: but its uſe requires caution,

 becauſe

becaufe it is acrid and heating; and it ought not to be given unlefs in a fmall dofe, efpecially to fenfible and irritable perfons.

Of the Biliary Calculi or Stones.

WHENEVER human bile is ftopt in its vefícula by any caufe, and particularly by fpafmodic contractions, as in the melancholia, hyfteria, long vexations, &c. it thickens and produces brown, light, inflammable concretions, of a very bitter tafte, which are called *biliary calculi*. Thefe concretions are frequently in great number; they diftend the veficula; they fometimes fill it entirely; they produce violent hepatic colics, vomitings, jaundice, &c.

Thefe calculi have been examined by M. Poulletier de la Salle. He obferved that they were foluble in ardent fpirit. Having fet thefe ftones to digeft in fpirit of wine, he remarked that this liquor was in a little time filled with minute, brilliant, and cryftallifed particles, having all the appearance of a falt. The experiments which he made upon this fubject fhowed him, that it was an oily falt, analogous in fome properties to the acid falt, which we have examined before under the article *Flowers of benzoin*. From the inveftigations of this philofopher, this falt is contained only in the biliary calculi of man: it was not found in thofe of the ox.

The difcovery of M. Poulletier de la Salle comes to be in part confirmed by facts collected by the Royal Society of Medicine, particularly upon the ftones of the veficula fellis. This company has received from feveral phyficians biliary calculi of a particular nature, which have not yet been defcribed. They are collections of tranfparent cryftalline laminæ, fimilar to mica or talc, which have abfolutely the fame form as the falt formed by M. Poulletier. It appears alfo, that human bile may yield a great quantity of thefe cryftals; fince the Society of Medicine has in its collection of calculi a veficula fellis, entirely filled with this tranfparent faline

con-

concretion. A full deſcription of theſe calculi may be found in the third volume of this company. It is to be wiſhed, that the nature of theſe calculi were examined: reſearches of this kind can be of uſe only to medicine.

From theſe details it is proper that two ſorts of biliary calculi be diſtinguiſhed: the one is opaque, fragile, inflammable, and truly bilious; it is a ſort of natural bilious extract. The other is tranſparent, cryſtalliſed in laminæ; and they appear to be a ſaline principle contained in the bile, which exiſts perhaps in greater quantity in certain morbific affections of this fluid, than in the natural ſtate; and which in this caſe is diſpoſed to precipitate and cryſtalliſe whenever the bile is ſtopt in great quantity in the veſicula,

Of the Saliva and Pancreatic Juice.

ANATOMISTS and phyſiologiſts have found a great analogy between the ſaliva and pancreatic juice. The ſalivary glands and the pancreas have indeed a ſtructure quite analogous; and the uſe of the humour which theſe organs prepare appears to be the ſame. Man and quadrupeds appear to be the only animals in which this humour is ſecreted; at leaſt ſalivary glands have not been found in moſt of the other animals.

Chemiſts have no exact knowledge on this ſubject. There is nothing the cauſe of this but the extreme difficulty of procuring even a very ſmall quantity. We only know, that the ſaliva is a very fluid juice, ſecreted by the parotid and ſeveral other glands, which runs continually into the mouth, but in greater abundance in maſtication. This humour ſeems to be ſoapy, impregnated with air, which renders it frothy; it leaves only very little reſiduum when it is evaporated to dryneſs. Salivary concretions, however, are ſometimes formed in the canals, which ſerve for the conveyance of this humour into the mouth. It appears to contain an ammoniacal ſalt, as lime and the cauſtic fixed alkalis diſengage from it a pungent and urinous odour. Dr Pringle's experiments ſhow that it is very ſeptic, and that it favours digeſtion by commencing a putridity in the aliments *.

Of

* All the fluids employed in the digeſtion of the food have been particularly mentioned by M. Fourcroy, except the gaſtric liquor, which poſſeſſes the moſt curious properties, though perhaps its chemical nature is the leaſt underſtood. A liquor of this kind, acting as a ſolvent, has long been known to exiſt in the ſtomachs of all animals with membranous ſtomachs, capable of diſſolving the food upon which the animals lived. M. Reaumur made a number of experiments upon the digeſtive powers of a hawk's ſtomach, by incloſing a variety of ſubſtances in a metallic tube perforated with holes: and the reſult of all his experiments was, that the gaſtric liquor of the hawk was capable of diſſolving fleſh, bones, and moſt animal ſubſtances; but that it had no effect upon the roots, leaves, and ſeeds of plants, or upon any of the vegetable matters upon which the herbivorous animals feed. Similar experiments have alſo been made upon dogs, and other animals, the ſtructure of whoſe ſtomachs reſembles the ſtructure of the human ſtomach, which lead to the ſame concluſion: ſo that there was every reaſon to infer from analogy, that the proceſs of digeſtion was conducted after the ſame manner in the human ſtomach. This conjecture was not however brought to the teſt of experiment, until Dr Edward Stevens got an opportunity to try the fact upon a Hungarian traveller, who could ſwallow large ſtones without hurting his health.

Dr Stevens gave him a variety of ſubſtances to ſwallow, incloſed within ſilver balls which had holes pierced in them. The balls paſſed through the whole courſe of the alimentary canal in the courſe of 30 hours; ſome days the time was longer, others ſhorter. Upon examination, the different ſubſtances were found to have been attacked by the gaſtric liquor, and to be partially diſſolved; at other times they were completely diſſolved, or reduced into the conſiſtence of a pulpy maſs; but at no time were they ever found to emit any thing of a fetid ſmell, or to be in the leaſt tainted with corruption. The fleſh of quadrupeds, birds, and fiſhes, apples, potatoes; all of them, both raw and cooked, were expoſed to the action of the gaſtric liquor, and all were diſſolved. The gaſtric liquor was not able to diſſolve bones; and ſome of the leguminous ſeeds were little affected by its action.

The reſt of theſe curious experiments will be met with in Dr Stevens's Theſis, publiſhed at Edinburgh in the year 1777.

It ſeems, however, moſt extraordinary, that ſo powerful a menſtruum of animal ſubſtances ſhould not act upon the ſtomach itſelf,

which

Of the Seminal Fluids.

The chemical nature of the ſeminal fluid is ſtill leſs known than that of the two preceding fluids. The few obſervations which it has been poſſible to make upon this humour, have ſhown, that it approached to animal mucilages; as it became fluid by cold and by heat, and as the action of the fire reduced it into a dry and friable ſubſtance.

Anatomical and microſcopical obſervations have gone farther than chemical experiments. They have demonſtrated, that the ſemen is an ocean in which ſmall round bodies ſwim, endued with a rapid motion, conſidered by ſome as living animals deſtined to reproduce the ſpecies; and by others as organic molecules, proper for forming by their approach a living animal. But we muſt aſſert, that theſe elegant experiments have contributed nothing to the advancement of ſcience, and that they have given riſe only to ſome ingenious conjectures.

LECTURE LXVI.

Of Urine.

URINE is a tranſparent excrementitious fluid, of a citron yellow, of a particular ſmell, and ſaline taſte, ſeparated from the blood by two glandular viſcera,

 called

which is continually expoſed to the action of the gaſtric liquor. This power of reſiſting ſolution has been by Mr Hunter aſcribed to the living principle of the ſtomach, which enables it to reſiſt digeſtion. Mr Hunter was led to this opinion, by finding the ſtomach of a perſon who died a violent death, without any previous complaint, partly diſſolved, and partly reduced into a pulpy maſs. The body was opened ſoon after death; and in ſimilar caſes the ſame appearance has occurred to other obſervers.

We are therefore, from conſideration of all the facts, to regard the gaſtric liquor as a very curious chemical ſolvent, diſtinguiſhed by peculiarities which we do not find to prevail in any other menſtruum we know.

called the *kidneys*, and conveyed from these organs into a reservoir, which is every where known by the name of *bladder*, where it remains some time: it is a kind of ley charged with the acrid matters contained in the animal humours; which if too long retained in the body, would produce disorder in the functions. Urine is a solution of a great number of salts, and particularly of extractive matters. When it is recent, it turns the syrup of violets neither green nor red: it varies in quantity and quality from several circumstances. That of man, which we propose to examine in particular, differs from that of quadrupeds. In the other classes of animals it exhibits still greater differences. The state of the stomach, and that of the humours in particular, produces an infinity of changes, which it will be impossible to appreciate till a set of experiments be completed, which have been only begun: we shall therefore confine ourselves to human urine in its state of health.

This fluid is distinguished by knowing physicians into two kinds: the one, called *urine of the drink*, or *crude urine*, flows a little time after repast; it is clear, almost tasteless and inodorous; it contains much fewer principles than the other, which is called *urine of the blood*, or *urine of coction:* this is secreted only when digestion is finished, and it is separated from the blood by the kidneys; whilst the former seems to filtrate partly from the stomach and intestines immediately into the bladder through the cellular texture.

The state of the health, and particularly the disposition of the nerves, singularly modify the urine. After hysterical or hypocondriacal paroxysms, it flows in considerable quantity: it has no smell, taste, nor colour. The diseases of the bones and articulations have singular influence upon the animal ley. It conveys often a great quantity of matter, apparently earthy, but which appears to be a calcareous phosphoric salt, as we shall take more notice afterwards: such is the deposition of the urine of gouty persons. Messrs Herissant and Morand, in partiticular, have obſerved, that when the bones are altered and ſoftened, the patients void a urine which depoſites a great quantity of this matter. It ſeems alſo, that in the ſtate of health, the urine conveys the quantity of this matter, the baſe of the bones, which is ſuperfluous in the nutrition and reparation of theſe organs.

Many aliments are ſuſceptible of communicating certain particular properties to the urine. Turpentine and aſparagus; the former gives it a violet, the ſecond a very fetid, colour. Perſons with weak ſtomachs void urine which retain the ſmell of the aliments taken in. Bread, garlic, onions, broth, and all vegetables, communicate to their urine a ſmell which diſtinguiſhes theſe ſubſtances. We have obſerved that this fluid does not alter the ſyrup of violets when it is freſh; kept ſome time, it turns it green inſtantly. Sometimes urine renders it green the moment it is voided; ſometimes it reddens it, ſince M. Macquer obſerved this property among the melancholici who had made uſe of greens or wine. From all theſe details, we conceive, that urine preſents phenomena to phyſicians, from which they may gain great advantage in practice. We muſt beware of imagining that we can judge, from the inſpection alone of the urine, of a diſeaſe, the ſex of a patient, and the proper remedies, as certain quacks pretend.

Human urine, conſidered anent its chemical properties, is a ſolution of a great number of different ſubſtances. Some are ſalts reſembling the mineral; and which, as M. Macquer thinks, come from the aliments, and have ſuffered no alteration. Others are ſubſtances analogous to the extractive principles of vegetables. Laſtly, there are ſome which appear particular to animals, and even to the urine, or at leaſt which have not been found in the products of the other kingdoms, nor even in the other animal ſubſtances, except urine. After having ſhown the means employed for the extraction of the different ſubſtances from urine, we ſhall give the hi-

hiſtory of theſe matters which are proper to this fluid, and of which we have as yet had no knowledge.

Recent urine, diſtilled in a balneum mariæ, yields a great quantity of phlegm, which is neither acid nor alkaline, but which readily putrefies. As this phlegm contains nothing particular, urine is generally evaporated in an open fire. In proportion as the water, which makes more than ſeven eighths of this animal humour, diſſipates, the urine aſſumes a brown colour; a powdery matter ſeparates, which has an earthy appearance, which has been taken for ſelenite, but which is a true ſalt, of little ſolubility, compoſed of phoſphoric acid and lime. This ſalt is of the ſame nature as the baſe of the bones; and we ſhall ſpeak of its properties in the hiſtory of theſe organs. When urine has acquired the conſiſtence of a clear ſyrup, it is filtrated and put into a cool place: it there depoſites ſaline cryſtals, which are made up of two particular ſaline ſubſtances, according to the diſcovery of M. Rouelle. Theſe cryſtals are known by the name of *fuſible ſalt*, *native ſalt of urine:* we ſhall examine its properties in a particular article. Theſe cryſtals may be collected at ſeveral times by repeated evaporations and cryſtalliſations. In theſe ſucceſſive evaporations, a certain quantity of marine and febrifugal ſalt cryſtalliſes: when the urine yields no more ſaline matter, it is in the ſtate of a very thick brown fluid, a kind of mother-water, and it holds diſſolved two particular extractive ſubſtances. By evaporation to the conſiſtence of a ſoft extract, and trituring the reſiduum with ſpirit of wine, M. Rouelle diſcovered, that a portion diſſolved in this menſtruum, and that another remained undiſſolved. The firſt has been called a *ſoapy matter*, and the ſecond an *extractive.*

The ſoapy ſubſtance is ſaline and ſuſceptible of cryſtalliſation. It dries only difficulty; and in this ſtate it attracts the air's humidity. It yields to the retort more than half of its weight of volatile alkali, a little oil, and ſome ſal ammoniac: its reſiduum turns the ſyrup of violets green.

The extractive ſubſtance, ſoluble in water and not in ſpirit of wine, dries eaſily in a balneum mariæ, like the extracts of plants: it is black, leſs deliqueſcent than the former: it yields to diſtillation all the products of animal matters. Such are the characteriſtic properties, according to M. Rouelle, that diſtinguiſh theſe two ſubſtances which form the extract of urine. To theſe details let us add, that this celebrated chemiſt extracted from an ounce to more than an ounce and a half of extract from a pint of urine voided after coction, whilſt a like quantity of crude urine gave only one, two, or three gros.

If, inſtead of ſeparating by means of ſpirit of wine this extract of urine into two diſtinct matters, it is fully diſtilled in a naked fire, it furniſhes a great deal of volatile alkali, a very fetid animal oil, ſome ſal ammoniac, and a little phoſphorus. Its coal contains a little common ſalt. This analyſis of urine ſhows, therefore, that this little is formed of a great quantity of water, a little marine ſalt, calcareous phoſphoric ſalt, or the baſe of the bones, a very great quantity of fuſible ſalt, and two particular extractive matters. Before we enter upon the examination of the fuſible ſalt, to which we ſhould direct a particular attention, let us proſecute the action of the different menſtrua upon urine.

Urine, expoſed to the air, is altered ſo much the faſter the hotter the atmoſphere: at firſt depoſitions are formed by ſimple cooling; ſeveral ſaline matters cryſtalliſe at its ſurface and bottom, and frequently a reddiſh ſalt called *gravel.* Nobody has paid more attention to the ſpontaneous alterations of this fluid than M. Hallé my colleague. He diſtinguiſhed, in the decompoſition of urine left to itſelf, ſeveral times, at which the nature of the ſediment, or of the cryſtals depoſited, and likewiſe the changes which they undergo, are evidently different. It is not our object to treat at full length of theſe changes, which will be found

found accurately described in an excellent paper written by the physician above mentioned, and which will be printed among those of the Royal Society of Medicine. We wish to show here only the great alterations which the urine suffers. Very soon after cooling, its smell alters, is exalted, and emits volatile alkali; its colouring part changes, and separates from the rest of the liquor; at last this alkaline odour is dissipated, and to it succeeds another, which is less pungent, but more disagreeable and nauseous; and the decomposition being complete terminates. M. Rouelle observed, that crude and serous urine did not putrefy so quickly; that its smell, when it altered, differed greatly from that of the urine of coction; and, lastly, that it was covered with mouldiness, like the juices of vegetables and the solutions of animal gellies. M. Hallé has seen certain specimens of urine pass to the acid before the putrid fermentation. Urine, putrefied for a year and more, subjected to evaporation, yields fusible salt just like recent urine; but it contains a portion of the acid of this salt uncombined; so that it raises an effervescence with the mild alkalis. The putrefaction volatilised a part of the volatile alkali. When it is evaporated, the salt deposited upon the sides of the bason is acid, and likewise effervesces with the oil of tartar. This observation is due to M. Rouelle junior.

Lime and the fixed alkalis instantly decompose the saline principles contained in the urine: it is sufficient for this purpose to pour some caustic fixed alkali, or to throw some quicklime into recent urine, and an insupportable putrid alkaline odour is evolved. It seems that it is from the decomposition of the fusible salt that these substances produce this odour. M. Bertholet, my colleague, discovered, that lime-water furnished a precipitate in recent urine, and that phosphorus might be extracted from this precipitate.

The acids have no action upon recent urine; but

they readily deſtroy the putrid ſmell of urine, and that of the depoſitions which it forms in this ſtate.

Urine decompoſes ſeveral mercurial ſolutions. Lemery exhibited, under the name of red precipitate, a magma of a roſy colour, which is formed when ſome nitrous mercurial ſolution is poured into urine. This precipitate is in part formed by the marine acid, and in part by the acid of the fuſible ſalt contained in this fluid. M. Brogniart obſerved, that this preparation ſometimes kindled by friction, and conſumes rapidly upon burning coals. This he attributed to a little phoſphorus.

Such is the preſent ſtate of our knowledge of the chemical properties of urine: there ſtill remains a great deal to be done, to complete what analyſis may diſcover upon this ſubject: it will be neceſſary to examine the different depoſitions obſerved in the urine, which are completely diſtinguiſhed by M. Hallé; the red or tranſparent ſaline concretions which are formed in it; the copious ſediment which the urine lets fall after gouty paroxyſms, and of that of patients attacked with the ſtone.

We are juſt now going to examine, in ſo many particular paragraphs, the ſaline products which are extracted from urine, and with whoſe properties it is neceſſary to be well acquainted.

§ I. *Of the fuſible or native Salt of Urine; or of the phoſphoric ammoniacal Salt.*

THE ſalt which is obtained by the cooling and repoſe of urine, evaporated to the conſiſtence of ſyrup, has been called *fuſible ſalt*, becauſe it fuſes with the fire, as we ſhall preſently ſee, *eſſential ſalt of urine*, and *microcoſmic ſalt*. The name of *phoſphoric ammoniacal ſalt* ſuits it beſt, becauſe it is formed of the particular acid of which phoſphorus is made, and of the volatile alkali.

Several authors, and among the reſt Meſſrs Margraaf, Rouelle, Pott, Schloſſer, and the Duc de Chaulnes, have

have begun to examine the properties of this salt: nevertheless, we are not acquainted with them all, as we shall see from the following observations. M. Proust has made a precious discovery upon the nature of this salt, as we shall presently observe.

This salt, obtained by the process just now described, is not pure; it is rendered impure by an extractive matter, and is frequently mixed with some marine salt: it is purified by solution in distilled water and evaporation, or, still better, by allowing this solution to evaporate in the open air. The form of its crystals has not yet been accurately described: sometimes they appear to be octagons, and in other circumstances it is obtained under the form of flat rhomboids, cut *en biseau* upon the sides, and placed obliquely, and as if covering one other. M. Romé de Lisle defines them regular four-sided figures, formed of four equilateral triangles. M. Rouelle has observed, that its form varied egregiously. Its taste is cool and pungent: this salt turns the syrup of violets green.

When it is put upon burning coals, it bubbles up, emits a volatile alkaline odour, and fuses. If if is distilled in a retort, very penetrating caustic volatile alkali is obtained. The residuum is a transparent glass, very fixed and fusible; which, according to M. Margraaf, is soluble in two or three parts of distilled water, and presents the characters of an acid. It has always been supposed, since the experiments of this chemist, that this glass was pure phosphoric acid; but M. Proust discovered in it a particular substance which is united with this acid, and which covers its properties. We shall treat of this substance in a separate article. It is sufficient to observe here, that the fusible salt seemed to be a compound of three bodies.

The fixed alkali, vegetable or mineral, and lime, decompose fusible salt, and disengage its volatile alkali. It decomposes mercurial marine salt, without touching mercurial nitre.

Lastly,

Laſtly, When a mixture of two parts of this ſalt to one of charcoal is ſtrongly heated in a retort, a ſolid product is obtained, combuſtible in the open air, which is known by the name of *phoſphorus*, whoſe properties ſhall be more fully noticed afterwards.

§ 2. *Of the Subſtance diſcovered by Meſſrs Pott, Margraaf, and Prouſt, in the fuſible Salt.* (See Phyſical Journal, vol. xvii. p. 145.)

M. Prouſt, juſt as M. Rouelle, was ſurpriſed at the ſmall quantity of phoſphorus which is obtained from fuſible ſalt treated with charcoal, and at the weak taſte of the vitreous reſiduum of this ſalt, when fuſed and decompoſed by the fire, and ſuſpected that this matter was not the pure phoſphoric acid, and that it contained ſome ſubſtance incapable of forming phoſphorus: conſequently, after having diſtilled a mixture of one part of fuſible ſalt and a half part of charcoal, and having obtained all the phoſphorus, he waſhed the reſiduum of the retort with diſtilled water; he filtrated and evaporated the ley in the open air; he got parallelogram cryſtals, an inch long, laid horizontally upon one another. Their quantity amounted from five to ſix gros in the ounce of phoſphoric glaſs. Meſſrs Pott and Margraaf obtained this ſalt by waſhing the reſiduum of the phoſphorus formed with fuſible ſalt and charcoal; but the former conſidered it as a ſelenitic earth; the latter does not explain its nature, and he ſuſpects that it contains ſtill a little phoſphoric acid. M. Prouſt has paid more attention to it; he has pointed out, by accurate experiments, ſeveral of its properties, and chiefly its ſtate in the fuſible ſalt. From this chemiſt we ſhall borrow the examination of this ſubſtance.

The ſalt, the baſe of the phoſphoric glaſs, cryſtalliſes regularly, as we have ſeen: its taſte is ſlightly alkaline: it turns the ſyrup of violets green.

Expoſed to the fire in a crucible, it bubbles up, loſes the water of its cryſtalliſation, reddens, and fuſes.

The glaſs which it forms is very fluid, and turns opaque upon cooling. Poured upon burning coals, when in fuſion, it gives a green colour to the flame.

Expoſed to the air, it effloreſces.

An ounce of hot water can diſſolve about five gros of this ſalt cryſtalliſed. This ſolution yields cryſtals by cooling.

The ſalt, the baſe of the phoſphoric glaſs, diſſolves the earthy ſubſtances by means of fuſion, and forms with them different glaſſes.

It unites with and ſaturates the alkalis. It is found in human urine, even combined with the mineral alkali; and forms a particular ſalt, to be examined afterwards.

It diſſolves in the mineral acids, and divides their water of ſolution, but ſpirit of wine precipitates it. It yields with the phoſphoric acid, obtained from the phoſphorus, a mixt ſubſtance, ſuſceptible of a vitreous form, and which makes part of the fuſible ſalt of urine, or of phoſphoric ammoniacal ſalt, as we have already ſaid.

It decompoſes nitre and marine ſalt, ſeparating their acids, and uniting with their baſe, as Margraaf has already obſerved.

Laſtly, it does not form phoſphorus when diſtilled with charcoal.

Theſe details demonſtrate, that the ſalt, the baſe of the phoſphoric ſalt, is a particular ſubſtance, owing, according to M. Prouſt, to the work of animaliſation. This chemiſt had obſerved, that this ſubſtance has great analogy with the ſedative ſalt, in its fuſibility, its property of tinging flame green, of decompoſing vitreous and marine ſalts, and of forming with the alkalis particular neutral ſalts. However, it widely differs in the form of its cryſtals, in its alkaline taſte, and property of turning ſyrup of violets green, in its effloreſcence, inſolubility in ſpirit of wine, &c.

As to the nature of this ſingular ſubſtance, the examinamination of which interests chemists and physicians, M. Proust says, that its answer is found; but that it so nearly relates to one of the most important problems in chemistry, that he imagines he should not give the one without the other *.

§ 3. *Of the fusible Salt with Base of Natrum, extracted from Urine.*

M. Margraaf has remarked, that fusible ammoniacal salt of urine was mixed with another sort of neutral salt, which greatly differs from it. M. Haupt has already made known several of the properties of this second native salt of urine, which he has called *mirabile perlatum*; but Mess. Rouelle and Proust have resumed the full examination of this salt, and to this last chemist in particular we owe an accurate knowledge of this salt.

When fusible or ammoniacal salt is purified, about the end of its crystallisation, and above it, there are formed fine saline groups of another nature, whose crystals are, according to M. Rouelle, flattened four-sided prisms, irregular, having one of the extremities two-sided, and composed of two rhomboids cut contrary-wise,

* M. Proust says, that it is astonishing M. Margraaf did not discover the substance united with the phosphoric acid of the fusible salt, and which deprives it of its acid properties in the glassy residuum of this salt, exposed to the action of heat. However, upon reading Margraaf's Dissertation, intitled, "Chemical Examination of a Salt of Urine, very remarkable," (first volume of his chemical works, Paris 1762, p. 123.) it is found, p. 172, 173, 174. that M. Margraaf, after having distilled an ounce of fusible salt, separated from its urinous part, with half an ounce of soot, and having extracted a gros of very fine phosphorus, he washed in boiling distilled water the caput mortuum of this operation; that this ley, filtrated and evaporated, produced seven gros of long crystals, which became dry in the air, but which was reduced into a farina by heat. These crystals, treated with charcoal, gave no phosphorus: dissolved in water, they precipitated metallic solutions, decomposed nitre and marine salt in small quantity. These crystals appear to be just the substance marked by M. Proust.

wiſe, and the other adheres to the baſe. The ſides of the priſm are alternate irregular priſms, and two long pyramids cut *en biſeau* *. It is this ſalt which M. Rouelle called *fuſible ſalt with baſe of natrum*. He ſuppoſed that it was formed of phoſphoric acid and mineral alkali. The details to be offered, according to M. Prouſt, will prove that it is not this acid which enters into its compoſition.

Fuſible ſalt with baſe of natrum deſerves this name, becauſe when it is expoſed to the fire in a crucible, it fuſes and yields a glaſs which becomes opaque upon cooling: heated in a retort, it yields only a phlegm, without any mark of its being acid or alkaline.

It diſſolves very well in diſtilled water, and cryſtalliſes by cooling: its ſolution renders the ſyrup of violets green.

The fuſible ſalt with baſe of natrum yields no phoſphorus with charcoal. Meſſrs Rouelle, Margraaf, and Prouſt, have proved this important fact; and it is undoubtedly this phenomenon which has induced this chemiſt to undertake a more complete examination of its nature.

Lime decompoſes this ſalt; and it has more affinity with the ſubſtance, which, as ſhall be more fully explained after, holds here the place of an acid, than with the mineral alkali. If ſome lime-water is poured into a ſolution of this ſalt, a precipitate forms, and the mineral alkali remains pure and cauſtic in ſolution.

The mineral acids, and even diſtilled vinegar, decompoſe it in an inverſe manner. By the action of the acids, M. Prouſt diſcovered the nature of this ſalt. M. Rouelle

* It were to be wiſhed that all cryſtals were deſcribed with the ſame accuracy with theſe. M. Rome de Liſle has a part of this deſcription in his excellent book, intitled *Chryſtallography*. The knowledge acquired theſe ſome years, leads us to preſume, that the ſecond edition, in which this learned man is now engaged, will contain many cryſtals, whoſe form was unknown at the time of the firſt publication.

Rouelle was of opinion, that the vitriolic and nitrous acids did not act upon this ſalt, becauſe they produced no ſelf-evident change. But M. Prouſt having mixed the vitriolic, nitrous, marine, and acetous acids, with a ſolution of fuſible ſalt with baſe of natrum, has obſerved, that though no precipitate fell in theſe mixtures, the liquors, evaporated and cooled, yielded Glauber's ſalt, cubic nitre, marine ſalt, and terra foliata mineralis; which proves, 1. That this ſalt was decompoſed by the acids; 2. That it contains mineral alkali, as M. Rouelle had already demonſtrated. As to the ſubſtance ſeparated, and which was united with the mineral alkali before, it is clear that it remains in ſolution in the liquors at the ſame time with the new-formed neutral ſalts. M. Prouſt has very fully examined it in the mother-water, which is obtained after the mixture of the vinegar and the cryſtalliſation of the terra foliata mineralis. By pouring upon this mother-water eight or ten times its bulk of hot ſpirit of wine, the laſt portions of the terra foliata diſſolve, and a magma is formed, which is waſhed with new ſpirit of wine, and then diſſolved in diſtilled water. This ſolution of the magma, evaporated in the open air, yields cryſtals like parallelograms, quite ſimilar to thoſe which are extracted from the waſhing of the reſiduum of the phoſphorus made with ammoniacal fuſible ſalt. It is, then, this particular ſubſtance, analagous to the ſedative ſalt, whoſe hiſtory we have given in the preceding article, which ſaturates the mineral alkali in the fuſible ſalt. This important diſcovery explains, why this ſalt gives no phoſphorus: it contains not an atom of phoſphoric acid; it has an analogy with borax, becauſe the ſubſtance which does the part of an acid in it reſembles ſedative ſalt.

Fuſible ſalt with baſe of natrum decompoſes the earthy neutral ſalts, and eſpecially calcareous nitre: it forms a precipitate, which is the combination of lime and the base of the phosphoric glass *: the supernatant liquor contains a neutral salt, formed by the acid of the calcareous salt, and the marine alkali of the fusible salt with base of natrum.

This salt is likewise decomposed by the metallic solutions: poured into a solution of mercurial nitre, it forms a white precipitate, which, distilled in a retort, yields a small quantity of a reddish sublimate, some flowing mercury; and at the bottom of the vessel remains an opaque white mass, adhering and combined with the glass. This is the base of the phosphoric glass, separated from the mercury and fused. This mercurial salt, boiled with a solution of salt of soda, reproduces the fusible salt with base of natrum, and leaves the mercury in the state of a red brick-coloured powder †.

LEC-

* It must be recapitulated, that this new substance, to which, for want of a better name, we give that of *base of phosphoric glass of the fusible salt*; which expresses, that it is found always in the glass which forms the ammoniacal fusible salt when exposed to the fire, constitutes with the mineral alkali the fusible salt with base of natrum, so called by M. Rouelle, or the sal mirabile perlatum of M. Haupt, which we have been examining in this paragraph.

† The microcosmic salt obtained by the first crystallisation proves extremely impure, from the adherence of a great deal of the mucilaginous part of the urine; and, what is exceedingly remarkable, we lose a large proportion of this salt in attempting to free it from those impurities by a subsequent crystallisation; which does not happen in the depuration of any other salt by solution and evaporation. M. Baumé first discovered this singular circumstance by accident, when he had the misfortune to lose almost the whole of the cargo which he had prepared to make Kunckel's phosphorus with. The Duke de Chaulnes afterwards found, that at least four-fifths of the salt were lost in the experiments which he made to ascertain the quantity which exhaled during the evaporation of the water; and the Duke, in order to prevent so immense a waste, tried several expedients to save the salt; and was at length persuaded that the following method answered the purpose best: He made some of the alt very hot in a glass vessel, and added one half the weight of warm water. He next poured the whole into a glass funnel lined with filtrating paper, and set the neck of the funnel into a phial full of boiling water, which stood in a sand-heat to support an equable warmth.

LECTURE LXVII.

§. 4. Kunckel's Phosphorus.

Phosphorus is the most combustible substance known. As it was first extracted from urine, and as the matter which yields the greatest quantity of it is

Ff 4 the

warmth. In this situation the salt runs all into the phial; which being set in a cool place to crystallise, afforded white and perfectly pure crystals, free from every impurity, and with the loss of no more than one fifth part of the original quantity. The advantage of this process seems to consist in making a strong solution in boiling water, which yields crystals by the mere exposure to cold, without being subjected to the expenditure of salt; which always happens in evaporating a weak solution down to a sufficient strength, by the application of heat alone.

M. Margraaf found, that a mixture of this fusible salt with filings of zinc, distilled in a retort, yielded a very pure phosphorus with the assistance of a much more moderate heat than what is necessary with any other inflammable addition. So that zinc seems to be the best substance to employ in preparing phosphorus.

With regard to the volatility of the phosphoric acid, M. Scheele makes a remark which deserves some attention. He placed some of the concrete phosphoric acid into a glass tube, and by approaching the flame of a candle to that side of the tube, the phosphorus sublimed over to the other; but by the addition of some water, the acid became so fixed as to bear a red heat without volatilising. So that in this case two substances, both volatile when separate, acquire an uncommon degree of fixity by mixture. The fact is indeed analogous to some others which occur in chemical researches. When the volatile vapours of the nitrous acid, for example, and of the volatile alkali, meet together in the air, they form a salt which is no way remarkable for volatility; and in some other cases the compound neither possesses the volatility nor the solubility of the ingredients when separate.

Since the publication of M. Fourcroy's lectures, the presence of phosphoric acid in the mineral kingdom has been announced to the world as a discovery made by Dr Gahn. The acid is mineralised with lead, from which the phosphorus is obtained pure, by dissolving the mineral salt completely in nitrous acid; then precipitating the whole lead by the addition of the vitriolic, pouring off the supernatural liquor, which consists of phosphoric and nitrous acids mixed together, and afterwards expelling the nitrous acid by heat when the phosphoric acid remains behind. This is the very same process which was originally employed to separate the phosphoric acid from animal bones, and is in every respect perfectly just.

the ammoniacal fusible salt, which we have just now been examining, it is our opinion that the history of this inflammable substance should be inserted here.

The discovery of phosphorus is owing to an alchemist called *Brandt*, a citizen of Hamburg, who discovered it in the year 1677. Kunckel associated with one called *Kraaft*, in order to gain the secret; but Kraaft not communicating it to him, Kunckel resolved to search for it; and after having undertaken an investigation of urine, from which he knew it was extracted, he succeeded in making phosphorus, of which he ought to be considered as the true inventor. Some ascribe the honour of this discovery also to Boyle, who indeed deposited a small quantity of it, 1680, in the hands of the secretary of the Royal Society of London: but Stahl says, that Kraaft told him, that he communicated the process to Boyle. This last physician gave his process to a German called *Godfreid Hankwitz*, who had an elegant laboratory in London, and was the only person who made phosphorus, and who sold it to all the philosophers in Europe. Although, from the year 1680 to the commencement of the present century, there appeared a great number of receipts for the preparation of phosphorus; and among others, those of Boyle, Kraaft, Brandt, Homberg, Teichmeyer, Frederic Hoffman, Niewentyt, and Wedelius; no chemists as yet prepared it; and this preparation was a real secret, when, in the year 1737, a foreigner offered at Paris a method of preparing phosphorus with success. The academy nominated four chemists, Messrs Hellot, Dufay, Geoffroy, and Duhamel, to investigate this operation in the laboratory of the royal gardens: this process succeeded very well. The minister recompensed the foreigner; and M. Hellot described it with accuracy, in a paper inserted among those of the academy *anno* 1737. This operation consists in evaporating five or six quarts of urine, till it be reduced to a grumous, consistent, black, and shining matter; in calcining this residuum in an iron laddle,

laddle, whoſe bottom is made red-hot, till it no longer fumes, and till it has the ſmell of flowers of peaches; in waſhing this calcined matter with the double, or leſs, of hot water; and in drying it after having decanted off the water. Three pounds of this matter are mixed with a pound and a half of groſs ſand or grated freeſtone, and four or five ounces of powder of charcoal of beech: this mixture is moiſtened with half a pound of water, and introduced into a Heſſian retort. We eſſay the matter by making it red in a crucible; when it emits a violet flame and a garlic ſmell, it will yield phoſphorus. The retort is placed in a furnace made fit for the purpoſe; and a large receiver, filled two-thirds with water, is adapted to it. This receiver muſt be pierced with a ſmall hole; and M. Hellot conſidered this as one of the moſt neceſſary ſteps to the ſucceſs of the operation. Three or four days after the apparatus has been ſet in order, the fire is applied with great moderation, in order to complete the exſiccation of the furnace and the lutes; it is gradually augmented to the greateſt force of heat, and in this ſtate ſupported for five or ſix hours. The phoſphorus diſtils only at the end of four hours after the commencement of the operation, which continues in the whole twenty-four; previouſly, a great quantity of concrete volatile ſalt riſes, which diſſolves in part in the water of the balloon. The volatile or aëriform phoſphorus paſſes over firſt in luminous vapours; the real phoſphorus next flows like an oil or fuſed wax. When no more paſſes over, the apparatus is left two days to cool; it is unluted, and ſome water added into the receiver, to detach the phoſphorus adhering to its ſides: it is made to melt in boiling water, and cut into ſmall pieces, which are introduced into a tapering glaſs tube, and then immerſed into boiling water. The phoſphorus fuſes, is purified, and becomes tranſparent, by the ſeparation of a black matter which is raiſed above it: then it is cooled in cold water, where it congeals; it is taken out of the tube by being puſhed from the

the ſmaller through the longer end. Such is a ſhort view of M. Hellot's deſcription; the length of the operation prevented chemiſts from repeating it, if we except M. Rouelle, who, in his chemical courſes, practiſed it with ſucceſs ſeveral times.

In the year 1743, M. Margraaf, in the Memoirs of the Berlin Academy, publiſhed a new method of making a great quantity of phoſphorus, with leſs trouble than had been done before. According to his proceſs, ſome plumbum corneum, the reſiduum of the diſtillation of four pounds of minium and two pounds of ſal ammoniac, is mixed with ten pounds of the extract of urine, in the conſiſtence of honey. Half a pound of powdered charcoal is added: the mixture is dried in an iron veſſel till it be reduced to a black powder; this is diſtilled in a retort, in order to extract by a well-regulated fire. the volatile alkali, fetid oil, and ſal ammoniac. The fire muſt not be made more violent than ſo as to make the bottom of the retort moderately red. The black and friable reſiduum of this operation is the matter from which the phoſphorus is extracted: it is aſſayed by laying a little of it on burning coals; if it emits a garlic ſmell and a blue phoſphoric flame, it is properly prepared. With this we fill three-fourths of a Heſſian or Picardy ſtone retort, well covered with a lute: this veſſel is placed in a reverberatory furnace, terminated by the head of a wind-furnace, and by an iron pipe ſix or eight feet high; a middle-ſized receiver, pierced with a ſmall hole, and half filled with water, is adapted to the retort; the joinings are ſecured with fat lute, covered with pieces of cloth lined with whites of eggs and lime: a brick-wall is raiſed between the furnace and receiver: the apparatus is allowed a day or two to dry; then the diſtillation commences by a gradual fire. This operation continues from ſix, eight, or nine hours, according to the quantity of matter employed. This phoſphorus is rectified by diſtilling it with a very gentle fire in a glaſs retort, with a recipient half full of water. Al-

moſt

moſt all the chemiſts have repeated M. Margraaf's proceſs with ſucceſs; and this alone was in uſe, till, ſeveral years ago, a method was diſcovered of ſeparating the phoſphoric acid from bones, as we ſhall remark upon ſpeaking of theſe organs.

It is evident that M. Margraaf's proceſs differs from that of M. Hellot only in the addition of plumbum corneum, the utility of which is not yet properly known, and in the diviſion of the operation into two. But what is of greateſt value in the labours of this celebrated Berlin chemiſt is, he has fully determined what is the ſubſtance contained in the urine which ſerves for the formation of the phoſphorus. By diſtilling a mixture of ammoniacal fuſible ſalt and charcoal, he obtained a very fine phoſphorus; and he obſerved, that the urine from whence this ſalt was got, yields almoſt no more of this combuſtible ſubſtance. It is therefore a conſtituent part of the ammoniacal fuſible ſalt which contributes to the formation of phoſphorus; and this ſubſtance is eaſily obtained by diſtilling two parts of the glaſs obtained from the ſalt decompoſed in a retort or crucible with one part of powdered charcoal. This operation requires much leſs time and much leſs fire than the methods hitherto deſcribed; ſince, according to M. Prouſt, the phoſphorus may come over in a quarter of an hour. It is without contradiction the beſt proceſs which we can adopt in the preparing of the phoſphorus of urine; but there are ſeveral obſervations to be made upon this point. 1. The vitreous reſiduum ariſing from the decompoſition by the fire of the fuſible ammoniacal ſalt, not being pure phoſphoric acid, but this combined with the ſubſtance diſcovered by M. Prouſt, and having in it only this acid which is capable of forming phoſphorus along with charcoal; only very little phoſphorus is obtained by employing this reſiduum, as an ounce yields only a gros, and often leſs. 2. When the fuſible ſalt is prepared in great quantity by evaporation and cooling, it is found to be mixed with a good part of fuſible ſalt with baſe of natrum; which, according to M. Prouſt's diſcovery, contains no phoſphoric acid, and cannot yield phoſphorus. It is conceivable then, from theſe two obſervations, why ſo little of this combuſtible body is obtained by the diſtillation of the fuſible ammoniacal ſalt with charcoal.

Phoſphorus obtained by all the methods now deſcribed is always the ſame. When it is very pure, it is tranſparent, of a conſiſtence ſimilar to that of wax; it cryſtalliſes in brilliant laminæ, as if micaceous, by cooling; it fuſes in warm water long before this fluid boils; it is very volatile, and turns into a thick fluid by means of heat: if it is in contact with the air, it emits a fume from the whole of its ſurface; this vapour, which diſcovers a ſtrong ſmell of garlic, ſeems white in the day-time, and very luminous in a dark place. This is then the ſlow inflammation of phoſphorus; indeed, if it is left any time thus expoſed to the air, it conſumes gradually, and leaves for a reſiduum a particular acid, whoſe properties we ſhall examine more fully after this. This ſlow combuſtion never goes on, except when the phoſphorus is in free contact with air; it even requires, in order to be very luminous, a heat from twelve to fifteen degrees, although it takes effect at an inferior temperature. This inflammation is produced without heat, and it kindles no combuſtible body. But when phoſphorus ſuffers a dry heat of twenty-four degrees, it kindles with decrepitation; it burns rapidly with a white flame mixed with yellow and green, very bright, and it deſtroys with great readineſs all the combuſtible bodies which it touches. The vapour which is then exhaled from it is very copious, white, and very luminous in the dark. It leaves a reſiduum, differing in the one and the other of theſe combuſtions. The firſt yields a liquor more than double the weight of the phoſphorus employed, which is called *phoſphoric acid*. The ſecond exhibits a thick matter, of a reddiſh white, which emits white vapours till it has attracted enough of the air's humidity to make it fluid; then it reſembles the acid and

and fluid residuum of the first combustion or slow inflammation. However, these acids present some differences in their combinations, as M. Margraaf has observed, and as M. Sage has shown in the Memoirs of the Academy 1777. We shall fully describe these differences when we come to treat of the phosphoric acid.

The combustion of phosphorus was considered by Stahl as the disengagement of phlogiston, which he imagined was combined with the marine acid * in this combustible body. M. Lavoisier, with a view of ascertaining what passes in this combustion, kindled, by means of a burning-glass, some phosphorus under a glass-vessel full of mercury. He has observed, that only a given quantity can burn in a determinate portion of air, and that this quantity amounts to a grain in sixteen or eighteen cubic inches of air: that after this combustion the phosphorus is extinguished, and that the air can no longer serve to support a new inflammation; that the volume of the air diminishes; and that the phosphorus is dissipated in white snowy flocci, which attach themselves to the sides of the vessel. These flocci have twice and a half the weight of the phosphorus employed; and this increase of weight corresponds exactly with that which the air lost, and depends solely upon the absorption of the pure air by the phosphorus. In fact, the white tufts are the concrete phosphoric acid, and formed by the combination of the phosphoric acid with the pure air contained in the atmospheric air, which has served for the combustion of this inflammable substance. The theory here is the same as that applied to sulphur; and it

* Stahl asserts in several of his works, that by combining marine acid with phlogiston, phosphorus may be formed. M. Margraaf consequently began some experiments with this view, by treating different combinations of marine acid with combustible substances; but he never produced an atom of phosphorus. He has also demonstrated, that the acid, the residuum of this combustible body, widely differs from muriatic acid; and all chemists are now convinced of this fact. We cannot say what should have led into such a mistake so able and so creditable a person as Stahl.

it would be uſeleſs to add what we have ſaid upon this ſubject in the mineral kingdom.

Phoſphorus liquifies in hot water. Although it is inſoluble in this fluid, it is, however, a little altered. It loſes its tranſparency; it turns yellow, and is covered with an efflореſcence or coloured powder. The water becomes acid; it appears luminous when ſhaken in the dark: phoſphorus then is ſlowly decompoſed in it.

The action of a great number of bodies upon phoſphorus is unknown. The knowledge we have of this ſubject is almoſt all owing to M. Margraaf, and from him we ſhall make the following brief obſervations.

Vitriolic acid diſtilled in a retort with phoſphorus, decompoſes it almoſt entirely, but without inflammation. The nitrous acid attacks it with violence, and ſuddenly inflames it. The marine acid makes no alteration.

Sulphur and phoſphorus combine by fuſion and diſtillation: a ſolid compound reſults of an hepatic ſmell, which burns with a yellow flame, and ſwells in water, to which it communicates acidity and the ſmell of liver of ſulphur, properties which certainly ſhow a particular reaction between theſe two bodies.

Phoſphorus does not unite ſo well with the metals as ſulphur does; although a very great number of analogous properties ſubſiſts between it and this laſt. M. Margraaf tried to make theſe combinations, by diſtilling each metallic ſubſtance with two parts of phoſphorus. Only arſenic, zinc, and copper, preſent particular phenomena: all the reſt of the metals were not altered by phoſphorus; and this matter was in part conſumed, and paſſed into the receiver without ſuffering any remarkable change. Phoſphorus, ſublimed with arſenic, preſented to him a matter of a beautiful red colour, ſimilar to realgar. Zinc, diſtilled ſucceſſively with this ſubſtance, yielded yellow, pointed, and very light flowers. Theſe flowers, expoſed to the fire under a red muffle, kindled, and yielded a tranſparent glaſs ſimilar to that of borax.

Copper,

Copper, treated in the same way with phosphorus, lost its brilliancy, and became very compact; it had acquired ten grains upon half a gros, and it kindled upon the approach of flame.

Phosphorus dissolves in all the oils, and renders them luminous. M. Spielman says, that he dissolved it in spirit of wine, and that this solution yielded sparks when poured into water.

Phosphorus is not yet of any use either in medicine or in the arts. Messrs Menzius, Morgenstern, Hartman, &c. say, that they have experienced good effects from it in malignant and bilious fevers, in debilities of the system, miliary fever. Others have recommended it in the measles, peripneumony, rheumatism, epilepsy, &c. But although several dissertations upon the virtues of phosphorus used internally have appeared already in Germany, we cannot as yet say any thing with certainty upon this subject.

§ 5. *Of the Acid of Phosphorus.*

The acid of phosphorus has been so named, because it was thought to exist already formed in phosphorus, from which we obtain it by combustion: but M. Lavoisier has proved, that this salt is a combination of phosphorus with pure air.

In general there are two processes for obtaining it. The one consists in burning phosphorus with rapidity, and by applying a brisk heat; the other in allowing it to consume and burn slowly by exposing it to the air. The first yields a concrete acid, which is twice and a half the weight of the phosphorus, and powerfully attracts humidity from the air; but it always retains some pieces of phosphorus undecomposed. The other method, after some time, affords a fluid without smell, of an acid taste, and not corrosive; which differs from the former obtained by deflagration in this, that with alkalis it is susceptible, according to M. Sage, of forming neutral salts different from those formed by the other

other acid. The differences, however, are but ſlight, at leaſt thoſe mentioned by M. Sage: and with ſtrictneſs we may look upon them as the ſame acids, notwithſtanding the different modifications of their properties.

In order to obtain the phoſphoric acid by deflagration, we muſt make uſe of Lavoiſier's proceſs; which conſiſts in ſetting fire to the phoſphorus by means of a lens, under a large glaſs bell immerſed in mercury, a little water being applied round the outſide of the bell where it ſinks in the mercury. This combuſtion is repeated according to the quantity of acid which we wiſh to have.

The phoſphoric acid *per deliquium*, or ſlow combuſtion, is obtained after *M. Sage's method*, by putting ſticks of phoſphorus upon the ſides of a glaſs funnel, the neck of which is inſerted into a bottle and baſe covered with a head; a glaſs tube being affixed to the neck of the funnel to retain the phoſphorus, and to give vent to the air in the bottle, which is diſplaced by the acid. After ſome time, for every ounce of the phoſphorus we obtain three of the acid, which runs into water deſignedly put into the bottle.

The phoſphoric acid obtained in this way is a tranſparent fluid very pure, has a ſour taſte, and reddens the ſyrup of violets, but is without ſmell. If we expoſe it to the fire in a retort, a pure phlegm comes over; the acid becomes concentrated, and weighs heavier than the vitriolic. By little and little it turns tenacious, and becomes white and ſoft like an extract. Laſtly, urged with a violent fire, it runs into a tranſparent glaſs. This glaſs differs from that of fuſible ſalt in this, that it has a ſour taſte, and attracts humidity from the air. Theſe properties are the conſequence of its being a pure acid, whilſt that of fuſible ſalt is combined with the ſubſtance diſcovered by M. Prouſt; of which we have treated in a particular article.

If we burn the phoſphoric acid *per deliquium* in an

 open

open vessel, it exhibits from time to time small flames, owing undoubtedly to a little phosphorus remaining, which has not been entirely consumed, accompanied with a smell of garlic: the remainder becomes concentrated, dries, and at last fuses in the same manner as when treated in a close vessel.

The concentrated vitriolic acid very quickly attracts humidity from the air. Its union with water is attended with heat. It combines with a great number of substances, as we shall soon see, and presents particular phenomena in its combination. Messrs Margraaf and Lavoisier have examined it when combined with alkaline and metallic substances; and upon their inquiries we shall found the history of the different phosphoric neutral salts.

The phosphoric acid seems to act on quartz and glass; but we have not yet examined exactly the nature of the alteration which it occasions on them.

It dissoves mild magnesia with effervescence: The salt resulting from this combination possesses but a small degree of solubility. A strong solution of it, set by for twenty-four hours, affords crystals like small needles, flattened very thin, several lines long, and cut obliquely at both ends. They are reduced into powder by exposure to a gentle heat. The vitriolic acid decomposes this salt, according to M. Lavoisier, who has the merit of this detail, as well as of several of the following details upon phosphoric alkaline salts.

This acid, mixed with lime-water, precipitates along with the lime in form of a very insoluble salt, which makes no effervescence with acids, and has always the acid to excess. It reddens blue paper; the mineral acids, and even the caustic fixed alkali, decompose it. From the solution of silver in aquafortis, it gives a precipitate of the colour of dregs of wine; and from that of mercury, in the same acid, a white powdery precipitate.

The phosphoric acid, saturated with fixed alkali,

 forms

forms a very soluble salt. Its solution evaporated, and set in the cold, affords crystals like four-sided prisms, terminated likewise by pyramids with four surfaces, corresponding with those of the prisms. This phosphoric tartar is acid. It dissolves much better in warm than in cold water. It puffs up when laid on burning coals: it fuses with great difficulty; and when once fused, it loses its saline taste. It throws down a white precipitate from silver in nitrous acid, and a whitish yellow from the solution of mercury.

The mineral alkali united with the phosphoric acid, affords a salt of an agreeable taste, analagous to that of marine salt. This salt does not crystallise at all. By evaporation it becomes like a gummy matter, ropy like turpentine, and deliquescent.

The salt which we mean here, M. Lavoisier prepared with the acid obtained by the deflagration of phosphorus; but M. Sage has asserted, that the acid *per deliquium* afforded crystals not deliquescent.

The ammoniacal phosphoric salt, formed by the union of this acid obtained by deflagration and the volatile alkali, is more soluble in warm than in cold water. In cold, it affords crystals, which, according to M. Lavoisier, have some resemblance to those of alum. This salt ought to differ from that formed by the evaporation of urine, in this respect, that the fusible salt contains an acid already neutralised, by the substance explained in § 2. and is composed of three different ingredients; whilst the true ammoniacal phosphoric salt, or the phosphoric acid united to the volatile alkali, contains only two.

The phosphoric acid decomposes nitre and sea-salt, and disengages their acids by distillation, not on account of its greater affinity with their bases, but on account of its fixity.

The fluid phosphoric acid acts on a small number of metallic substances only. It dissolves zinc, iron, and copper well. Their solutions evaporated do not crystallise,

ſtalliſe, that of iron excepted, which ſeems ſuſceptible of cryſtalliſation. The ſolutions of zinc and copper become ductile and ſoft maſſes, reſembling extracts. If we urge them with heat, they throw out ſparks, and appear formed of true phoſphorus. M. Margraaf and the academicians of Dijon have examined in detail the action of this acid on the metals and ſemimetals. It precipitates ſome metallic ſolutions likewiſe; ſuch as that of mercury in the nitrous acid; the precipitate of which is white, and is rediſſolved when the mixture is expoſed to cold.

The ſolution of lead in the ſame acid is in like manner precipitated by the phoſphoric acid. It acts on the oils; it exalts their flavour, and imparts to thoſe that have none a ſweet odour like that of ether. It thickens others. When diſtilled in its ſtate of dryneſs with charcoal, it forms phoſphorus. Heated in a retort with ſpirit of wine, it afforded to the academicians of Dijon a very acid liquor, which burnt with a little ſmoke, and exhibited ſeveral of the properties of ether. This experiment ſhows that the phoſphoric acid had acquired volatility, ſince the product was acid. Laſtly, it diſſolves the ſubſtance remaining after the formation of phoſphorus from the fuſible ſalt, and forms by fuſion with it a hard glaſs, inſipid, inſoluble, not deliqueſcent; in a word, the ſame reſult as that left after the fuſible ſalt has been urged by fire. According to M. Prouſt, it can even take up an exceſs of this ſubſtance.

This acid is as yet of no uſe. In the mean time, it merits all the attention of phyſicians, ſince it makes part of ſeveral animal ſubſtances, and is ſecreted in greater or ſmaller quantities by the urinary veſſels, in the different diſeaſes which attack the bones, articulations, &c.

It is proper to mention here what Margraaf imagines, that this acid can very well exiſt in the aliment, and paſs from vegetables to animals; ſince he ſays he ob-

obtained phosphorus by heating charcoal of mustard, wheat, &c. with a great fire. Some chemists even believe the phosphoric acid to exist in minerals; but such conjectures are not to be regarded, except so far as they are demonstrated by accurate experiments.

§ 6. *Of the urinary Calculi.*

THE calculi or stones which form in the human bladder, constitute one of the most terrible diseases which man has to encounter. Physicians have more knowledge of the origin and formation of these concretions than chemists have of their nature. There are some analyses in the retort, which indicate that the calculi yield an alkaline phlegm, a little oil, and much concrete volatile alkali: it is known, that in general the acids dissolve them, and that the alkalis have also a remarkable action upon these substances; but they have not been completely enough examined to form an accurate judgment of their nature. This examination will undoubtedly furnish interesting facts; particularly at this time, when so valuable knowledge is acquired of the nature of the bones, with the base of which the calculous matter has been long known to have the greatest analogy. It is very probable, that the same salt as in the bones will be found in the calculi of the bladder; and the analysis of the bones may, to a certain degree, be referred to that of these concretions.

LECTURE LXVIII.

Of the INSENSIBLE PERSPIRATION *and* SWEAT.

PHYSICIANS have discovered a great analogy between the urine and the fluid which issues through the skin; they know that these secretions supply one another's place in different circumstances, and they are naturally led to consider the vapoury fluid of transpiration as of the same nature with urine. However, it has been

been hitherto impossible to collect a sufficiently great quantity of this humour for chemical investigation. The practice of medicine has shown, that its qualities vary; that its smell is insipid, aromatic, alkaline, or sour; that its consistence is sometimes glutinous, thick, tenacious, and that it leaves a residuum upon the skin; that it frequently tinges linen yellow and of different shades; lastly, M. le Count de Milly has observed, that through the pores of the skin a gaseous fluid issues, which he took to be aërial acid, or fixed air. Therefore a great number of experiments remains to be made, which particular circumstances alone will be able to give physicians the opportunity of undertaking and pursuing.

Of the solid Excrements of Animals.

The aliments with which animals are nourished contain a great quantity of matter which is incapable of affording them any support, and which is thrown out of the intestines in a solid form. The excrements are coloured by a portion of bile which they carry along with them; their fetid smell is owing to the commencement of putrefaction which they undergo in the long tract thro' which they pass in the intestines. Homberg is the only chemist who has examined these matters. He has observed, that the phlegm furnished from the excrements, distilled in a balneum mariæ, was of a disagreeable smell: he extracted from it, by washing and evaporation, a salt, which fuses like nitre, and inflames in close vessels. This matter, distilled in a retort, yielded him the same products with the other animal-matters. Putrefied excrements furnished him an oil without colour or smell, which did not fix mercury into silver, as he had been told.

It must be observed, that the feculent matter which Homberg examined was furnished from men who were nourished with bread of Gonesse and Champagne wine.

No doubt, the kind of nouriſhment muſt influence the fæces, as they are only the reſiduum of the aliments.

Of the ſoft and white Parts of Animals.

ALTHOUGH the analyſis of the ſoft parts of animals be leſs advanced than that of their fluids, we are now coming to perceive the different matters of which they are compoſed. We mean to talk here only of the ſoft and white parts, becauſe the muſcles and the bones will be treated of in a ſeparate article.

All the ſoft and white parts, ſuch as the membranes and ligaments, the tendons and cartilages, in general contain a mucous ſubſtance, very ſoluble in water, and inſoluble in ſpirit of wine, which is known by the name of *gelly*. This gelly is extracted by boiling theſe animal parts in water, and by evaporating this decoction to a ſolid tremulous maſs, and by cooling. If it is more ſtrongly evaporated, a dry ſubſtance is obtained, which is brittle and tranſparent, and known by the name of *glue*. This laſt is prepared with all the white parts of animals. With the ſkin, cartilages, oxen feet, the ſtrong Engliſh, Flemiſh, and Dutch glue is prepared; with ſerpent's ſkin the glue for gilding is made; with the parings of gloves and parchment a glue is made, which is uſed by the painters, &c.; and, laſtly, there are almoſt no animals whoſe tendons, cartilages, nerves, and particularly ſkin, cannot ſerve for the preparation of theſe kinds of glue. It muſt, however, be obſerved upon this ſubject, that glues differ from one another in conſiſtence, colour, taſte, ſmell, and ſolubility. Some ſoften very well in cold water; others diſſolve only in boiling water. The beſt of theſe ſubſtances ought to be tranſparent, of a yellow colour inclining to brown, without ſmell and taſte; it ought to diſſolve entirely in water, and form a viſcous uniform fluid, which dries, preſerving a tenacity and tranſparency equal throughout.

Animal

Animal gelly differs from *glue*, properly so called, only in its having less viscosity and consistence.

The first is extracted especially from the white and soft parts of young animals: it is found also in their flesh or muscles, in their skin and bones. Glue is obtained only from more aged animals, whose fibre is stronger and more dry. However it be, as these two substances present the same chemical phenomena, we shall for the example which we are going to take, pitch upon the gelly which is yielded by the cartilages or membranes of a calf.

This matter has almost no smell in its natural state. Distilled in a balneum mariæ, it yields an insipid phlegm without smell, and susceptible of putrefying. In proportion as it loses its water, it acquires the consistence of glue; and when fully dried, it resembles horn. Exposed to a stronger fire and air, it swells, bubbles up, and liquefies; it blackens, and emits a copious fume of a fetid smell; it kindles with a violent heat only, and still difficultly. Distilled in a retort, it yields an alkaline phlegm, an empyreumatic oil, and concrete volatile alkali, or mild sal ammoniac. It leaves a large coal very difficult of incineration, and containing only very little fixed alkali and febrifugal or marine salt.

The gelly, exposed to a moist and hot air, passes very soon to acidity, and very soon after that putrefies.

Water dissolves it in all proportions. The acids, and particularly the alkalis, dissolve it. The oils and spirit of wine have no action upon it. The most part of these properties approximate the gelly to the insipid vegetable juices, if we except that of giving volatile alkali in its analysis. But might not this last property be attributed to a portion of lymphatic matter, which water extracts at the same time with the gelatinous substance, particularly when the gellies or glues have been prepared by a strong and long decoction?

Of the Flesh or Muscles of Animals.

The muscles of animals, chemically considered, are formed of a parenchymatous and cellular substance, in which different humours are contained, partly concrete, and partly fluid. These humours are composed, 1. Of a red and white lymph. 2. Of a gelatinous substance. 3. Of a sweet oil of the nature of fat. 4. Of a particular extractive substance. Lastly, of a saline matter, whose nature is still little known. As the analysis of entire flesh, which yields in a balneum mariæ a vapid water, in the retort an alkaline phlegm, an empyreumatic oil, and concrete volatile alkali, which leaves a coal from which a little fixed alkali and febrifugal or marine salt is extracted by incineration, affords no accurate proof of the nature of these different principles, recourse must be had to means by which these substances may be extracted without altering them, and which allow their properties to be examined apart.

In order to obtain and separate these different substances investigated by M. Thouvenel *, different methods may be employed. This physician used expression in order to obtain the fluids in spongy muscles; the action of heat to coagulate the lymph, and to obtain the salt by evaporation; water to dissolve the gelatinous mucilage; salt, and extract, and spirit of wine, to carry off these two last principles without the gelly. It is in ge-

* In this place the medico-chemical paper of this physician should be read, which treats of the virtues and principles of the animal medicated substances, which won the prize of the Bourdeaux Academy 1778. M. Thouvenel has paid great attention to the analysis of animal matters: to him we owe several discoveries; but we would do well to keep in sight, that as his intention was only to appreciate, by chemical analysis and observations the medicinal properties of these maters; what he has said upon their nature is not so full, and perhaps not so clear, especially with regard to the flesh of animals. as could have been wished in inquires of philosophical chemistry. He himself has made the same reflection in his above-mentioned work.

general very difficult to ſeparate completely theſe different matters, becauſe all are ſoluble in water, and ſpirit of wine diſſolves the ſoapy extract at the ſame time and a part of the ſalt. The proceſs which ſucceeds beſt, appears to be that which conſiſts, firſt in waſhing the fleſh in cold water, which carries off the colouring lymph with a part of the ſalt; next, in digeſting the reſiduum of this waſhing in ſpirit of wine, which diſſolves the extractive part and a portion of the ſalt; laſtly, in boiling in water the fleſh treated by theſe two proceſſes. This fluid diſſolves the gelatinous part by ebullition, and it carries off alſo the portions of extract and ſalt which have eſcaped the action of the firſt menſtrua.

By ſlowly evaporating the firſt water employed cold, the lymph coagulates; it is ſeparated by filtration; and the ſlow evaporation of the filtrated liquor furniſhes the ſaline matter. Upon evaporating likewiſe the ſpirit of wine, the colouring extractive matter is obtained: and, laſtly, decoction furniſhes the gelly and fat oil, which ſwims upon the ſurface, and fixes by cooling. After the extraction of theſe different ſubſtances, no more remains than the fibrous texture; it is white, inſipid, ſoluble in water; it contracts when burning; it yields a great deal of volatile alkali and very fetid oil to the retort; laſtly, it has all the characters of the fibrous part of the blood. Hence it ſeems to be demonſtrated, that the muſcular organ is the reſervoir in which the action of life depoſites the fibrous matter, which becomes concrete by repoſe, and which appears to be the focus or the baſe of the animal property called *irritability* by phyſiologiſts.

No more remains for us in knowing completely the fleſh of animals, but to examine the properties of each of the ſubſtances of which it is compoſed.

The lymph, the gelly, and the fat part, are already known to us: the firſt perfectly reſembles that of the blood. We muſt obſerve, that it is it which upon coagulation by the heat of water in which the meat is boiled to make

make broth, produces the ſcum, which is taken off with care. This ſcum is of a dirty brown red, becauſe the red lymph is altered by the boiling heat. The gelly, extracted from the fleſh, generally converts into a tremulous maſs the broths prepared with the fleſh of young animals, which contains much more of it than that of old ones; it is entirely ſimilar to that which conſtitutes the white and ſoft parts of animals, whoſe properties we have deſcribed in the preceding article The fat matter which forms the flat round drops which ſwims upon the ſurface of broth, and which becomes ſolid by cooling, preſents all the characters of fat. We have then only to examine the extractive matter and the ſalt which is obtained in the analyſis of the muſcles.

The ſubſtance, which M. Thouvenel calls *mucous extractive*, is ſoluble in water and ſpirit of wine; it has a remarkable taſte, whilſt the gelly has none. When it is very concentrated, it forms an acrid and bitter matter; it has a particular aromatic ſmell, which heat evolves; it is it which colours the broth, and which gives it the agreeable taſte and ſmell, which is peculiar to it. When it is evaporated too much, or when a great quantity of meat is put in to that of the water, the broth is more or leſs coloured, and more or leſs acrid. Laſtly, the action of heat evolves and extracts the taſte of this extractive matter almoſt to the point of communicating the taſte of ſugar or caramel, as is obſerved at the ſurface of roaſted meat. If we examine farther the properties of this extractive ſubſtance evaporated to a dry conſiſtence, it is obſerved, that its taſte is acrid, bitter, and ſalt; when put upon a burning coal, it bubbles up and liquefies, exhaling a pungent acid ſmell ſimilar to that of burnt ſugar; expoſed to the air, it attracts its humidity, and forms a ſaline effloreſcence at its ſurface; it turns ſour and putrid in a warm air, when it is diluted in a certain quantity of water; and, laſtly, it is ſoluble in ſpirit of wine. All theſe characters approximate this ſubſtance to the ſoapy extracts, and to the ſaccharine matter of vegetables.

As to the salt which crystallises in the slow evaporation of the decoction of flesh, its nature is not yet fully known. M. Thouvenel obtained it in the form of down, or in that of ill-formed crystals. This chemist thinks that it is a perfect neutral salt, formed by the vegetable fixed alkali and an acid which has the character of the microcosmic acid in frugivorous quadrupeds, and the marine in carnivorous animals: but this salt should be considered as unknown, till a sufficiently great quantity shall have been collected to afford a complete examination, especially by means of the tests. It is necessary to observe, that a long troublesome labour is requisite to free it of the extractive, gelatinous, and lymphatic matters, which blunt its properties.

Of the Bones of Animals.

THE bones are the support of all the other organs of animals, and the base upon which all the soft parts rest. These hard parts ought not to be considered as passive in the animal-œconomy; they are real secretory organs, which separate from the blood and other humours a particular saline matter, of which they are the deposition or the reservoir.

The bones, considered in all animals from man down to insects and worms, differ in their texture, solidity, position relatively to the muscles, and probably in their nature. Chemical analysis has not yet decided this last point: but we cannot refuse to believe that the bones of man and quadrupeds are of a different form from the soft and flexible bones of fishes, reptiles, and especially the horny skeleton of insects, and likewise from the calcareous shell of shell-worms. The point of view under which we must examine animals bones here, does not allow us to insist upon these differences which chemists have not yet accurately investigated, so as to fix the opinion of physiologists.

The bones of man and quadrupeds, which alone have been hitherto examined by chemists, are not earthy matters,

ters, as was formerly imagined; they contain a certain quantity of gelatinous matter diſperſed in the ſmall cavities formed by the ſeparation of the ſolid laminæ, which compoſe their texture; and theſe ſolid laminæ themſelves, which in their inſolubility and conſiſtence ſeem to approximate to earthy matters, have been theſe ſeveral years known to be a true neutral ſalt formed of the phoſphoric acid and lime.

Bones expoſed to the fire with the contact of air, kindle by means of a certain quantity of medullary fat which they contain. If they are diſtilled in a retort, they yield an alkaline phlegm, a fetid empyreumatic oil, and a great deal of concrete volatile alkali. Their coal is compact; it incinerates with very great difficulty; it leaves a white reſiduum, which furniſhes by waſhing with cold water a ſmall quantity of ſoda or natrum.

Hot water next carries off a ſmall quantity of ſelenite. What remains after theſe ablutions is inſoluble in water; it is a calcareous phoſphoric ſalt, which the vitriolic and nitrous acids are ſuſceptible of decompoſing.

The water in which we boil bones, reduced into ſmall pieces or grated down, is mixed with a ſubſtance which gives it viſcidity, and is a true gelatinous matter.

The alkalis are capable of decompoſing the calcareous phoſphoric ſalt, which forms the baſe of the bones. This decompoſition is announced by the Dijon Academicians. They ſay, that they effected it by fuſing a mixture of calcined bones and fixed alkali.

The acids act upon bones, and decompoſe the phoſphoric ſalt which they contain. By means of them Meſſrs Scheele and Gahn, about ſix years ago, extracted from them the phoſphoric acid. Theſe chemiſts at firſt employed the nitrous acid, which completely diſſolved the bones. This acid ſeized upon the lime, with which it formed calcareous nitre, which remains diſſolved at the ſame time with the phoſphoric acid. Into this mixture they poured ſome vitriolic acid, which carried off the lime of the calcareous nitre, and formed ſelenite: they

they ſeparated this laſt, which precipitated on account of its inſoluhility by filtration: laſtly, they diſtilled in a retort the filtrated liquor, which was a mixture of nitrous and phoſphoric acids; and they obtained this laſt in the ſtate of phoſphoric glaſs. By heating it in a retort with ſome powdered charcoal, they procured phoſphorus much more readily than had been done from the reſiduum of urine. Meſſrs Poulletier de la Salle and Macquer are the firſt who repeated theſe excellent experiments at Paris. Then the Dijon Academicians, Rouëlle, Prouſt, and Nicolas, communicated their reſearches and proceſſes. Several other chemiſts examined the different ſolid matters of animals; and among theſe laſt, M. Berniard extracted phoſphoric acid from foſſil bones, from thoſe of the whale, porpoiſe, elk, ox, man, from the teeth of the ſea-cow, the elephant's grinders; and he obſerved that all theſe bones yielded the ſame ſubſtances, and contained the phoſphoric acid in different quantities. See *Phyſical Journal, October* 1781. M. le Marquis de Bullion alſo extracted ſome phoſphoric glaſs from ivory.

The acid of phoſphorus is extracted from bones in the following manner: The bones are calcined to whiteneſs, reduced to a powder, paſſed through a ſearch, mixed in an earthen veſſel with an equal part of oil of vitriol. and a ſufficient quantity of water added, ſo as to make the whole of a milky conſiſtence: the mixture is left to reſt for ſome hours: after the decompoſition is complete, the liquor is put through a double cloth; the precipitate is waſhed with water till this liquor which paſſes through the cloth is taſteleſs, and no longer precipitates lime-water. Then we are certain that the reſiduum contains no more free phoſphoric acid: the water of the different ablutions is evaporated; it gradually depoſites a white matter, which is ſelenite, and which is ſeparated by the filtre; this ſelenite muſt be carefully waſhed in order to get all the phoſphoric acid: theſe filtrations are repeated till the liquor

depoſites

depofites no more. Then the liquor is evaporated till it put on the confiftence of honey or of a foft extract; it then acquires a brown colour and a fat appearance. It is put into a crucible, and heated by degrees, till it ceafe to emit any fulphureous, as if aromatic, odour, and till it no longer boils. In this ftate this matter has a femivitreous confiftence and acid tafte: it attracts humidity from air. If it undergoes a ftronger heat, it fufes into a tranfparent glafs, hard, infipid, infoluble, fhowing no mark of acidity any longer. When we want to obtain phofphorus from it, we ought not to wait till the refiduum of the evaporated acid liquor be in this ftate of infoluble glafs, becaufe it then yields it only by means of a ftrong heat, and much more flowly than when it is foft and deliquefcent. It is reduced into phofphorus by reducing it to powder: mixing it with half its weight of very dry powder of charcoal, it is introduced into a retort of freeftone, to which a recipient half filled with water is adapted, and pierced with a very fmall hole, or terminated by a fyphon, with M. Woulfe's apparatus: a gradual heat is applied, till the retort be of a white heat; then the phofphorus runs in drops; and the operation continues upon the whole from five to feven or eight hours, according to the quantity of matter which we are diftilling, and the force of heat which the furnace cangive. From fix pounds of bones twenty ounces, or a little more, of vitreous refiduum, are obtained; and this refiduum yields from three to four ounces of very fine phofphorus, and a few gros of phofphorus half decompofed.

It is the cafe here with this product, extracted from bones by the vitriolic acid, as with the refiduum of the fufible falt of urine decompofed by the fire. This product is not pure phofphoric acid, fince it yields upon the whole only a fifth of its weight of phofphorus; it appears to contain a certain quantity of calcareous earth, becaufe the felenite, which cannot be taken away entirely, is decompofed by the concentrated phofphoric

phoric acid, and reforms calcareous phosphoric salt, which enters into fusion by means of the excess of vitrescible acid. Besides that, the osseous phosphoric glass is partly formed of the particular substance discovered by M. Proust, which seems constantly to accompany the phosphoric acid combined in the humours of animals. This chemist has not yet shown the process for extracting this substance from the osseous phosphoric glass; but the examinations which he has promised of this important subject will surely indicate this process, and elucidate in a great degree the history of this product of art. Lastly, some modern chemists have thought, that the phosphoric acid is not contained in bones, and that it is formed by the vitriolic acid employed in its extraction. M. Berniard is at present studying to decide this question; and we may expect great satisfaction from this chemist's investigations.

LECTURE LXIX.

Of the different SUBSTANCES *useful in* MEDICINE *or the* ARTS, *which are extracted from different* ANIMALS.

IF we proposed to give an accurate and complete account of all the substances which animals furnish to medicine and the arts, we shall have more to say upon this subject alone than has yet been delivered upon the animal-kingdom, especially if we speak of the different animal-substances, which empiricism or blind credulity formerly introduced into medicine as famous remedies, and which are now happily regarded as entirely useless. Our design is to show only the principal of these substances; those in which chemical and medicinal experience has owned very remarkable virtues, or which are of great use in the arts. The number of these matters is not very considerable; and the most part of them is so fully known, and so excellently handled by medical authors, that we shall offer, in a few words, only their

their principal properties; with ſo much the more reaſon, that they almoſt all relate to the fluid or ſolid ſubſtances of animals, which we have been juſt now examining.

Among the matters which quadrupeds furniſh, we ſhall chooſe caſtor, muſk, and cornu cervi or hartſhorn. Spermaceti, produced by a cetacea, merits a particular examination. Among the products of birds, we ſhall relate the analyſis of eggs; the amphibia, the tortoiſe, frog, and viper, ſhall have a ſeparate article. Ichthyocolla alone among the fiſhes will take up our attention. The claſs of inſects will furniſh a greater number of ſubjects: we ſhall take notice of cantharides, ants, millepedes; honey and wax, reſin, lac, kermes, cochineal, and lapides cancrorum. As for the worms, among which we comprehend the lumbrici, oyſterſhells, mother-of-pearl, the bone of the ſound, we ſhall not make any remark about them here; either becauſe ſeveral of theſe ſubſtances have not been analiſed, or becauſe ſome of them entirely reſemble thoſe which we have examined before. Laſtly, we ſhall terminate our examination of the products of the animal-kingdom with coral and coralline. From this ſhort enumeration it is evident, that we paſs over in ſilence a great number of other matters which were formerly employed in medicine: ſuch are, among the reſt, ivory, unicorn, the teeth of the hippopotamus, thoſe of the beaver and wild boar, the bones of the ſtag's heart, elk's foot, bezoar ſtones, civet, the blood of the wild goat, in quadrupeds; the ſwallow's neſt, gooſe's greaſe, peacock's dung, the membrane of the chick's ſtomach, among birds; the toad, among reptiles; the gall and ſtones of the carp, ſerpent's liver, perch-ſtones, the jaws of the pike, among fiſhes; ſilk, ſpider's web, crab's claws, among inſects; laſtly, ſnails and the dentales, among the naked or covered worms. Of all theſe ſubſtances, ſome have only ſuch virtues as exalted imagination has beſtowed upon them;

them; and others are very well ſupplied by thoſe which we have ſelected, and of which we are going to give the particular examination.

Caſtor.

The name of *caſtor* has been given to two bags, ſituated in the inguinal region of the male or female beaver, which contain a matter very odorous, ſoft, and almoſt fluid, when they are recently taken from the animal; but which dry, and put on the conſiſtence of a reſin, through length of time. This ſubſtance has an acrid, bitter, nauſeous taſte: its ſmell is ſtrong, aromatic, and even fetid: it is formed of a coloured reſinous matter, which ſpirit of wine and ether diſſolve; of a mucilage, gelatinous and partly extractive, which water carries off; and a ſalt, which cryſtalliſes in the aqueous ſolution when evaporated, but whoſe nature is as yet a ſecret. The reſin of caſtor, in which all its virtue reſides, ſeems to be very analogous to that of the bile. All the ſubſtance of this animal product is contained in membranous cells, which riſe from the internal tunic of the bag which contains them. There is no accurate analyſis of caſtor yet: we only know, that it gives a little eſſential oil and concrete volatile alkali to diſtillation; and that by means of ſpirit of wine, ether, and water, the different matters of which it is compoſed are ſeparated.

In medicine, it is employed as a powerful antiſpaſmodic in hyſteric and hypochondriacal fits, and in convulſions which depend upon the ſame affection. The moſt happy and ready effects are frequently produced by it; but it happens ſometimes that it irritates inſtead of calming, according to an unknown diſpoſition of the nervous and ſenſible ſyſtem: it therefore ſhould be given only in a very ſmall quantity at firſt. It has been given with ſucceſs alſo in the epilepſy, tetanus, &c. Its doſe is from a few grains to half a gros in ſubſtance: it is given in the form of boluſes: it is frequently mixed, and almoſt always with advantage, with opium and all the calming or narcotic extracts. Its ſpirituous and etherial tincture is alſo made uſe of, which is preſcribed from a few drops to twenty-four or thirty-ſix grains, in convenient drinks.

Muſk.

Musk, a ſubſtance known to every body by its ſtrong tenacious ſmell, is contained in a bag ſituated near the umbilical region of a ruminant quadruped analogous to the antelope and kid, and which differs from them ſufficiently to make a particular genus. This matter is ſimilar to caſtor in its chemical properties. It is a reſin, united with a certain quantity of mucilage, bitter extract, and ſalt. It is often adulterated. Its virtues are more exalted than thoſe of caſtor; it is more active; therefore it is employed only in the moſt preſſing caſes. It is given as a powerful antiſpaſmodic in convulſive diſeaſes, in the hydrophobia, &c. It is looked upon, too, as a violent aphdrodiſiac. We ought to be very careful in its uſe, becauſe it frequently excites, inſtead of allaying, nervous diſorders.

Hartſhorn.

Hartshorn is one of the animal-ſubſtances which is moſt employed in medicine. It is an oſſeous ſubſtance, which differs in no reſpect from bones. It contains abundance of ſoft gelly, very light and nouriſhing, which is extracted from it by boiling it, when reduced into very ſmall pieces, in eight or ten times its weight of water. If it is diſtilled in a retort, it yields a reddiſh and alkaline phlegm called *ſpirit*, an oil more or leſs empyreumatic, and a great quantity of mild volatile alkali. In the operation an enormous quantity of gas, in great part inflammable, is diſengaged. As the volatile ſalt is coloured, it is digeſted in a little ſpirit of wine, which carries off the oil that renders it impure. The charry reſiduum incinerated, contains a little natrum, ſelenite, and a great quantity of calcareous

calcareous phoſphoric ſalt, which is decompoſed by oil of vitriol, as has been remarked in treating of the bones. The ſpirit and ſalt of hartſhorn are employed in medicine as very good antiſpaſmodics. The former, ſaturated with the acid ſalt of amber, forms the liquor cornu cervi ſuccinatus. Oil of hartſhorn, rectified with a gentle heat, becomes very white, odorous, very volatile, and almoſt as inflammable as ether: it is known by the name of *Dippel's animal-oil*, a German chemiſt, who firſt prepared it. Formerly a great number of rectifications were employed to obtain the oil very white and fluid. It is ſince diſcovered, that two or three ſuffice, provided they have been done with the precaution, 1. Of introducing the oil to be rectified into a retort by means of a long funnel, ſo that the neck of the veſſel may be very clean; for a ſingle drop of coloured oil is ſufficient to communicate colour to all that is diſtilling. 2. Of taking only the firſt moſt volatile and white portions: Meſſrs Model and Baumé made theſe obſervations. M. Rouelle has given alſo a very good proceſs for obtaining this oil; it conſiſts in diſtilling it with water. As it is only the moſt volatile and moſt really ethereal portion contained even in the oil of the firſt diſtillation which the heat of boiling water can raiſe, we are certain by this means of having only the moſt ſubtile and penetrating portion. This oil has a briſk ſmell, and a ſingular lightneſs and volatility; it preſents all the characters of vegetable eſſential oils; it ſeems to differ from them only in containing volatile alkali, ſince it turns the ſyrup of violets green, according to M. Parmentier.

This oil is employed in drops in nervous affections, epilepſy, &c.

Spermaceti.

SPERMACETI is an oily, concrete, cryſtalline, ſemitranſparent matter, of a particular wild ſmell, which is obtained from the cavity of the cranium of the whale, and which is purified by liquefaction and ſeparation from another fluid inconcrescible oil, which is mixed with it. This substance presents very singular chemical properties, which assimilate it on the one side to the fat oils, and on the other to the essential oils.

Spermaceti, heated with the contact of air, kindles and burns uniformly, without emitting a disagreeable smell; very fine candles on that account are made with it in the countries where it is manufactured, at Bayonne, Saint-Jean de Luz, &c.

If it is distilled in a naked fire, it gives no acid phlegm like the fat oils, according to M. Thouvenel; but it all passes over almost quite unaltered into the recipient when it begins to boil, and it leaves in the retort a charry mark. By repeating this operation, it loses its solid form, and remains fluid, without being more volatile.

Spermaceti exposed to hot air turns yellow, and becomes rancid, but not so easily as the other concrete fat oils. The water in which it is boiled yields by evaporation only a small mucoso-unctuous residuum.

The caustic alkali dissolves spermaceti, and forms with it a soap, which gradually acquires solidity till it become friable.

The nitrous and marine acids have no action upon it. The concentrated vitriolic acid dissolves it, and alters its colour. This solution is precipitated by water like oil of camphor.

Spermaceti unites with sulphur like the fat oils.

The fat and essential oils dissolve it by means of heat: hot spirit of wine also dissolves it, and allows it to precipitate by cooling. Ether dissolves it in the cold, or by the heat of the hand alone.

Might spermaceti be to the fat oils what camphor is to the essential?

Medicine formerly made very extensive use of this substance; a great number of properties was ascribed to it. It was used especially in catarrhal cases, erosions, ulcers of the lungs, kidneys, &c. At present it

is

is ſeldom employed except as an emollient, and even in a ſmall quantity, as it is found to oppreſs the ſtomach, and to occaſion nauſea and vomiting.

Of Eggs.

THE eggs of birds, and in particular thoſe of hens, are compoſed, 1. of an oſſeous ſhell, which contains a gelly and ſome calcareous phoſphoric ſalt, demonſtrated by M. Bernſard ; 2. of a membranous pellicle, placed below the ſhell, which ſeems to be a texture of fibrous matter ; 3. of the white; 4. of the yolk contained and ſuſpended in the middle of the white. Upon this laſt ſubſtance the germ is ſupported.

The white of an egg is abſolutely ſimilar to the lymph of the blood : it is viſcous, gluey, turns the ſyrup of violets green, and contains free fixed mineral alkali. Expoſed to a gentle heat, it coagulates into a white opaque maſs, which emits an hepatic ſmell and gas. This white, coagulated and dried in a balneum mariæ, yields an inſipid phlegm, which putrefies, and acquires the hardneſs and reddiſh tranſparency of horn. Diſtilled in a retort, it yields concrete volatile alkali and empyreumatic oil : its coal contains a little mineral alkali. The white of an egg, expoſed to the air in ſmall layers, dries rather than corrupts, and forms a ſort of tranſparent varniſh. It diſſolves in water in all proportions. The acids coagulate it : if this coagulum, diſſolved in water, is filtrated, the fluid which paſſes over yields by evaporation the neutral ſalt, which ſhould be formed with the acid employed and the fixed mineral alkali contained in this liquor. Spirit of wine alſo coagulates the white of an egg.

The yolk is formed in great part of a lymphatic matter, but is mixed with a certain quantity of ſweet oil ; ſo that this mixture diſſolves in water, and forms a kind of animal emulſion. If it is expoſed to the fire, it turns into a maſs leſs ſolid than the white ; when it is dry, it becomes ſomewhat ſofter, from the diſengagement of its oil which issues out at its surface. If in this state it is submitted to the press, this oil is obtained, which is sweet and fat, smelling and tasting somewhat of roast meat, or it is empyreumatic.

The yolk of an egg distilled after the extraction of the oil, yields the same products as all animal matters.

The acids and spirit of wine coagulate it. The sweet oil which it contains establishes a striking analogy between the eggs of animals and the grain of vegetables, as these last also contain an oil, which is at the same time covered with a mucilage, and reduced to the state of an emulsion.

Eggs are of very extensive use as an aliment. Its different parts are used in pharmacy and medicine. The shell calcined is used as an absorbent. The oil of an egg is an emollient: it is used externally in burns, chaps, &c. The yolk of an egg renders oils soluble in water, and forms lohochs. It is triturated with resins. The whites of eggs are employed with success in pharmacy and butteries, to clarify the juices of plants, weak milk, syrups, liquors, &c. It is also employed to preserve the surface of tables, upon which it forms a transparent varnish.

Of the Tortoise, Frog, and Viper.

These three animals are employed in medicine: broths are made with their flesh and their bones, to which particular virtues are ascribed. It would indeed appear that the zoophyte animals, whose humours are more attenuated than those of most quadrupeds, whose parts have in general a stronger and quite particular smell, and appear to contain more saline matter, as they furnish much volatile alkali upon distilling them with a gentle heat, after they have been triturated with oil of tartar; these animals, I say, seem to enjoy more powerful and more numerous virtues. However, many physicians doubt of their energy, and ascribe them to the other animals. In ſpite of this opinion, it is ſtill a practice to preſcribe broths of tortoiſe and frog fleſh, in languid habits, conſumptions without any apparent cauſe, to convaleſcents from acute diſeaſes; and good effects are experienced from them frequently. It ſeems that their decoctions are more nouriſhing, more light, and endued at the ſame time with a certain activity, which their ſtrong ſmell and particular taſte ſufficiently demonſtrate.

Vipers are accounted more active. The ancients boaſted greatly of their virtues in diſeaſes of the ſkin, breaſt, chronic affections, where the lymph is vitiated. We cannot be hindered from believing, that their broths are better reſtoratives than aliments, and ought to produce depurations by the ſkin, by means of their exalted ſpiritus rector. They are alſo adminiſtered entire, and as aliments, and with ſucceſs in the ſame diſeaſes.

The chemical analyſis of theſe animals demonſtrated to M. Thouvenel a gelly lighter or heavier, conſiſtent or viſcous; an acrid extract, bitter and deliqueſcent; a concreſcible aluminous matter; an ammoniacal ſalt, and an oily ſubſtance of a particular taſte and ſmell, ſometimes ſoluble in ſpirit of wine, &c. See this author's Memoirs upon the Medicinal Animal Subſtances, p. 6. to 21.

Ichthyocolla.

ICHTHYOCOLLA, or fiſh-glue, is a ſubſtance in part gelatinous and in part fibrous, which is prepared by rolling the membranes which form the vas natatorius of the ſturgeon, and of ſeveral other fiſhes, and by drying them in the air, after forming them into a cord twiſted into a heart. This matter gives a viſcid gelly by ebullition in water. When it is macerated ſome time in this fluid, it cannot be folded and extended into a kind of membrane. It is never brittle, as the glues properly ſo called; but it is flexible, on account of its fibrous and elaſtic texture. A kind of it is alſo prepared by the decoction of the ſkin and inteſtines of fiſhes; but it has not the ſame properties in the arts. Ichthyocolla yields all the products of other animal ſubſtances. It may be employed in medicine as an emollient in diſeaſes of the throat, inteſtines, &c.; but ſeveral other vegetable ſubſtances are generally preferred which enjoy the ſame properties. In the arts, it is uſed to clarify liquors, wine, coffee, &c. It forms a filtre, which carries from their bottom up to their ſurface the ſtrange matters which alter their tranſparency.

LECTURE LXX.

CANTHARIDES, ANTS, MILLEPEDES.

INSECTS have alſo been very little examined by chemiſts: however, the ſingularities which ſome have preſented in their anlyſis, ſhow that inquiries of this kind will be intereſting to medicine and the arts. We are going to give here the reſult of the experiments made upon three inſects; two of which are employed in medicine, and the other preſents chemical facts, too intereſting to be paſſed over in ſilence.

1. Spaniſh flies, or cantharides, a remedy ſo important on account of its corroſive and epiſpaſtic qualities, are formed, according to M. Thouvenel, 1. of a parenchyma, whoſe nature has not been demonſtrated, and which makes the half of the weight of theſe inſects dried; 2. of three gros in the ounce of a reddiſh yellow extractive matter, very bitter, which yields an acid to diſtillation; 3. of twelve grains in the ounce of a yellow and waxen matter, to which theſe animals owe their golden yellow colour; 4. of ſixty grains of an oily green ſubſtance, analogous to wax, of an acrid taſte, in which principally the odour of cantharides reſides. To diſtillation this ſubſtance yields a very pungent acid, and a concrete oil like wax. Water diſſolves the extract;

tract, the yellow oil, and even a little of the green oil; but ether alone attacks this ſubſtance, and may be employed with ſucceſs to ſeparate it from the others. Upon this kind of green wax depends the virtue of the cantharides. This laſt ſubſtance is extracted at the ſame time with the extractive matter, and a tincture is formed very copious in theſe inſects, by employing a mixture of equal parts of water and ſpirit of wine. By diſtilling this mixt tincture, a ſpirit of wine is extracted, which preſerves a ſlight colour of the cantharides; and the different matters which it kept in ſolution, ſeparate from one another as the evaporation proceeds.

2. The ants have been analyſed by Meſſrs Margraaf and Thouvenel. Theſe chemiſts found that they contained a particular acid, a fat oil, and an extractive matter. The ſecond puſhed his inquiries much farther. Upon diſtilling theſe inſects with water, the fluid which was obtained is but ſlightly acid. With a view of collecting this ſaline principle, M. Thouvenel extended pieces of linen impregnated with oil of tartar upon ants with the ſkins taken off. The ants emitted their acid, and the odorous principle of the ſame nature, which they emit in ſo great abundance as to ſaturate the fixed alkali ſpread upon the cloth. The ley of theſe ſalts evaporated yielded a neutral ſalt, cryſtalliſed in flat parallelograms, or in priſmatic columns, not deliqueſcent, whoſe acid appeared to this chemiſt to have all the characters of that of the microcoſmic or fuſible ſalt. Spirit of wine digeſted upon ants extracted a little eſſential oil from them, which conſtitutes with this fluid the ſpirit of magnanimity. If theſe inſects are boiled in water, and then preſſed through cloth, a fat oil is got, which amounts to thirteen gros in the pound. This oil is of a greeniſh yellow; it congeals in a temperature much leſs cool than that in which oil of olives congeals, and it is very analogous to wax. The water of the decoction evaporated, yields a reddiſh brown extract of a fetid ſmell, acidulous, and like cheeſe;

cheese; of a bitter, nauseous, and acid taste. This extract separates into two substances by the successive application of water and spirit of wine. The parenchyma of ants, deprived of these different substances, amounts to three ounces two gros in the pound.

3. The palmers, millepedes, aselli, porcelli, onisci, &c. showed to M. Thouvenel some differences in their analysis. Distilled in a balneum mariæ without addition, they yielded an insipid alkaline phlegm, sometimes effervescing with acids, and turning syrup of violets green. In this operation they lost five-eighths of their weight; then treated by water and spirit of wine, they furnished in the ounce two gros of soluble matter, of which more than two-thirds were extractive matter, and the rest an oily or waxen substance. These two substances are separated by ether, which dissolves the last without touching the extract. These matters differ from those of cantharides and ants, in their yielding more concrete volatile alkali, and no acid in their distillation. M. Thouvenel upon this subject has observed, that in insects the millepedes seem to be to the cantharides and ants, what the reptiles are relatively to quadrupeds.

As to the neutral salts contained in insects, they are in greater quantity, and very difficult of extraction. M. Thouvenel says, that the millipedes and earth-worms, lumbrici, constantly yielded marine salt with earthy and vegetable alkaline base; whilst in ants and cantharides these two bases, the first of which always seemed to him the most abundant, are united with an acid, which has the character of the microcosmic acid. It is necessary to observe that this chemist has given in his dissertation neither the means of extracting these salts, nor the processes which he employed in the investigation of their nature.

They are seldom employed in medicine, except cantharides and millepedes. These last appear to act only as stimulants and slight diuretics; and, from the experiments of M. Thouvenel, they ought not to be employed in a stronger dose than what is generally employed.

The

The expreſſed juice of forty or fifty living millepedes given in an emollient drink, or mixed with the juice of certain aperient plants, may be employed with ſucceſs in the jaundice, ſerous maladies, *à ſeroſa colluvie*, milky collections, &c. As to the cantharides, it is one of the moſt powerful medicines which medicine poſſeſſes. M. Thouvenel himſelf experienced the effects of the green wax matter, in which the virtue of theſe inſects reſides: applied to the ſkin in the doſe of nine grains, it raiſed a bliſter full of ſeroſity, as the cantharides in powder do. But what is moſt valuable in theſe experiments upon this powerful remedy, is what this phyſician obſerved from the effects of the ſpirituous tincture of the flies. He employed it with great ſucceſs externally, in the doſe of two gros to two ounces and a half, in rheumatic and ſciatic pains, irregular gout. It heats the parts, accelerates the motion of circulation, excites the evacuation by ſweat, urine, ſtool, according to the parts upon which it is applied. He alſo relates ſome good effects from its internal uſe by foreign phyſicians. But young phyſicians ought to be very cautious in the internal uſe of this medicine; it has been ſeen to occaſion frettings of the ſkin, inflammations, ſpittings of blood, pains in the kidneys, bladder, dyſuries, &c, Vid. *Diſſert. cit. M. Thouvenel*, p. 46.—50.

Of the Honey and Wax.

THESE two matters, prepared by the bees, ſeem to belong to the vegetable kingdom, as theſe inſects go to collect the firſt in the nectaria of the flowers, and the ſecond in the anthera of their ſtamina. However, they have undergone a particular elaboration; and beſides, as they are extracted after the labour of the bees, their properties ſhould be examined in the hiſtory of inſects.

Honey is a matter perfectly ſimilar to the ſaccharine juices, which we have examined in vegetables. It has a white or yellow colour, a ſoft and granulated conſiſtence, a ſaccharine and aromatic taſte. A true ſugar may be ex-

extracted with ſpirit of wine and water, by means of certain manipulations. It yields an acid phlegm and an oil to the retort; and its coal is rare and ſpongy, like that of the mucilages of plants. The nitrous acid extracts from it an acid entirely analogous to that of ſugar. It is very ſoluble in water; it forms a ſyrup, and it paſſes as ſugar does to the ſpirituous fermentation. It is a very good aliment, and an emollient, pectoral, and ſlightly aperient medicine. It is given diſſolved in water and mixed with vinegar, called *oxymel;* it is often combined with ſome acrid plants, as when it forms the oxymel ſcilliticum, colchicum. It forms the baſe of ſeveral medicines, which bear its name, as mel-roſarum, roſe-water, mel mercuriale, &c.

Wax is a concrete oily juice, analogous to the ſolid fat oils, ſuch as the butter of Cacao; and ſtill more ſo to the royal pigment of China. Although it cannot be doubted that this ſubſtance comes from the ſtamina of flowers, it is, however, demonſtrated, that it receives in the body of the animal a particular elaboration; ſince, according to the eſſays of M. de Reaumur, a flexible powder cannot be made with the powder of the antheræ. The wax, which compoſes the cells of the bees, is yellow, of an inſipid taſte; it is whitened by expoſure to the dew and the air, after it has been reduced into minute laminæ; it ſoftens when heated with a gentle fire, fuſes, and forms a tranſparent oily fluid; it becomes ſolid and opaque by cooling. When it is heated with the contact of air, it kindles as ſoon as it is volatiliſed; ſuch is the effect produced by the wick of candles. If it is diſtilled in a retort, it yields an acid phlegm of a ſtrong pungent ſmell, an oil at firſt fluid, which afterwards fixes in the receiver, and has the conſiſtence of butter; it leaves only a very ſmall quantity of coal, very difficult of incineration. By ſeveral times rectifying the butter of wax, it becomes fluid and volatile. Wax is not alterable by the air; it becomes coloured by it in a certain time; it diſſolves in the oils, to which

which it gives conſiſtence. By melting it in theſe fluids with a gentle heat, it forms medicines, called *cerates*. Spirit of wine has no action upon wax. The acids blacken it; the alkalis combine with it, and convert it into a ſoapy ſtate.

Wax is employed in a great number of arts; it is employed in pharmacy to make liniments, ointments, and plaſters.

Reſin Lac.

THE improper name of gum lac has been given to a reſinous ſubſtance of a deep red, which is depoſited upon the branches of trees by a kind of ant particular to the Eaſt Indies. This ſubſtance appeared to M. Geoffroy, Memoirs of the Academy 1714, a kind of hive in which the ants depoſite their eggs. Indeed, if the lac is bruiſed into ſticks, it is found filled with ſmall cavities or regular cells, in which ſmall oblong bodies are placed, which Geoffroy conſidered as the embryos of the ants. This chemiſt thinks that the lac owes its colour to this animal matter. He conſiders this laſt as a true wax: however, its dryneſs, the aromatic ſmell which it emits upon burning, and its ſolubility in ſpirit of wine, ſeem to approximate it to the reſins; it yields to diſtillation a kind of butter, according to the ſame author. In commerce, we diſtinguiſh lac in ſticks, lac in grains, and flat lac. It muſt be obſerved, that many other colouring ſubſtances, and in particular the red animal or vegetable feculæ, prepared in a particular manner, are called in dyeing by the name of *lacs*. Reſin lac is uſed in the Levant for painting cloths and ſkins. It makes the baſe of ſealing-wax. A tincture of it is made with ſpiritus cochleariæ. It enters the troches of karabé, dentifrice powders and opiates, ſmelling paſtes, &c.

Of Kermes and Cochineal.

KERMES, coccus infectorius, was conſidered by the firſt naturaliſts as a tubercle or an excreſcence of plants. More accurate obſervation has ſhown, that it is the female

male of an insect, ranked among the hemipteræ by M. Geoffroy. This female insect fixes to the leaves of the green oak; after having been fecundated, it turns larger, grows up, and soon assumes the form of an insect; it represents a round brown shell, under which are comprehended eggs in very great number. This shell was formerly employed in dyeing; its use was given up after cochineal was known. Kermes presents the same chemical properties as this last.

There was the same mistake about cochineal as about kermes: it was a long time considered as a grain. Father Plumier is one of the first who discovered this error. Indeed this substance is the female of an hemipterous insect, which differs from the kermes in this, that it preserves its form although fixed upon plants. Cochineal employed in dyeing grows upon the opuntia, the Indian fig-tree. It is cultivated in great quantities in South America. Geoffroy, who analysed it, found in it the same principles as in the kermes; he got from it some volatile alkali. The form of this insect may be seen by macerating this insect in water. Cochineal is employed to make carmine, and in dyeing. A crimson or scarlet colour is extracted from it, according to the manner in which it is used. As it is an extractive colouring matter, it cannot be fixed upon the substances to be dyed without the assistance of some active substance. It acts easily upon wool, and dyes it scarlet by means of the solution of tin in aqua regia, which decomposes the colouring extract, and singularly exalts the colour. This colour could not be given to silk before the discovery of M. Macquer. This chemist found the method of fixing it upon this substance, by impregnating the silk with the solution of tin before plunging it into the bath of cochineal, instead of mixing this solution in the bath, as is done to wool.

Of Crab's Eyes and Coral.

The stony concretions, falsely called *crab's eyes*, *occuli cran-*

cancrorum, are found two in number in the interior and inferior part of the ſtomach of theſe inſects. They are round, convex on one ſide, concave on the other, and ſituated between the two membranes of their ſtomach. As they are found only when the crabs are changing their ſkin and ſtomach, and as they are gradually deſtroyed in proportion as their new covering acquires conſiſtence, it is believed with great probability, that they ſerve for the reproduction of the calcareous ſubſtance which makes the baſe of their ſcales.

Theſe ſtones have no taſte; and contain a ſmall quantity of gelatinous matter. They are prepared by waſhing them ſeveral times, and porphyriſing them with a little water to reduce them into a ſoft paſte, which are moulded into troches and dried. The water employed in waſhing carries off what animal gelly theſe ſtones contain, and no more than the earthy ſubſtance remains. When prepared in this way, they make a briſk effervefcence with all acids, and are abſolutely of the ſame nature with chalk. They have no other virtue than that of abſorbing acidities in the primæ viæ. From the very doubtful opinions about all theſe animal ſubſtances in general, they have been accounted aperient, diuretic, and even cordial remedies.

Coral is quite of a ſimilar nature; it is a kind of calcareous white, roſy, or red ramification, which forms the baſe of the habitation of the polypi. It is prepared in the ſame way as crab's eyes; it enters the confection alkermes, *la poudre de guttete*, troches of karabé. Numberleſs properties are aſcribed to it; but it has abſolutely no virtue but that of a pure abſorbent, unleſs combined with the acids. It is frequently employed, juſt as crab's eyes, in the ſtate of a neutral ſalt, formed with vinegar or citron juice, as an aperient, diuretic, &c.

Of Coralline.

CORALLINE, called *ſea-moſs*, is a habitation particular to the polypi. It yields to the retort the ſame principles as animal ſubſtances; it has a ſaline taste, bitter and disagreeable; it is successfully employed as a vermifuge; it is given in powder in the dose of twenty-four grains to children, and increased to two gros or more to adults. An anthelmintic syrup is made of it; it enters the compositon of the worm-powder.

Of the Chemical Analysis of Animals and Vegetables.

If naturalists and anatomists have found very great analogies between the structure and functions of vegetables and animals, with a view of uniting these two classes of bodies into one kingdom alone, called *organic;* chemists have not found less striking ones between these substances in the nature of their principles, and their researches are proper to confirm the analogies which approximate the nature of these bodies to each other. In fact, upon comparing the properties of the substances which analysis demonstrates in the vegetable and animal kingdoms, striking relations are to be found. The mucilages, extracts, spiritus rector, saccharine matter, sweet and essential oils, resins, and colouring parts, are found in both kindoms: the fermenting property is also found in them, and these two classes of bodies contain matters equally susceptible of passing to the spirituous, acid, and putrefactive fermentations. It is unnecessary to give proofs of this assertion; they are to be found in the history of each of the substances of these two kingdoms, which we have examined one after the other: but it is necessary to fix the differences which distinguish their products.

It is a long time since physicians were convinced that animals procured from vegetables the most part of the matter proper to make up their substance; but the fluids of the first being exposed to more various and more rapid motions than those of vegetables, it is easy to conceive that they ought to be susceptible of greater alteration, and that their combinations are more numerous. Indeed the mucilages of animals are more attenuated; the extracts

tracts are found only in ſmall quantity; the ſpiritus rector is more penetrating, more active, volatile, and tenacious; the reſins are found only in a few individuals; their colouring parts are more fine and alterable; laſtly, they contain ſaline matters of a quite different nature; and the lymphatic fluid, concreſcible by heat, juſt like the fibrous ſubſtance concreſcible by ſimple repoſe, are two new ſubſtances, owing to the motion from animal life, to which nothing has been found analogous in the vegetable kingdom, although the glutinous matter ſeems to produce ſome reſemblance between them.

From theſe obſervations it follows, that the phenomena which animal ſubſtances preſent in chemical operations, ſhould differ from thoſe which vegetable ſubſtances preſent when treated in the ſame way. On this account, the former yield in general to the retort a great quantity of volatile alkali, fetid oils, a coal of difficult incineration; and, eſpecially, they are much more alterable by heat, repoſe, and humidity.

Of the Putrefaction of Animal Subſtances.

Although vegetable ſubſtances be ſuſceptible of being decompoſed and entirely deſtroyed by the putrid fermentation, as we have related, they are however very far from being ſo proper to undergo this inteſtine motion as animal matters. The putrefaction of theſe laſt is much more rapid, and its phenomena are different: all the fluid and ſolid parts of animals are equally expoſed to it, whilſt ſeveral vegetable matters ſeem to be exempt from it, or at leaſt to undergo it with great difficulty and ſlowneſs.

The putrefaction of animals, which we muſt conſider with Boerhaave as a true fermentation, is one of the moſt important phenomena, and at the ſame time very difficult to be underſtood.

All the works of the philoſophers, from Bacon of Verulam, who had fully felt the importance of the inquiries upon this ſubject, down to our days, have elucidated only a few points, and hinted at the general phenomena presented by putrefying substances. Becher, Hales, Stahl, Drs Pringle and Macbride, Gaber, Baumé, the author of the essays upon putrefaction, and those of the dissertations upon antiseptics, presented with the prize 1767 by the Dijon academy, have observed and carefully described the facts which the putrid alteration exhibits: but it will be seen by the relation we are going to give, that a great number of experiments still remains to be made, to understand fully the phenomena of this natural operation.

Every fluid or soft substance, extracted from the body of an animal, exposed to the air at a temperature of ten degrees or above it, suffers more or less rapidly the following alterations. Its colour turns pale; its consistence diminishes: if it is a solid part like meat, it softens, and allows a serosity to issue out, whose colour very soon alters: its texture is relaxed and disorganised; its smell becomes disagreeable; it gradually thickens and diminishes in size; its smell is exalted, and becomes alkaline. In this state, if it is contained in a close vessel, the progress of the putrefaction seems to relent; only an alkaline and pungent smell is felt. The substance effervesces with the acids, and turns the syrup of violets green: but by opening a communication with the air, the urinous exhalation is dissipated, and, with a kind of impetuosity, a particular putrid insupportable smell is emitted, which continues a long time, diffuses itself every where, affects animal bodies as a ferment capable of changing their fluids. This smell is corrected, and as if stopt, by the volatile alkali. When this last substance is volatilised, putrefaction turns active again, the putrefying mass swells presently, the colour alters, the fibrous texture of the flesh is almost no longer distinguishable; it is changed into a soft brown or yellow matter; its smell is nauseous, and very active upon the bodies of animals. This odorous principle gradually loses its force: the fluid portion of the flesh acquires a sort of consistence; its colour deepens, and it finishes by being reduced in-

to

to a friable matter, half dry and a little deliquefcent; which being rubbed between the fingers, is reduced into a grofs powder like earth. Such is the laft ftate which terminates the putrefaction of animal fubftances: they take a longer or fhorter time in arriving at this point. Eighteen months, two and even three years, fcarcely fuffice for the entire deftruction of animal bodies when expofed to the air; nor has the total deftruction of dead bodies interred, relatively to the time which elapfes before this change takes place, been yet accurately determined.

From this relation it follows, 1. That the conditions proper for evolving and fupporting the putrefaction of animal matters are, the contact of the air, heat, humidity, and repofe, or the inertia of bodies: 2. That the volatile alkali is lefs a product of putrefaction than a matter difengaged, like the oils, during this natural operation, as it exifts only for a certain time, beyond and within which it never appears: 3. That putrefaction, effected by an inteftine motion proper to organifed fubftances, may be compared to the action of the fire, as M. Godard has remarked; and may be confidered as a fpontaneous decompofition, as M. Baumé has thought; and that it differs from both only in its flownefs: 4. That in this operation of nature, the immediate principles of animals act upon one another by means of the water which favour their contact, and of the heat which the inteftine motion produces; that thus the volatile matters are gradually diffipated in the order of their volatility, and that nothing more remains after putrefaction, except an infipid, as if earthy, refiduum: 5. That the putrid exhalation, fo well characterifed and diftinguifhed by the olfactory nerves, and whofe action is fo ftrong upon the animal œconomy, fhould be confidered as the true product of putrefaction, as this exhalation conftantly exifts in all fubftances which are putrefying, and throughout the whole time of putrefaction; fince it is proper to this operation, and is found in no other natural phenomenon:

 laftly,

laftly, As it evolves the putrefactive motion in all animal bodies which are expofed to its action. As to the nature of this fubftance, it is particularly this point which is fo incompletely underftood, and requires a farther inveftigation. What we know of it, fhows us that it is extremely volatile, attenuated, penetrating; that pure air, water in great quantity, and the acid gafes, are fufceptible of moderating its effects. Altho' it is no miftake to confound it with the fixed air, which is difengaged in great quantity from putrefying fubftances, to the difengagement of which Dr Macbride entirely attributed the caufe of this natural phenomenon; although we ought no longer to compare it either with inflammable gas difengaged from putrefcent bodies, or to the luminous matter which fhines at the furface of the putrefying folids of animals, and which makes thefe fubftances appear as fo many phofphori; we cannot, however, be hindered from believing, that it has fome very direct relation with thefe fubftances, as it conftantly accompanies them, and is as volatile, as fubtile, as they, and as it acts with fo much energy upon the organs of animals.

We may, with M. de Boiffieu, diftinguifh four ftages in the putrid fermentation of animal-fubftances. The firft, called by this phyfician a *tendency to putrefaction*, confifts in a fmall alteration, which is manifefted by an infipid or very flightly mufty fmell, and in the foftening of thefe fubftances. The fecond ftage, that of commencing putrefaction, is fometimes indicated by marks of acidity. The fubftances which are fubject to it lofe their weight, affume a fetid fmell, are foftened, and allow the ferofity to get free when they are in clofe veffels; or they dry, and put on a deep colour, if they are expofed to the free air. In the third ftage, or advanced putrefaction, the putrefcent matters emit an alkaline fmell, mixed with the putrid and naufeous odour: they become as if diffolved; their colour alters more and more; and at the fame time they lofe their weight and volume. Laftly, the fourth degree, that of complete

plete putrefaction, manifests itself by the entire dissipation of the volatile alkali, of which no traces remain: the fetid odour loses its force; the volume and the weight of the putrefied substances are considerably diminished; a gelatinous mucosity separates; they gradually dry; and are at last reduced to an earthy and friable substance.

Such are the general phenomena which are observed in the putrefaction of animal-substances; but they are far from being the same in all putrescent substances. There is at first a great distinction to be made between the putrefaction of the parts of animals when alive and that of their organs when dead. The motion which exists in the first, singularly modifies the phenomena of this alteration; and physicians have frequent opportunities of seeing the differences which exist between these two states with regard to putrefaction. Besides that, every humour, every solid part, separated from a dead animal, has also its proper manner of putrefying: the muscular, membranous, or parenchymatous, and more or less compact texture of the organs; the oily, the mucilaginous, or lymphatic nature of the humours; their consistence; their state, relative to that of the animal which produced them; influence the putrefactive motion, and modify it in a thousand ways, which are perhaps inappretiable. Lastly, How shall we enter into a discussion of the state of the air, its temperature, elasticity, weight, dryness, or humidity of the exposure of the putrefying substance in different places, even of the form of the vessels which contain it? circumstances, all of which produce alteration in the phenomena of spontaneous putrefaction. It must therefore be agreed, that the history of animal-putrefaction is only begun, and that the investigations and experiments of the physicians and chemists of several ages, all united, are requisite to render this subject complete.

F I N I S.

BOOKS

IN THE

DIFFERENT BRANCHES OF MEDICINE,

Printed for and fold by C. ELLIOT, Edinburgh; T. CADELL, and G. ROBINSON, London.

1. FIRST LINES OF THE PRACTICE OF PHYSIC, by Dr WILLIAM CULLEN, Profeffor of the practice of Phyfic in the Univerfity of Edinburgh. Corrected, much enlarged, and now firft completed, in 4 vols 8vo; with a large Index. Price 1l. 4s. in boards, and 1l. 8s. bound.—Volume IV. may be had feparately. Price 6s. in boards.

2. An INQUIRY into the NATURE and CAUSES of FEVER; with a review of the feveral opinions concerning its Proximate Caufe, as advanced by different Authors, and particularly as delivered from the Practical Chair in the Univerfity of Edinburgh. Including fome Obfervations on the exiftence of Putrefaction in the living Body, and the proper method of Cure to be purfued in Fever. By Caleb Dickinfon, M. D. Price 3s.

3. A SYSTEM of SURGERY. By Benjamin Bell, Member of the Royal College of Surgeons, one of the Surgeons to the Royal Infirmary, and Fellow of the Royal Society of Edinburgh. Illuftrated with Copperplates, Vol. I. II. and III. 8vo. 6s. each, boards. This work, when completed, with a former volume on Ulcers, &c. by the fame Author, will comprehend a full fyftem of Modern Surgery. The whole to be contained in other two vols 8vo. Vol. IV. and V. will be publifhed as foon as poffible.

4. A TREATISE on the THEORY and MANAGEMENT of ULCERS; with a Differtation on White Swellings of the Joints. To which is prefixed, An Effay on the Chirurgical Treatment of Inflammation and its Confequences. By the fame Author. A new edition, being the third, confiderably improved and enlarged, 6s. in boards.

5. The third Edition, corrected, of MEDICAL CASES, felected from the Records of the Public Difpenfary at Edinburgh; with Remarks and Obfervations. By Andrew Duncan, M. D. F. R. and A. S. Edin. Phyfician to the Prince of Wales for Scotland, &c. Price 6 s. bound.

☞ A very fine Print of Dr Duncan, painted by Weir, and engraved by Trotter. Price 2 s. 6d.

6. EXPERIMENTS on the RED and QUILL PERUVIAN BARK: with Obfervations on its Hiftory, mode of Operation, and Ufes; and on fome other Subjects connected with the Phenomena and Doctrines of Vegetable Aftringents. Being a Differtation which gained the firft Prize given by the Harveian Society of Edinburgh for the year 1784. By Ralph Irvine. One vol. 8vo. Price 3s.

7. Out

7. OUTLINES of the Theory and Practice of MIDWIFERY, by Alexander Hamilton, M. D. F. R. S. Edin. Professor of Midwifery in the University, and Member of the Royal College of Surgeons, Edinburgh. Price 6s. bound; or, with Dr Smellie's 40 Tables and Explanations, 11s. boards, and 12s. bound.

8. A SYSTEM of ANATOMY, from Monro, Winslow, Innes, Hewson, and the latest authors.—Arranged, as nearly as the nature of the work would admit, in the order of the Lectures delivered by the Professor of Anatomy in the University of Edinburgh.—In Two Volumes Octavo, illustrated with 16 Copperplates. Price 13s. boards.

This System will be found very useful to the young as well as the advanced Students of Anatomy, as it comprehends the whole of Dr Monro on the Bones; Innes on the Muscles, his Explanations of the Skeleton and Muscles, and his Eight Anatomical Tables; as also Dr Winslow, Mr Hewson, and the latest authors, on the other parts of the body, with many alterations, corrections, and additions by the Editor, so as to form a complete SYSTEM of ANATOMY, in a very commodious size, and at a very moderate price.—In addition to Mr Innes's Tables are given, two views of the Viscera, two views (a fore and back) of the Veins and Arteries, two plates of five figures of the Lymphatic Vessels from Hewson, and two views (a fore and back) of the Nerves, all accurately engraved.

9. The WORKS of ALEXANDER MONRO, M. D. F. R. S. Fellow of the Royal College of Physicians, and late Professor of Medicine and Anatomy in the University of Edinburgh. Published by his Son ALEXANDER MONRO, M. D. President of the Royal College of Physicians, Professor of Medicine, Anatomy, and Surgery, in the University of Edinburgh. To which is prefixed, the LIFE of the AUTHOR. In one very large volume in quarto, elegantly printed upon a royal paper, and ornamented with a capital Engraving of the Author by Mr Basire, from a painting by Allan Ramsay, Esq; besides several Copperplates illustrative of the subjects. Price 1l. 5s. in boards.—The fine Engraving of the Doctor, by itself, at 5s. for proof impressions. The simple engraving cost 40 guineas.

10. A TREATISE on COMPARATIVE ANATOMY, by Alexander Monro, M. D. F. R. S. &c. &c. Published by his Son, Alexander Monro junior, M. D. A new Edition, with considerable Improvements and Additions by *other hands*, 12mo. Price 2s. in boards.

11. A TREATISE on the Theory and Practice of MIDWIFERY; by W. Smellie, M. D. To which is now added, his Set of Anatomical Tables, exhibiting the various cases that occur in practice; accurately reduced and engraved by A. Bell, on 40 cop-

copperplates, (including an additional plate of instruments, by the late Dr Thomas Young), with explanations. A new edition, on fine paper, in 3 vols 12mo. Price 10s. 6d. in boards, or 12s. bound.

12 Dr. SMELLIE's set of Anatomical TABLES, and an Abridgement of the Practice of Midwifery, with a view to illustrate his Treatise on that subject and Collection of Cases, 8vo size 6s. 12mo size 5s. in boards.

13. INNES's Eight Anatomical TABLES of the Human Body, containing the principal parts of the Skeleton and Muscles represented in the large Tables of Albinus; to which are added, Concise Explanations. New edition, with an Account of the Author. Neatly half-bound, quarto, price 6s 6d.

14. INNES's short description of the Human MUSCLES, chiefly as they appear on dissection, together with their several uses, and the synonyma of the best authors; a new edition, greatly improved by Alex. Monro, M. D. 2s. 6d. in boards.

15. The LONDON MEDICAL JOURNAL, from January 1781 to the end of 1783, in 4 vols 8vo. price 1 l. 8s. bound.

The same for 1784 in single numbers, as well as any of the former at 1s. 6d. each.

16. PRACTICAL OBSERVATIONS on the MORE OBSTINATE and INVETERATE VENEREAL COMPLAINTS, by J. Schwediauer, M. D. 8vo. price 3s. 6d. sewed.

In the press, and shortly will be published by C. Elliot,

I. The STRUCTURE and PHYSIOLOGY of FISHES explained and compared with those of Man, illustrated with 46 large Copperplates, in one very large volume in folio, English and Latin, by Alexander Monro, M. D. Fellow of the Royal College of Physicians, and Professor of Physic, Anatomy, and Surgery, in the University of Edinburgh, price 2l. 2s. boards.

II. The NEW DISPENSATORY, on the plan of the late Dr Lewis, by a Physician in Edinburgh, large 8vo.

III. THESAURUS MEDICUS, or a selection of the best medical theses, selected and approved of by the R. Med. Society, from the 1758 to the 1785; Vol. III. IV. which complete the Work.

IV. Dr Alexander Hamilton's TREATISE on MIDWIFERY and FEMALE COMPLAINTS, with the Treatment of Lying-in Women, and the management of new born children, for the use of female practitioners and private families. It will be had with Dr Smellie's 40 plates and explanations at 10s. in boards, or without the same at 4s. only.

V. Baron Haller's FIRST LINES of PHYSIOLOGY, translated from the correct Latin copy printed under the inspection of William Cullen, M. D. To which is added, a translation of the laborious Index composed for that edition. The present edition has been compared with the last published at Gottingen by Professor Wristberg; and includes a translation of the whole of his notes added to that edition.

www.ingramcontent.com/pod-product-compliance
Lightning Source LLC
Chambersburg PA
CBHW020858210326
41598CB00018B/1704
* 9 7 8 3 7 4 3 4 3 2 9 2 5 *